IPM in Practice

Principles and Methods of Integrated Pest Management

Second Edition

IPM in Practice

Principles and Methods of Integrated Pest Management

Second Edition

Mary Louise Flint
UC Statewide Integrated Pest Management Program
and Department of Entomology
University of California, Davis

University of California
Agriculture and Natural Resources
Oakland, California
Publication 3418

To order or obtain ANR publications and other products, visit the ANR Communication Services online catalog at *http://anrcatalog.ucdavis.edu* or phone 1-800-994-8849. You can also place orders by mail or FAX, or request a printed catalog of our products from

University of California
Agriculture and Natural Resources
Communication Services
1301 S. 46th Street
Building 478 - MC 3580
Richmond, CA 94804-4600

Telephone 1-800-994-8849
(510) 665-2195, FAX (510) 665-3427
E-mail: anrcatalog@ucdavis.edu

Second edition, 2012

Publication 3418
ISBN-13: 978-1-60107-785-1

Photographs by Jack Kelly Clark, except as noted.
Cover photo by Michael L. Poe; design by Celeste A. Rusconi

Library of Congress Cataloging-in-Publication Data

Flint, Mary Louise, 1949-
IPM in practice: principles and methods of integrated pest management / Mary Louise Flint. — 2nd ed.
p. cm. — (Publication; 3418)
ISBN 978-1-60107-785-1

1. Pests—Integrated control. 2. Garden pests—Integrated control. I. University of California (System). Division of Agriculture and Natural Resources. II. Title. III. Title: Principles and methods of integrated pest management. IV. Series: Publication (University of California (System). Division of Agriculture and Natural Resources); 3418.

SB950.F565 2012
363.7'8–dc23
2012029754

This publication has been anonymously peer reviewed for technical accuracy by University of California scientists and other qualified professionals. This review process was managed by ANR Associate Editor Joseph Grant.

 Printed in Canada on recycled paper.
5m-pr-8/12-SB/CR

PRECAUTIONS FOR USING PESTICIDES

Pesticides are poisonous and must be used with caution. READ THE LABEL CAREFULLY BEFORE OPENING A PESTICIDE CONTAINER. Follow all label precautions and directions, including requirements for protective equipment. Use a pesticide only on crops specified on the label. Apply pesticides at the rates specified on the label or at lower rates if suggested in this publication. In California, all agricultural uses of pesticides must be reported. Contact your county agricultural commissioner for details. Laws, regulations, and information concerning pesticides change frequently, so be sure the publication you are using is up to date.

Legal Responsibility. The user is legally responsible for any damage due to misuse of pesticides. Responsibility extends to effects caused by drift, runoff, or residues.

Transportation. Do not ship or carry pesticides together with foods or feeds in a way that allows contamination of the edible items. Never transport pesticides in a closed passenger vehicle or in a closed cab.

Storage. Keep pesticides in original containers until used. Store them in a locked cabinet, building, or fenced area where they are not accessible to children, unauthorized persons, pets, or livestock. DO NOT store pesticides with foods, feeds, fertilizers, or other materials that may become contaminated by the pesticides.

Container Disposal. Dispose of empty containers carefully. Never reuse them. Make sure empty containers are not accessible to children or animals. Never dispose of containers where they may contaminate water supplies or natural waterways. Consult your county agricultural commissioner for correct procedures for handling and disposal of large quantities of empty containers.

Protection of Nonpest Animals and Plants. Many pesticides are toxic to useful or desirable animals, including honey bees, natural enemies, fish, domestic animals, and birds. Crops and other plants may also be damaged by misapplied pesticides. Take precautions to protect nonpest species from direct exposure to pesticides and from contamination due to drift, runoff, or residues. Certain rodenticides may pose a special hazard to animals that eat poisoned rodents.

Posting Treated Fields. For some materials, reentry intervals are established to protect fieldworkers. Keep workers out of the field for the required time after application and, when required by regulations, post the treated areas with signs indicating the safe reentry date.

Harvest Intervals. Some materials or rates cannot be used in certain crops within a specific time before harvest. Follow pesticide label instructions and allow the required time between application and harvest.

Permit Requirements. Many pesticides require a permit from the county agricultural commissioner before possession or use. When such materials are recommended in this publication, they are marked with an asterisk (*).

Processed Crops. Some processors will not accept a crop treated with certain chemicals. If your crop is going to a processor, be sure to check with the processor before applying a pesticide.

Crop Injury. Certain chemicals may cause injury to crops (phytotoxicity) under certain conditions. Always consult the label for limitations. Before applying any pesticide, take into account the stage of plant development, the soil type and condition, the temperature, moisture, and wind direction. Injury may also result from the use of incompatible materials.

Personal Safety. Follow label directions carefully. Avoid splashing, spilling, leaks, spray drift, and contamination of clothing. NEVER eat, smoke, drink, or chew while using pesticides. Provide for emergency medical care IN ADVANCE as required by regulation.

AUTHORS

Mary Louise Flint, Associate Director for Urban and Community IPM, UC Statewide IPM Program, and Extension Entomologist, Department of Entomology, UC Davis

Patricia Gouveia, former Senior Writer, UC Statewide IPM Program, was coauthor of the first edition.

ACKNOWLEDGMENTS

This book was produced under the auspices of the University of California Statewide IPM Program.

Many people contributed to this book. For the First Edition, an Ad Hoc PCA Study Materials and Examination Advisory Committee advised on book development and the Knowledge Expectations (KEs) that define what PCAs need to know to make effective IPM decisions. A second group advised on updating the book and the KEs for the 2012 edition and suggested new content.

Tunyalee Martin, Content Supervisor for the UC Statewide IPM Program, served as Technical Editor for the Second Edition. Steve H. Dreistadt also contributed significantly to the revision.

CONTRIBUTORS AND REVIEWERS FOR THE SECOND EDITION

Kassim Al-Khatib, UC Statewide IPM Program, UC Davis
Roger Baldwin, UC Statewide IPM Program, Kearney Agricultural Center
Walt Bentley, UC Statewide IPM Program, Kearney Agricultural Center
Jianlong Bi, UC Cooperative Extension, Monterey County
Devon Carroll, PCA, Fresno, CA
Nita Davidson, California Department of Pesticide Regulation, Sacramento
R. Michael Davis, Department of Plant Pathology, UC Davis
Steve H. Dreistadt, UC Statewide IPM Program, UC Davis
Peter B. Goodell, UC Statewide IPM Program, Kearney Agricultural Center
David Haviland, UC Cooperative Extension, Kern County
Richard Melnicoe, Western Regional IPM Center, UC Davis
Cliff Ohmart, PCA, Davis, CA
Carolyn Pickel, UC Statewide IPM Program, Sacramento Valley
Renee Rianda, PCA, Woodland, CA
John Roncoroni, UC Cooperative Extension, Napa County
Jay Rosenheim, Department of Entomology, UC Davis
Steve Swain, UC Cooperative Extension, Marin and Sonoma Counties
Lucia Varela, UC Statewide IPM Program, North Coast
Patrick Weddle, PCA, Placerville, CA
Cheryl A. Wilen, UC Statewide IPM Program, South Coast
Frank G. Zalom, Department of Entomology, UC Davis

CONTRIBUTORS TO THE FIRST EDITION

Syed Ali, California Statewide Water Resources Control Board, Sacramento
Michael Baefsky, PCAS, Baefsky & Associates, Orinda, CA
Nancy Brownfield, East Bay Regional Parks District, Berkeley, CA
Tim Butler, California Association of Pest Control Advisers, Sacramento
Edward Caswell-Chen, Department of Nematology, UC Davis
Art Craigmill, Department of Environmental Toxicology, UC Davis
Kim Crum, California Association of Pest Control Advisers, Sacramento
R. Michael Davis, Department of Plant Pathology, UC Davis
Joseph Ditomaso, Plant Sciences Department, UC Davis
Steve H. Dreistadt, UC Statewide IPM Program, UC Davis
Les Ehler, Department of Entomology, UC Davis
Peter B. Goodell, UC Statewide IPM Program, Kearney Agricultural Center
Elizabeth Grafton-Cardwell, Department of Entomology, UC Riverside
Lyndon Hawkins, Calfiornia Department of Pesticide Regulation, Sacramento
Steve Koike, UC Cooperative Extension, Monterey County
Tom Lanini, Department of Plant Sciences, UC Davis
Marshall Lee, California Department of Pesticide Regulation, Sacramento
James MacDonald, Department of Plant Pathology, UC Davis
Mark Mayse, Departments of Entomology and Plant Sciences, CSU Fresno
Richard Melnicoe, Department of Environmental Toxicology, UC Davis
Chris Morgner, PCA, Agri-Valley, Merced, CA
Robert Norris, Department of Plant Sciences, UC Davis
Patrick O'Conner-Marer, UC Statewide IPM Program, UC Davis
Michael Parrella, Department of Entomology, UC Davis
Sandy Ratliff, California Department of Pesticide Regulation, Sacramento
Gary Ritenour, Department of Plant Science, CSU Fresno
Jay Rosenheim, Department of Entomology, UC Davis
Kay Rudolph, U.S. EPA, Region 9, San Francisco
Judy Stewart Leslie, PCA, Pest Management Associates, Exeter, CA
Ron Thomsen, PCA, California Association of Pest Control Advisers, Sacramento
Patrick Weddle, PCA, Hansen & Associates, Placerville, CA
Desley Whisson, Department of Wildlife, Fish and Conservation, UC Davis
Cheryl Wilen, UC Statewide IPM Program, South Coast
Muffet Wilkerson, California Department of Pesticide Regulation, Sacramento
Frank G. Zalom, Department of Entomology, UC Davis

Contents

1 Introduction

The purpose of this book is to help pest management professionals apply the principles of *integrated pest management* (IPM) to the many different environments in which they work, such as agricultural crops, urban landscapes, greenhouses, forests, rights-of-way, and other managed *ecosystems*. The emphasis is on practical application of IPM in the field. After studying this book, readers should be able to adapt IPM techniques and strategies to fit their specific pest management situations. This book has been written as an IPM study guide for those preparing for California's written licensing exams for *pest control adviser* (PCA). However, it is also intended as a resource for others seeking practical IPM information, including students, growers, landscape and property managers, *pesticide* applicators, and pest control consultants throughout North America.

SIDEBAR 1-1

What Is IPM?

Integrated pest management (IPM) is an ecosystem-based strategy that focuses on long-term prevention of pests or their damage through a combination of techniques such as biological control, habitat manipulation, modification of cultural practices, and use of resistant varieties. Pesticides are used only after monitoring indicates that they are needed according to established guidelines, and treatments are made with the goal of removing only the target organisms. Pest control tools are selected and applied in a manner that minimizes risks to human health, to beneficial and nontarget organisms, and to the environment. For more information on the concept of IPM, refer to Chapter 3.

THE IMPORTANCE OF INTEGRATED PEST MANAGEMENT

Integrated pest management programs (see Sidebar 1-1) emphasize ecosystem-based strategies that provide economical, long-term solutions to *pest* problems. Pesticides are used only when they are necessary to prevent imminent loss or damage to the managed resource; IPM strategies thus minimize hazards to human health, the environment, and *nontarget organisms*. Many pest management professionals regularly employ the concepts of integrated pest management in their daily work. For example, to avoid pests or reduce pests to noneconomic levels, pest managers often recommend using pest-resistant plant varieties, a change in planting or harvest dates, or a system of *crop rotation*. IPM recommendations may include managing water and fertilizer, adjusting cultivation techniques, the use of mating disruption techniques, or the destruction of pest habitats. Also, strategies that encourage natural enemies such as *parasites*, *predators*, *pathogens*, or competitors for food can help limit pest numbers. Most important, in IPM programs some or all of these techniques are combined or integrated into an overall strategy to protect crops.

Pesticides are an integral component of many integrated pest management programs. In many situations their use is required. However, in IPM programs, pesticides are chosen and used in ways that enhance their effectiveness and attempt to minimize negative impacts on workers and people at the site and on the ecosystem. The need for application, choice of material, and spray timing and rate are carefully determined based on pest identification, *monitoring* (Figure 1-1), and consideration of factors such as past crop history, previous experience, and research guidelines.

Prior to any control action, pest density and stage of development are verified. The presence of natural enemies and their capacity to exert control, along with any environmental factors that might impact the pest, natural enemies, or the crop, are carefully noted. The growth characteristics of the crops are observed and the amount of nutrients available is monitored. This information, combined with an understanding of the interrelationship of pests in the ecosystem, presents the challenge of pest control: using all available management methods to create an unfavorable environment for the pest while maximizing favorable conditions for the plant or crop.

As more variables are woven into a total-systems approach to managing the ecosystem and pest problems, the breadth and scope of knowledge that a pest manager requires continue to grow. Pest management has become an interdisciplinary science requiring managers to integrate their knowledge of entomology, *weed* science, plant pathology, nematology, vertebrate pests, crop development, environmental quality, and other sciences. Integrated pest management combines a complex mix of interdisciplinary practices and technologies specific to each locale to manage pest *populations* efficiently. IPM is a flexible, evolving strategy that is continually updated as new information becomes available.

FIGURE 1-1

PCAs regularly monitor crops using various techniques such as shake or beat sampling to support pest management decision making.

Later chapters in this book provide the fundamentals on which to build an IPM program and the resources needed to stay up to date.

WHAT IS A PEST CONTROL ADVISER?

In California, any person who offers a recommendation on any agricultural use of a pest control product or technique, presents themselves as an authority on any agricultural use, or solicits services or sales for any agricultural use is a pest control adviser (PCA). In many parts of the United States, PCAs are often called pest management consultants; however, we will use the term *pest control adviser* throughout this book to include anyone who recommends specific pest management practices. PCAs can be distinguished from pesticide applicators, who are licensed in California to apply pesticides but do not make recommendations about which pesticide to use.

According to California law, all PCAs must be licensed by the California Department of Pesticide Regulation (DPR). California law defines *agricultural use* broadly. The term refers to the use of any pesticide, method, or device for the control of plant or animal pests, or any other pests, or the use of any pesticide for the regulation of plant growth or defoliation of plants. Agricultural use includes but is not limited to commercial production of plants or animals and applications in parks, golf courses, waterways, forests, roadsides, cemeteries, and other similar areas. Pesticides that are intended for home use, structural pest control, industrial or institutional use, or for the control of an animal pest under the written prescription of a veterinarian are not defined as agricultural use.

California law does not require a written recommendation from a licensed PCA when pesticides are applied for residential, industrial, institutional, structural, or veterinary use. Structural pest control operators are licensed by DPR but make pest control decisions without involvement of PCAs. Veterinarians are licensed by the state Veterinary Medicine Board and also are regulated by the California Department of Consumer Affairs. Officials of federal, state, and county departments of agriculture, and University of California personnel engaged in official duties related to agricultural use, are also exempt from the licensing requirement. However, their agricultural use recommendations must be in writing.

California has the oldest licensing program for PCAs in the United States. Its strong education, examination, and continuing education requirements have allowed California to develop the largest and best-trained group of professional pest management advisers in the nation (Figure 1-2). However, PCAs (sometimes called pest management consultants, crop consultants, or integrated pest management consultants) are active throughout the country and are represented by many professional organizations. They are trained professionals and work in various types of pest management situations, including urban and landscape, rangeland, right-of-way, forest, aquatic, and agricultural areas. PCAs work for pest control companies, public agencies, farming operations, or other companies or may work as independent consultants hired by growers or others to provide advice.

Growers may legally apply many pesticides without the advice or written recommendation of a PCA. Most growers, however, often consult PCAs because they recognize that PCAs have a high level of pest management expertise. PCAs keep up with the latest developments in pest management to a degree that most growers are not able to achieve. Further, a PCA can improve reliability of a pest management program and focus greater attention on avoiding damage to human health and the environment.

FIGURE 1-2

Through continuing education, PCAs are constantly updating their expertise. Here PCAs are learning how to identify pests and beneficials in grapes.

HOW TO BECOME LICENSED AS A CALIFORNIA PEST CONTROL ADVISER

To become a PCA in California, an applicant must pass examinations administered by DPR in laws and regulations and in at least one pest control category. The Laws and Regulations exam can be taken concurrently with the pest control categories, which are listed below. A person who has never held a California agricultural PCA license must meet the minimum qualifications established by DPR, which include educational and core course requirements. Educational and apprenticeship requirements change from time to time, so check with DPR, Cooperative Extension, or the county agricultural commissioner for up-to-date information. The licensing section of the DPR website, www.cdpr.ca.gov, is a good resource.

ADVISER LICENSING CATEGORIES

California PCAs must show expertise in identifying and managing different types of pests, such as insects, weeds, plant pathogens, *nematodes*, and vertebrates. For this reason, licensing categories follow pest types. The seven California PCA categories are insects, mites, and other *invertebrates*, with subcategories in plants and animals; plant pathogens; nematodes; vertebrate pests; weed control; defoliation; and plant growth regulators. This differs from the situation with pesticide applicators, who are licensed according to the type of area managed, such as agricultural, right-of-way, forest, landscape, or structural/institutional pest control. PCAs may make recommendations only in those categories in which they have been issued a license. All PCAs are expected to demonstrate knowledge of IPM, regardless of their license category.

KNOWLEDGE EXPECTATIONS

Knowledge expectations define the basic background information PCAs need when starting their first job. Meeting knowledge expectations will require recalling facts, comprehending information, applying knowledge, and analyzing situations. Specific knowledge expectations have been developed for each license category. In addition, all PCAs are expected to meet the knowledge expectations for integrated pest management listed on the DPR website.

Use knowledge expectations to study and prepare for the licensing exams. If you can answer all the questions listed in the knowledge expectations, you are probably ready for the exam. Information required to meet the IPM knowledge expectations (see the DPR website, http://www.cdpr.ca.gov/docs/license/adviserke.htm) is covered in this manual. Additional information needed to meet the knowledge expectations for each subject area licensing category is contained in study materials recommended by DPR. Knowledge expectations and study materials for other licensing categories can be found on the DPR website.

In addition to the more general information PCAs must know to meet knowledge expectations, some of the recommended study materials contain additional in-depth information not covered in the examination process. Applicants are not expected to memorize this information to pass exams. However, PCAs often need in-depth material on new pests or management techniques and will find these texts useful references after they pass their exam. In all licensing categories, a basic knowledge expectation is that PCAs will know how to use reference materials to identify pests and disorders, find biological information, and adapt and evaluate new management strategies.

As the science of pest management evolves, knowledge expectations for PCAs will change, too. A committee of PCAs, university experts, and regulatory specialists established by DPR and the University of California keeps knowledge expectations up to date. Once licensed as a PCA, acquisition of knowledge does not end—in fact, it is just beginning. Keep in touch with changes in each of the licensing areas and be sure to cover these areas in meeting continuing education requirements.

CONTINUING EDUCATION

The science of pest management is never static. New pests arrive every year, new pest management tools regularly become available, and pesticide regulations change frequently. To serve their clientele properly, PCAs must continually add to their pest management knowledge. California recognizes the need for continuing education in its license renewal regulations. Because of the demand created by these requirements, California PCAs have access to a large variety of pest management courses (Figure 1-3). It is essential that PCAs choose courses that are practical and can be used to improve the services and advice they offer.

California PCA licenses must be renewed every 2 years. Within each 2-year renewal period, the adviser must complete a minimum number of hours of pest management and pesticide laws and regulations educa-

FIGURE 1-3

Looking at the relationship of soil to turf problems in this hands-on workshop helps these landscape PCAs complete their continuing education requirement.

tion. Instruction approved by the director of DPR includes college-accredited courses in pest management, professional or technical seminars, certain online courses, and *field trial* tours or demonstrations relating to pest management and pesticide use. Advisers must maintain a record of their continuing education credits. Continuing education requirements may change over time, so check with the county agricultural commissioner or DPR for the latest information.

STANDARD SERVICES OFFERED BY PEST CONTROL ADVISERS

Growers, public agencies, and others employ PCAs because of their expertise. PCAs and pest management consultants provide up-to-date pest management information, perform monitoring services, and provide written recommendations for a variety of pest management situations.

Written Recommendation

The basic service that PCAs offer is advice on pest management practices or products that can be used to prevent, control, or curtail pest presence or damage. By law, no one except a licensed PCA (and certain government officials and university experts) can provide such information for agricultural use in California. When advice is given for a pest that is site and time specific, the recommendation must be made in writing following guidelines established by DPR. Growers cannot hire commercial applicators to make applications of restricted-use pesticides unless a written recommendation from a PCA is in hand.

Pest Management Expertise

By virtue of their license and training, PCAs are recognized as experts in the field of pest management. PCAs bring specialized expertise and also have access to the latest pest management information. PCAs have knowledge of pest identification, monitoring methods, information on new pest management tools, and skills such as an ability to use computerized pest prediction models. They are expected to be aware of potential environmental, health, or plantback risks associated with pesticide applications, and they are expected to warn their clients of possible problems as well as regulatory restrictions. A good PCA will often save a grower, public agency, or other resource manager time and money by avoiding unnecessary pesticide applications.

Pest Monitoring

Pest management decisions must be based on actual field conditions. PCAs are expected to know how to sample or monitor fields to determine whether pests pose a threat. Generally, monitoring is done on a regular basis and in a systematic manner so that pest counts or damage symptoms can be compared over the course of a season. Monitoring often involves counting pest numbers or the incidence of damage symptoms, but it also can involve estimating natural enemies, measuring crop growth, and monitoring *weather* and field conditions (Figure 1-4). For some pests, particularly weeds, surveys are carried out during the season before planting. A monitoring program includes a good record-keeping program so that problems can be mapped and analyzed over time.

FIGURE 1-4

Keeping monitoring records helps PCAs recognize trends in pest numbers or incidence of disease so they can be noted and analyzed over time. A hands-free magnifier makes scouting more efficient.

PCAs must be able to interpret the results of their monitoring and relate them to the need to take pest management actions.

ADDITIONAL SERVICES OFFERED BY PEST CONTROL ADVISERS

PCAs often become involved in other aspects of resource management. In their day-to-day activities, PCAs collect field data that can help them make other types of crop production decisions. For instance, they may suggest modifications of irrigation or fertilization practices to enhance plant growth and development or to reduce pest populations. Also, because of their expertise in pest management and pesticides, PCAs may keep records of pest management activities and pesticide used, or they may document worker training activities for their employers or clients.

100% Pesticide Use Reporting

Users of *agricultural pesticides* in California are required to provide complete, site-specific documentation of every pesticide application, including use in parks, golf courses, cemeteries and other public agency sites, rangeland, pastures, along roadside and railroad rights-of-way, as well as on agricultural crops. Pesticide use reports (Figure 1-5) must include the name of the pesticide, the amount applied, the date of application, the application site, and the name of the applicator. Growers and PCAs must remember that use of all registered pesticides must be reported, including use of organically acceptable materials such as soaps, microbials, and mating disruptants.

The data from use reports provide valuable information for PCAs, growers, and site managers. Use reports document efforts at reduced-risk pest management practices and provide a complete history of pesticide use in a given area from which relevant pest management strategies can be planned. Many PCAs record this information for their own use and may also provide it as a service to their clients. However, by law, it is the user of the pesticide (e.g., the grower or public agency) who is responsible for filing use reports, not their PCAs.

Worker Training

Employers must maintain a written training program describing the materials and information that will be used to train employees who will handle pesticides or who may be exposed to pesticides in the course of their work. The training program must address the immediate and long-term hazards involved, the safety procedures to be followed, the procedures for handling emergency situations, applicable laws and regulations, and the requirement for medical supervision if workers handle organophosphate, carbamate, or pesticides with the signal words *Danger* or *Warning*. PCAs, in the course of reading the label and writing the recommendation, have reviewed much, if not all, safety and precautionary information and can assist clients in organizing necessary

FIGURE 1-5

Users of agricultural pesticides in California are required to file Pesticide Use Reports. PCAs sometimes perform this requirement.

STATE OF CALIFORNIA

MONTHLY SUMMARY PESTICIDE USE REPORT

DEPARTMENT OF PESTICIDE REGULATION
ENFORCEMENT BRANCH

PR-ENF-060 (REV. 9/07)

INSTRUCTIONS FOR COMPLETING THIS FORM ARE INDICATED BELOW AND ON THE REVERSE SIDE

OPERATOR (FIRM NAME)	ADDRESS	CITY	ZIP CODE	PHONE NUMBER

OPERATOR ID/PERMIT NUMBER	LICENSE NUMBER	COUNTY WHERE APPLIED	COUNTY NUMBER	MONTH/YEAR OF USE	TOTAL NUMBER OF APPLICATIONS

1. Complete Columns A, B, C, and D for All Users
2. Complete Column E by using one of the following codes:
Code 10 - Structural Pest Control.......... includes any pest control work performed within or on buildings and other structures.
Code 30 - Landscape Maintenance Pest Control.......... includes any pest control work performed on landscape plantings around residences or other buildings, golf courses, parks, cemeteries, etc.
Code 40 - Right-of-Way Pest Control.......... includes any pest control work performed along roadsides, power lines, median strips, ditch banks, and similar sites.
Code 50 - Public Health Pest Control.......... includes any pest control work performed by or under contract with State or local public health or vector control agencies.
Code 80 - Vertebrate Pest Control.......... includes any vertebrate pest control work performed by public agencies or work under the supervision of the State or county agricultural commissioner.
Code 91 - Commodity Fumigation (Nonfood/Nonfeed)..... includes fumigation of nonfood/nonfeed commodities such as pallets, dunnage, furniture, burlap bags, etc.
Code 100 - Regulatory Pest Control.......... includes any pest control work performed by public employees or contractors in the control of regulated pests.
3. Complete Columns F and G, if use does not fit one of the above codes

A	B	C	D	E	F	G
MANUFACTURER AND NAME OF PRODUCT APPLIED	EPA/CALIFORNIA REGISTRATION NUMBER FROM LABEL INCLUDE ALPHA CODE	TOTAL PRODUCT USED (Check One Unit of Measure)	NUMBER OF APPLICATIONS	CODE	COMMODITY OR SITE TREATED	ACRES/UNITS TREATED
		□ LB □ OZ □ PT □ QT □ GA				
		□ LB □ OZ □ PT □ QT □ GA				
		□ LB □ OZ □ PT □ QT □ GA				
		□ LB □ OZ □ PT □ QT □ GA				
		□ LB □ OZ □ PT □ QT □ GA				
		□ LB □ OZ □ PT □ QT □ GA				
		□ LB □ OZ □ PT □ QT □ GA				
		□ LB □ OZ □ PT □ QT □ GA				
		□ LB □ OZ □ PT □ QT □ GA				
		□ LB □ OZ □ PT □ QT □ GA				

REPORT PREPARED BY ______________________ DATE ____________

training content. Additionally, because of education and licensing requirements, PCAs are considered to be qualified trainers for pesticide handlers, fieldworkers, and others who work around pesticides, and this is a service many PCAs offer.

Calibration

When a pesticide application is recommended, it is critical that the application equipment be calibrated correctly to deliver the recommended rate (Figure 1-6). PCAs often assist in the *calibration* of equipment because incorrect or inaccurate delivery of a material may cause plant injury or inadequate control, or lead to environmental contamination.

Fertilizer Use and Other Crop Production Advice

Fertilizer recommendations are a service many PCAs offer. The formulation of the fertilizer depends on complex relationships among the plant's nutritional needs, soil type, available water, and other environmental factors. Fertilizer applied at the correct time and in the correct amounts can aid overall plant health; an incorrect application can injure or kill plants. Pest management considerations may also influence fertilizer recommendations. For instance, certain weed species exhibit rapid growth after fertilizer applications, and sucking insects such as aphids and psyllids are attracted to lush, new growth promoted by nitrogen application in landscaped areas.

Cultural and mechanical practices are part of an IPM approach. PCAs often make recommendations regarding tillage operations, irrigation practices, and crop rotation schedules. Likewise, PCAs consider the impact of past and future crops and areas adjacent to the site. Because PCAs must be knowledgeable about all aspects of crop or landscape management, they are often consulted about decisions that may be only indirectly related to pest management. Some PCAs also become certified as a Certified Crop Adviser (CCA) by the American Society of

FIGURE 1-6

PCAs often assist in the calibration of equipment to help make sure that the proper amount of pesticide is applied.

Agronomy to demonstrate their expertise in other aspects of crop production.

THE CHALLENGES OF PEST MANAGEMENT

Informed decision making requires constant vigilance and attention to detail. PCAs must understand the interrelationships among environmental conditions, cultural practices, crop biology, and pests and beneficials. Pest problems do not arise in a vacuum; they occur because a combination of factors in the environment favors the growth of the pest population.

Likewise, pest management actions should not arise out of a vacuum. Pest management decisions must be based on systematic, timely information using the latest techniques to give the most up-to-date protection possible. Consultation with local university and extension experts, growers, and fellow PCAs is necessary to make sound decisions.

PCAs must stay abreast of the latest developments in pest management and employ the techniques applicable to each field situation. PCAs must also, in their decision making, keep their eyes and ears tuned to the regulatory requirements as they affect pest management. The public's perception of pesticides, the various pest control strategies, and the regulatory impetus toward reduced-risk pesticide use are issues that must be factored into every pest management decision.

Education

To obtain their licenses, California PCAs must meet certain educational requirements as stated by law and listed on the DPR website, www.cdpr.ca.gov. The interdisciplinary emphasis of IPM requires a well-rounded background in plant biology, crop production or animal science, entomology, plant pathology, weed science, soil and water science, *ecology*, economics, and business. PCAs must understand the marketing of the crop or the use of the resource. Good record-keeping and communication skills are necessary to keep the client as well as the public informed.

The PCA must be an educator as well as educated. The day-to-day work of examining fields, parks, urban landscapes, rights-of-way, forests, or other managed systems and discussing problems with clients and recommending pest control measures is best accomplished by PCAs who are familiar with all aspects of crop and plant protection and know when to call on a specialist for advice. Keeping growers up to date, through discussion, reports, and literature, educates and reassures them that they have access to the best pest management strategies available.

Research

Keeping up with progress in research should be a priority for every PCA. Advances in the development of pest-resistant crops, *transgenic* crops, pest-detection tools, manipulation of cultural practices, monitoring practices, treatment thresholds, and *biological control* agents, as well as selective, rapidly degradable, and environmentally safe pesticides, are continually changing and improving IPM programs. Knowledge of new research assists PCAs in making up-to-date decisions and providing the best service possible (Figure 1-7).

PCAs should take every opportunity to play an active role in pest management research and its implementation. At the same time, PCAs can keep up to date by reading technical journals and attending educational seminars. As new technologies develop, PCAs are often in a unique position to evaluate their effectiveness and their economic and environmental benefits. PCAs offer advice on how new tools and techniques are adapted for best use in the field. They also cooperate with researchers in field trials. The practical experience of PCAs can help direct research to areas of the most concern in agriculture and the landscape.

Technology

Technological advances are constantly improving pest management practices. New products are available every season. Every PCA must make a special effort to keep up with the new products that are available. Often there may be little university research to guide PCAs on the effectiveness of these products, especially in local areas. Manufacturers may provide research data, but PCAs must be able to interpret these research results for their own situation.

FIGURE 1-7

The science of pest management is constantly changing. To make up-to-date decisions, PCAs must stay abreast of recent developments in research.

Changing technologies may include new formulations of pesticides, new application equipment, the use of biological control agents or mating disruption, new devices for monitoring pests or weather, or new computer programs. A PCA should use reliable methods for comparing new products before adopting or recommending them. Methods for conducting field trials can be found in Chapter 7.

Regulation

Pesticide regulation in California is a three-tiered system—the federal and state governments make and adopt laws, regulations, and policies, and county agricultural commissioners are involved in enforcement and may adopt supplemental regulations for enforcement in their counties (Figure 1-8). State laws regulating pesticide use and pest control are part of the California Food and Agricultural Code. These laws are made by the California legislature in response to needs arising in the state or from federal requirements. Federal pesticide laws are contained in the Federal Insecticide, Fungicide, and Rodenticide Act (FIFRA). Any pesticide used in California must first be registered by both the U.S. Environmental Protection Agency (EPA) and DPR.

Regulations are the working rules developed by the responsible agency to interpret and carry out laws. At the national level, the U.S. EPA is the primary agency regulating pesticides. At the state level, regulations pertaining to pest control and pesticide use are under the jurisdiction of DPR. Several other state agencies as well as federal agencies—the U.S. EPA and the U.S. Food and Drug Administration—are involved in cooperative enforcement with DPR. The county agricultural commissioners, under the direction of DPR, enforce most of the regulations governing pesticide use and worker safety on a day-to-day basis. Agricultural commissioners develop pesticide use policies or conditions specific to the needs of their counties. Counties and local governments may also

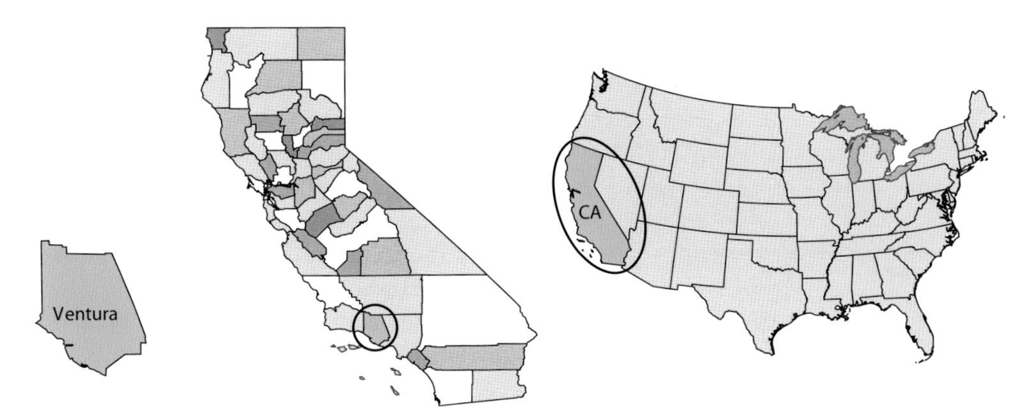

COUNTY AGRICULTURAL COMMISSIONER'S OFFICE
County agricultural commissioners enforce laws in 58 counties. They may develop local policies on pesticide use.

CALIFORNIA DEPARTMENT OF PESTICIDE REGULATION
California has its own laws to make pesticide use safer in our area. These laws are part of the California Food and Agricultural Code.

U.S. ENVIRONMENTAL PROTECTION AGENCY
Pesticide applicators in all states must obey the Federal Insecticide, Fungicide and Rodenticide Act (FIFRA).

FIGURE 1-8

Federal, state, and county governments are involved in the regulation of pesticides.

create laws and ordinances governing the use and storage of pesticides.

Laws and regulations have a direct impact on pest management practices. Regulations affect control methods, pest *eradication* and plant *quarantine* efforts, pest control or abatement districts, the pesticide registration process, and the licensing and training of PCAs, applicators, and dealers. For up-to-date information on laws and regulations, refer to the DPR *Laws and Regulations Study Guide*. Once you are licensed, DPR will send you regulatory updates each time you renew.

Laws and regulations frequently change or are amended in response to new situations. For instance, identification of new needs regarding worker safety, *groundwater* protection, endangered species protection, and restricted-use materials have resulted in new regulations over the last decade. Keep up to date by attending continuing education programs and reviewing DPR publications, news releases, or the DPR website for changes in laws and regulations.

In some cases, the regulatory process can be used to control and abate pests. Grower-initiated pest management projects, such as the pink bollworm project and the Beet Curly Top Virus Control Program, are state run but grower funded. The state monitors for the presence of the pest and mandates specific actions to the growers. State law has also facilitated the formation of grower-run pest management districts, such as several county Citrus Pest Control Districts working to detect and eradicate tristeza virus. The state collects funds and ensures that the district complies with the laws.

Regulatory actions can be used to ensure the judicious use of pesticides. For instance, by changing water-holding requirements, the flow of a rice *herbicide* into the Sacramento River was reduced significantly. Off-site movement and decreased risk to nontarget organisms were accomplished without eliminating the use of an effective tool.

Personal Liability

PCAs assume legal responsibility for the recommendations they make. By virtue of being licensed and providing a service on which growers rely, PCAs are considered to be experts in pest management. When pest control practices fail and pests damage crops or when pesticides or other pest control practices injure crops or property, the PCA may be subject to damage or liability claims in court. PCAs must be aware of their responsibility to follow practices that limit risks to people and the environment. PCAs must keep all parties well informed about the risks and benefits of suggested practices. Record keeping and documentation of written communications are necessary to limit liability.

Public Perception

The public has become increasingly concerned about pesticides in food, water, and the environment. Public perception presents many challenges for the PCA. Yet, the general public's fear of pests and demand for blemish-free produce and aesthetically pleasing landscapes often lead to increased pesticide use. Consumers should be presented with a balanced and unbiased view of pest management options and considerations. PCAs should also take a leading role in educating the public about integrated pest management. Great strides can be made toward reassuring the public about efforts to reduce pesticide risk by demonstrating strategies that manage pests with lower pesticide inputs without sacrificing quality, yield, or consumer costs.

2 Ecological Principles as They Apply to Pest Management

Ecology is the study of the interrelationships among organisms and their surrounding environment. The *community* of organisms in an area and their nonliving environment function together as an ecosystem. Organisms that become pests are often integral components of managed ecosystems. They are defined as pests because they interfere with people's management of a resource, whether it is a crop or animal grown for food, a forest, or an ornamental plant in the landscape. When pest management actions are taken against a pest, those actions can have an impact on the entire ecosystem. The interactions among the pest, its natural enemies, a plant *host*, other organisms, and the environment must be understood before making a pest management decision. Understanding the ecological factors that allow pests to adapt and thrive in a particular ecosystem will help identify weak links in the pests' life cycle and factors that can be manipulated to manage them.

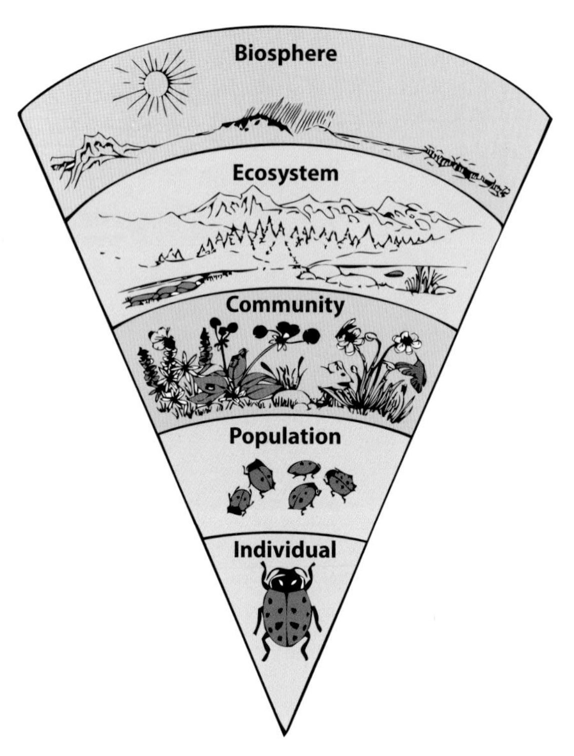

FIGURE 2-1

The levels of ecological organization in an ecosystem. The individual in this case, the convergent lady beetle, is not a pest but rather an important predator of several insect pests. Source: Flint and van den Bosch 1981.

LEVELS OF ECOLOGICAL ORGANIZATION

The first level of organization in an ecosystem is the individual. Locally interacting, interbreeding groups of individuals of the same species form populations. Populations of interacting plants, animals, and microorganisms form communities. Communities of plants and animals interacting in a physically defined area form ecosystems (Figure 2-1).

The Individual

To understand an ecosystem, it is necessary to recognize the role of the individual. An individual's response to members of the same species, other organisms, environmental conditions, and resource availability underlies the more complex interactions that occur at the population and community levels. Environmental and ecological forces that favor survival and reproduction of certain individuals relative to others shape population characteristics. If some individuals in a population carry *genes* that allow them to better tolerate a new, otherwise lethal set of conditions, then those individuals are more likely to survive and reproduce. This process, whereby some individuals carrying alternative inherited traits survive and reproduce under stressful conditions, is called *natural selection*. Over succeeding generations, genetic traits from the adapted individuals are passed to their *progeny*, changing the genetic composition of the population and leading to a population with an enhanced ability to survive the new set of conditions. This type of evolution has allowed many pest species to develop strains that are resistant to pesticides. Although individuals share many morphological, physiological, genetic, and behavioral characteristics, no two individuals are exactly the same. Variation among individuals can be attributed to heredity, the environment, or a combination of the two. The set of genes an organism carries is referred to as its *genotype*. Variations in genotype may produce individuals with differences in appearance, physiology, or behavior (e.g., foraging strategy).

The interaction between the genotype and the environment produces an individual's observable or expressed characteristics, which comprise its *phenotype*. Under different environmental conditions, such as growing in full sun versus growing in the shade, a single genotype may give rise to an array of phenotypes. Many perennial weeds provide good examples of phenotypic expression. For example, if local populations of plants with identical genotypes (also called clones) are exposed to different levels of light, nutrition, and moisture, they will look different after several months of growth. Although identical genetically, the plants are likely to grow larger or smaller and perhaps produce different-shaped leaves or varying numbers of flowers or fruit, depending on environmental conditions (Figure 2-2).

The Habitat and Ecological Niche. The environment in which an individual or species population lives is its *habitat*. Habitat includes the biological and *physical environment* that surrounds an individual organism or population, including the population of other organisms and the *abiotic* environment. For example, a mosquito's habitat must include water but also includes the surrounding plants and organisms, which provide food, shelter, cover or *prey* for predators, and competitors for mosquitoes (Figure 2-3). The place occupied by an entire community of individual organisms is also referred to as the community's habitat. The natural communities of California—such as chaparral, coastal scrub, and mixed evergreen forest—are examples of diverse habitats that have specialized groups of species associated with them.

Ecological niche refers to all the components of the habitat with which an organism or population interacts. An individual's ecological niche includes the role it plays in the community as well as the combination of conditions and resources that allow a viable population to exist. A description

FIGURE 2-2

The same species of weed may display different phenotypes when growing under varying conditions of shade, temperature, moisture, and nutrients. The three plants shown here are all mature annual sowthistle, Sonchus oleraceus, *growing in different habitats: from left, a dry, hot sunny tomato field; a shady, well-watered citrus orchard; and a backyard situation.*

FIGURE 2-3

Standing water provides suitable habitat for mosquitoes. The mosquitoes' habitat also includes the surrounding plants and organisms.

of an individual's niche would include its sources of energy, its nutrition, how it affects other organisms (e.g., as a predator, prey, or competitor), and the extent to which it is capable of modifying the ecosystem.

For the pest manager, habitat provides information on where to find a pest organism. The ecological niche may reveal insights into the pest's role in the community. Some weed species, for instance, occupy specific niches in an area. Nutsedges thrive in sunny areas with waterlogged soil; their presence often indicates a low area in a field, overwatering, or a break in the irrigation system. Nutsedges often outcompete other weeds or crop plants in these areas.

A given species can occupy different niches in different regions. Likewise, different life stages of some organisms may occupy different niches. For instance, the larvae of Lepidoptera usually feed and live on a few plants in a limited area, while the adult moth or butterfly may not feed or may feed only on nectar and may travel relatively great distances.

The ecological niche of an insect pest may also include various plant associations. Some insect pests may move to different plants, such as weed hosts, or switch to other foods during certain times of the year. Some pathogens, such as the organism that causes stem rust of wheat, require two different or *alternate hosts* (wheat and barberry) to complete their life cycle.

Populations

A population is a group of individuals of the same species that occupies a distinct space and possesses characteristics that are unique to the group. The genetic makeup of the population together with local environmental and ecological factors determine the population's success (how well it survives and how fast it grows). Characteristics of local populations can evolve through natural selection. For instance, certain populations of many invertebrates and some plant pathogens and weeds have developed resistance to commonly used pesticides. Once resistance has developed, these populations can no longer be controlled with those pesticides and alternative management strategies must be developed. Chapter 5 provides details on managing pesticide-resistant pest populations.

Population Density. The number of individuals of a species in a defined area is a measure of its *population density*. Population density is variable; that is, population size increases and decreases over time. There are factors, however, that usually operate in an ecosystem to keep populations within certain boundaries. For instance, as population density increases, there is likely to be increased competition for limited resources—such as nutrients, water, and sunlight in the case of plants or food, water, and shelter for many animals—until the population ceases to grow, declines, or migrates. Conversely, as population density decreases, necessary resources may again become abundant, allowing the population to recover. Many other factors, such as predators, disease, or climate, can impact population density.

Age Distribution. Populations consist of individuals of different ages. The proportion of individuals in each age group defines a population's *age distribution*. Birth and death rates are important factors determining age distribution. *Immigration* and *emigration* can also be sources of population increase or decline.

The proportion of individuals of different ages in the population can indicate whether the population is expanding, declining, or remaining stable. Expanding populations characteristically have a large percentage of young individuals, declining populations have a large percentage of old individuals, and stable populations have a relatively even distribution among age groups. For example, the population of a vertebrate pest, such as meadow voles, can expand rapidly in a new environment due to high birth rates but then stabilize over time as population density rises and *mortality* increases among its young due to factors such as predation or food shortages (Figure 2-4). Population fluctuations, however, are normal, and population stability can be rare, as in vole populations in agricultural situations, where management is frequently necessary.

Many of the organisms in managed systems are short lived, with life cycles that are well synchronized with the culture of the crop. Often the entire population begins the season at the same stage—for example, egg, overwintered larva, seed, vegetative offshoot, or *spore*. For organisms that live only one season and have just one generation per year, most individuals are about the same age at the same time during the season. Likewise, many annual weed species in a field will germinate at about the same time (winter or summer *annuals*) when first irrigated, resulting in all newly germinated individuals being approximately the same age. Even organisms with several generations per year, like codling moth, can maintain brood or *cohort* effects where most individuals in the population are of similar age at any one time (Figure 2-5). These generational broods can be important in timing management practices. Very rapidly reproducing species, like aphids, which have many generations per season, tend to show fewer brood effects and have individuals at different ages present at any time. In a population of perennial weed species or vertebrates, several overlapping generations are also likely to be present.

Population Growth and Regulation. Population growth occurs when birth rates exceed death rates or immigration exceeds emigration. Population size may be affected by physical factors such as weather, water availability, and nutrient availability. Populations may also be biologically controlled by factors such as predators, competitors, disease, or food availability.

Factors that affect population density can be categorized as density dependent or den-

FIGURE 2-4

Age pyramids for vole populations. In an unlimiting environment, the population expands at an exponential rate, as depicted by the pyramid on the left. But when birth rates and death rates are equal, the age distribution in the population becomes more stable, as on the right side of the graph.

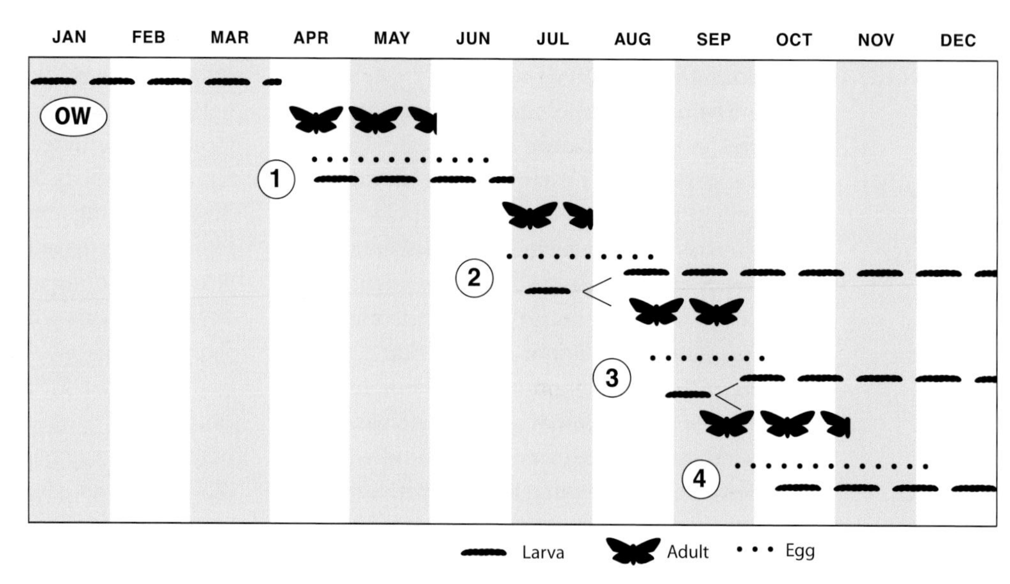

FIGURE 2-5

Generalized description of codling moth seasonal development. With the emergence of the overwintering population (OW), up to four generations (or broods) are possible. Moth flights and developmental times vary with temperature and location, but adult moths of each generation tend to appear at about the same time at a specific location.

sity independent. *Density-independent factors* affect populations regardless of population density and include disturbances such as floods, drought, fire, other unpredictable environmental conditions, and most pest control actions. In each case, a similar percentage of the population is killed regardless of whether the density is high or low.

Density-dependent factors have a different effect when populations are high compared with when they are low. Competition for resources, and predation, parasitism, and disease are examples of factors that can limit population growth in a density-dependent way. For example, areas of high prey density attract certain predators and allow them to more easily find prey, so that a higher portion of the population is killed by predation than when numbers are low and prey are harder to find. Likewise, disease-causing pathogens spread more readily in areas of high population density, increasing the mortality rate. Density-dependent factors are important in regulating populations, and they help keep populations at equilibrium as opposed to the often-catastrophic effects of density-independent factors (Figure 2-6).

Sometimes density-dependent and density-independent factors work together to help control a pest. For example, when hot, dry temperatures occur when citrus red mite populations are high, virus *epidemics* often decimate the mite populations. Similarly, fungal diseases of aphids are favored by warm and humid conditions but move through populations and exert their greatest impact when aphids are abundant.

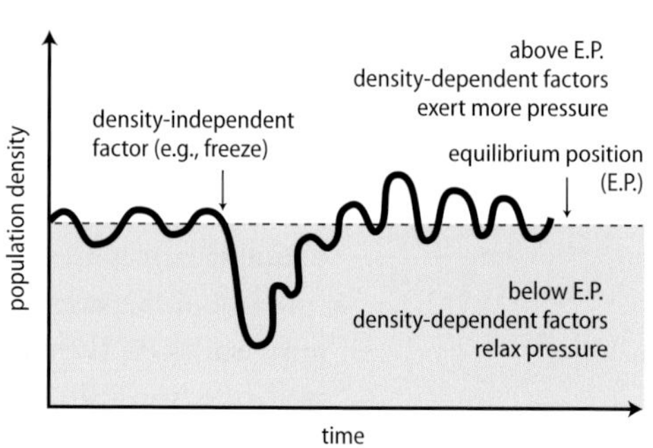

FIGURE 2-6

The effect of density-dependent and density-independent regulating factors on the equilibrium position of a population. When populations are high, density-dependent (E.P.) factors exert more pressure helping to keep the population at equilibrium. Density-independent factors, however, can cause sudden, drastic changes in population as noted in the sudden decline in population as the result of a freeze. Source: *Flint and van den Bosch 1981.*

Dispersal. The movement of individuals or their offspring (e.g., seeds, spores, or larvae) into or out of an area is called *dispersal*. Dispersal allows individuals to colonize new areas. Along with birth rates and death rates, dispersal regulates populations. Dispersal also plays an important role in the evolution of populations because it allows the mixing of genes between populations.

Dispersal is accomplished through immigration, emigration, or *migration*.

Immigration is movement into a population. When populations are at low densities, immigration accelerates population growth and in extreme cases prevents extinction. Emigration is movement out of the population. When populations become extremely abundant in a given area, emigration reduces population density and often helps in reestablishing an equilibrium. Migration is the frequent movement into and out of a population area. Migration may involve the mass movement of populations. Occupation of areas that are sometimes unfavorable is made possible by seasonal and diurnal migration. Birds flying south for the winter is an example of seasonal migration. Many pests migrate as well. For instance, grasshoppers and lygus bugs migrate from uncultivated foothill areas to crop fields in the late spring or early summer.

Ecotypes. An ecotype is a locally adapted population of a species that allows it to live in a specific habitat. Ecotypes are genetically determined and evolve through natural selection. For instance, a hot-weather ecotype of a species may evolve in an area characterized by high temperatures that other populations of the same species may not tolerate. Among plant pathogens, pathotypes could be considered as ecotypes. Pathotypes are strains of a plant-pathogenic fungus adapted to particular cultivars of a host plant. New ecotypes of pest species may allow the species to expand its range and become a more serious threat to certain managed systems. For instance, purple nutsedge, *Cyperus rotundus*, has been traditionally suppressed in many areas of the world by flooding fields for lengthy periods, often in rice rotations; however, recently, new ecotypes of purple nutsedge have evolved that tolerate flooding and have even become a major problem in flooded rice fields in southeast Asia.

Community

A community consists of all the populations of plants, animals, and microorganisms that share the same habitat and interact directly or indirectly with one another. Communities can be of any size and are often defined by the environment in which they occur or by the dominant species in the community. Not all organisms in a community have an equal effect on the community. Generally, a relatively few species, such as the large species of trees in a forest, exert a major controlling influence on the entire community.

Interactions among populations of different species in a community can be complex and may be sensitive to disturbances or fluctuations. Efforts to control one component of the community can have a positive or negative impact on other organisms in the community. For instance, the application of broad-spectrum insecticides may eliminate or reduce one species, such as a pest. However, as a consequence of this application, many other species in the community may be affected, either directly by the toxicity of the pesticide or indirectly because the host, prey, or competition has been eliminated.

Species Richness and Diversity. Species richness is a measure of the number of different species found in a community. Species diversity is a measure of the num-

FIGURE 2-7

Ecosystems are not isolated from one another; they blend into one another through a transitional zone called an ecotone.

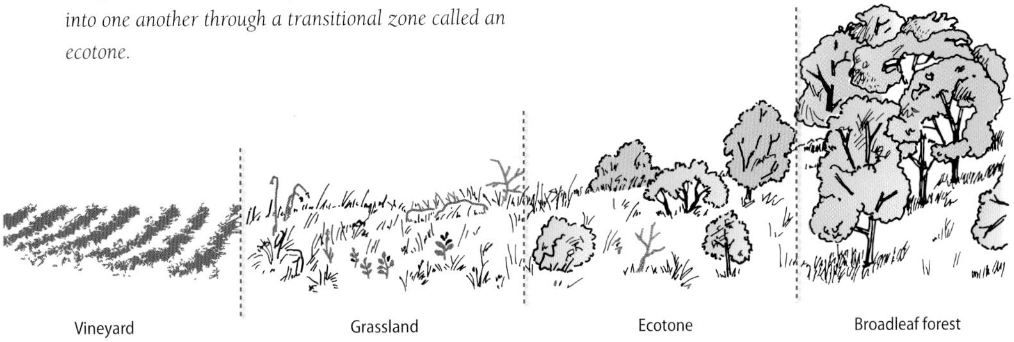

ber of species in an area and their relative abundance. Species diversity is often higher in biologically controlled (natural) ecosystems and lower in physically controlled (managed) ecosystems. Large increases in population size of individual species and more frequent fluctuations between high and low population densities may be more likely in communities with relatively low diversity.

Ecotones and the Edge Effect. Communities do not have impermeable barriers or simple boundaries constructed around them. A transitional zone called an *ecotone* exists between communities (Figure 2-7). Ecotones support organisms from each community and often contain habitats and organisms not found in either community. The tendency for the variety or number of species (species richness) to increase in an ecotone is sometimes called the *edge effect*. The nature and intensity of the edge effect is influenced by the complexity of the communities that overlap, by the differing environmental conditions at the overlap, and the size of the overlap. Destruction of nearby alternate hosts for pests, planting or treating trap plants around the edge of a field, or installation of hedgerows around the edges of fields to provide habitat for natural enemies are three examples of managing ecotones to improve pest management.

Ecosystem

An ecosystem consists of a community of interacting and interdependent populations of plants, animals, and microorganisms and their physical environment. The biotic community and the nonliving environment function as a unit that is essentially self-sustaining, assuming that the essential inputs and the basic structure remain relatively stable. Principally the physical environment determines the type and number of organisms that can successfully survive and reproduce.

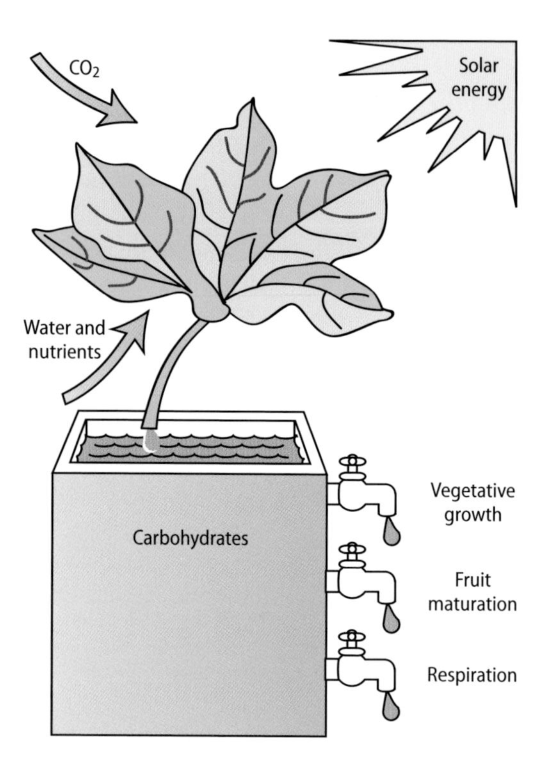

FIGURE 2-8

Carbohydrates produced from photosynthesis form an energy pool that supports all plant functions. As demands drain energy, vegetative growth is the first function to shut off; fruit maturation is next, followed by respiration.

THE ECOSYSTEM CONCEPT

The concept of communities of organisms interacting with their physical environment is the basis of ecology. Organisms interact and depend on other organisms, the biotic components of the ecosystem, as well as the abiotic (nonliving) components such as dead matter, minerals, water, and energy.

Energy flow drives the ecosystem, determining limits to the food supply and the production of all biological resources. Light energy from the sun is captured by green plants and converted to chemical energy. Energy is stored in plants as carbohydrates and used by the plant to support all functions (Figure 2-8). Other organisms use and convert this chemical energy to various forms through food chains. At each step, some of the chemical energy is assimilated and used by the organism and the rest is released in respiration and waste products. The goal of crop production is to maximize

FIGURE 2-9

In photosynthesis, plants use energy from the sun to convert carbon dioxide from the air and water from the soil into carbohydrates (sugar and starches), their primary food source. In the process, oxygen is released into the air.

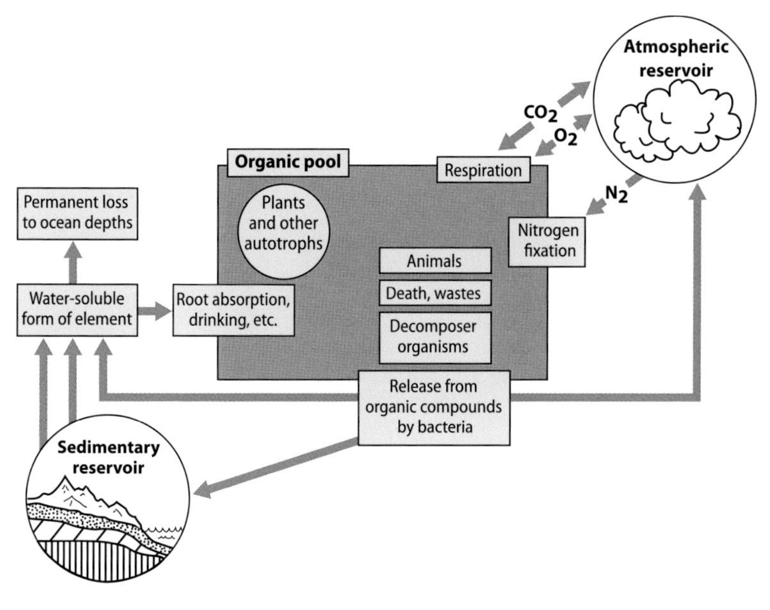

FIGURE 2-10

In a typical biogeochemical cycle, the minerals or inorganic elements required for the growth and development of living organisms circulate from the nonliving to the living and back to the nonliving components of the ecosystem. Source: *Flint and van den Bosch 1981.*

ecosystem energy into a harvestable product; use of plant energy by pests and competition by weed pests are undesirable as both take away from crop production.

Photosynthesis

The process by which green plants use energy from the sun to convert carbon dioxide and water into carbohydrates is called *photosynthesis* (Figure 2-9). Nearly all plant growth depends on the capture of solar energy, which occurs in the green plant parts containing chlorophyll. Chlorophyll molecules capture sunlight and convert it into chemical energy. The importance of photosynthesis to an ecosystem is paramount because, as primary producers, plants are the principal source of energy and organic material for the ecosystem. A reduction in the photosynthetic capacity of the ecosystem will limit the survival of other organisms.

Abiotic Components

Abiotic, or nonliving, factors in ecosystems influence the growth, development, and ability of organisms to survive. They also influence the interactions between organisms. The abiotic components that most commonly affect organisms in managed situations include minerals, soil, water, temperature, climate, light, and gases.

Biogeochemical Cycles. Many mineral or inorganic elements are required for the growth and development of living organisms. These inorganic elements and compounds, which include carbon, hydrogen, oxygen, and nitrogen, circulate through the ecosystem from the nonliving to the living and back to the nonliving parts of the biosphere. This circular path is known as a biogeochemical cycle (Figure 2-10). During this process, some nutrients are returned immediately to the environment as quickly as they are removed. Other nutrients are stored in the tissues of plants and animals, while still others are chemically bound or stored in the soil surface for a long time before becoming available to living organisms. A slow and steady exchange takes place between the

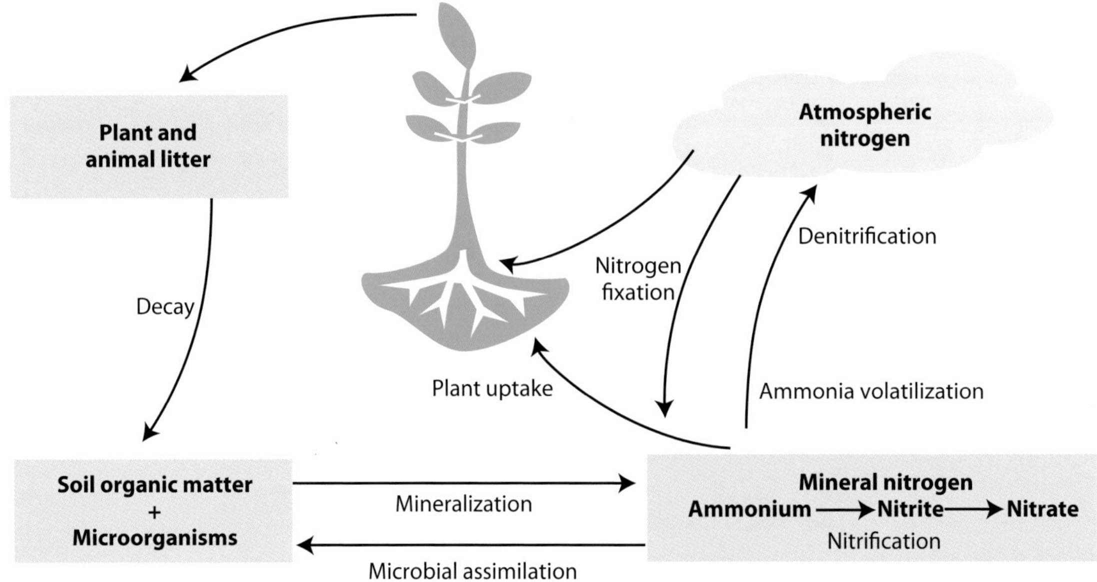

FIGURE 2-11

A simplified nitrogen cycle. Nitrogen circulates from a reservoir in the atmosphere through microorganisms, plants, and animals and into soils through leaching, decomposition of organic matter, and mineralization. Nitrogen returns to the atmosphere through volatilization and denitrification by bacteria.

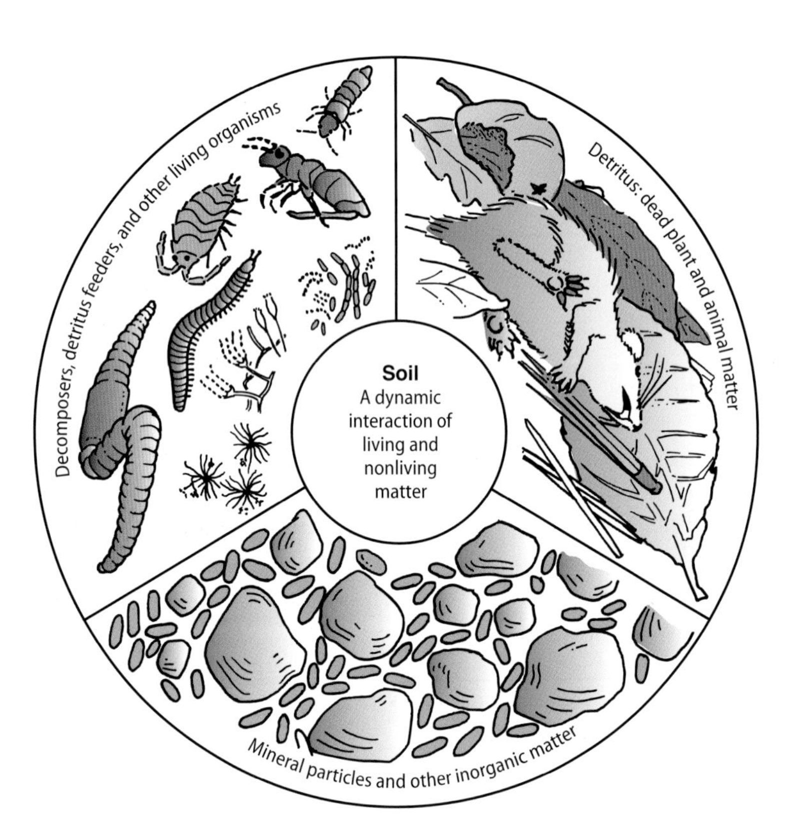

FIGURE 2-12

A productive soil is composed of a mixture of organic matter, inorganic minerals, soil organisms, microorganisms, gases, and moisture. Altering the mix of any of these components changes the soil composition.

easily available nutrients and the relatively unavailable nutrients.

In a typical biogeochemical cycle, plants take up nutrients from the soil and organize them into biologically useful compounds. Many other organisms take the nutrients into their bodies by feeding on plants or herbivores. Organisms of decomposition, such as some *fungi*, return the nutrients to their elemental state. Air and water move the nutrients between the organic and inorganic components of the ecosystem. Each of these factors is essential for the cyclic flow of nutrients to occur. There are two basic types of biogeochemical cycles: gaseous and sedimentary. In gaseous cycles, such as the nitrogen cycle (Figure 2-11), the atmosphere acts as the nutrient reservoir. In sedimentary cycles, such as the phosphorus cycle, the earth is the reservoir.

A variety of biogeochemical cycles exist for the cycling of nutrients such as nitrogen, hydrogen, phosphorus, carbon, and sulfur. Trace elements or micronutrients also enter biogeochemical cycles, but plants and animals require them in much lower quantities. These elements include iron, copper, boron, zinc, iodine, silicon, and nickel.

Soil. Soil is a mixture of inorganic minerals, *organic matter*, soil organisms and microorganisms, gases, and water (Figure 2-12).

TABLE 2-1.

Relationship between soil texture and physical and chemical properties.

	Physical and chemical properties					
Soil type	Particle size (mm)	Water infiltration	Water-holding capacity	Nutrient- holding capacity	Aeration	Workability
sand	2–0.05	good	poor	poor	good	good
silt	0.05–0.002	medium	medium	medium	medium	medium
clay	0.002 and less	poor	good	good	poor	poor
loam	mix of sizes	medium	medium	medium	medium	medium

It is the planting medium, the reservoir for nutrients, the site of decomposition, and the area where nutrients are recycled. Soil is the foundation upon which all terrestrial life depends. Soils are important in determining the species composition of an ecosystem. In turn, the ecosystem's inhabitants help determine the makeup of the soil.

Important attributes of soil are texture, organic matter, and exchange capacity. The proportion of different-sized soil particles, which are classified as sand (large), silt (medium), and clay (small), determine texture. For instance, loam is a mixture of different-sized soil particles (such as 35% sand, 45% silt, and 20% clay). Soil texture directly affects water-holding capacity, infiltration, nutrient-holding capacity, aeration, and workability. Loam soils generally provide the best balance for aeration and water-holding capacity. The relationships between soil texture and physical and chemical properties are summarized in Table 2-1.

The organic component of the soil is made up of a complex of living soil organisms that include numerous species of *bacteria*, fungi, protozoans, algae, nematodes, mites, insects, millipedes, spiders, centipedes, earthworms, snails, slugs, moles, and other burrowing animals, plus organic matter such as decaying plant, microbial, and animal matter (detritus). Certain constituents of organic matter are converted into a substance called humus. Humus, to a large extent, is made up of microbial products and the residue of decaying organic matter found mainly in the topsoil. Although organic matter generally makes up only a small percentage of total soil mass (typically 0.1–1% in conventional farms and 1–2% in organic farms), it plays an important part in soil structure and fertility.

Soil cation exchange capacity is a measure of the soil's electrical properties. Clay particles and humus particles are negatively charged (anions). Most minerals in fertilizer products such as ammonium (NH_4^+), hydrogen (H^+), calcium (Ca^{++}), magnesium (Mg^{++}), potassium (K^+), and sodium (Na^+) are positively charged (cations). Without the negatively charged clay and humus particles, the soil would not have the ability to retain and store the positively charged nutrient cations.

The capacity of soil to bind and hold nutrients or other positively charged chemicals so that they are not washed away through *leaching* is important when using fertilizers and pesticides. The propensity for chemicals to be removed through leaching contributes to a loss of soil fertility and is a major factor in determining the potential for groundwater contamination, or pesticide or fertilizer *runoff*. This is one reason why there are application restrictions for certain very leachable pesticides such as triazine herbicides on sandy soils. Sandy soils have low levels of organic matter and a low cation exchange capacity, allowing triazine to move easily through the soil.

Water. Water plays a fundamental role in all ecosystems. It is a major factor in

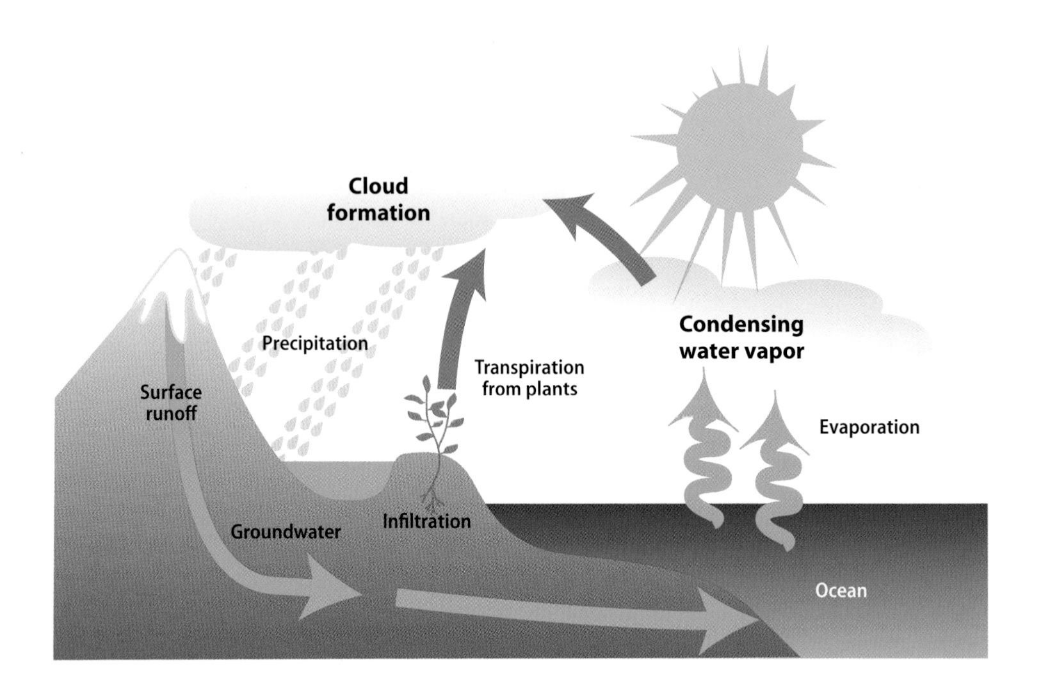

FIGURE 2-13

Water enters the ecosystem primarily from rainfall; it is held in the soil, drains into the groundwater, or runs off into surface water. It is recycled back into the atmosphere mainly through evaporation and transpiration.

weathering, leaching, and erosion, and it controls to a great extent the introduction and loss of nutrients from the ecosystem. Water is continuously recycled throughout the ecosystem, entering the atmosphere through evaporation and returning to the earth through condensation and precipitation (Figure 2-13).

The primary means by which water enters many ecosystems is from rainfall. For some ecosystems, water enters primarily through irrigation or runoff from snow. The amount of rainfall (or water) in an area determines the number and size of organisms that inhabit a particular ecosystem. Some organisms, for example, are able to adapt to conditions of drought, while others adapt to conditions of moderate to heavy rainfall.

Water loss occurs primarily through evaporation, transpiration, drainage (infiltration) to groundwater, and runoff to streams. Plants and animals in ecosystems react differently to evaporation. Animals often hide to avoid dehydration. Water that enters plants from the soil is lost mostly through transpiration. Transpiration is necessary to supply water to the leaves for photosynthesis, to carry nutrients up from roots and through the plant, and to cool the plant. Transpiration efficiency is an important factor in the growth pattern and survivability of plants. To improve plant growth and crop quality, you can schedule irrigations based on water evaporation from soil surfaces and transpiration from plants. Combined, these water processes are called *evapotranspiration* (see the section "Irrigation and Water Management" in Chapter 5).

Temperature. Temperature is important in determining which organisms will inhabit an ecosystem. The developmental rate of plants, microorganisms, and cold-blooded animals is directly related to the temperature of their environments. Where temperatures are warmer rather than cooler, more generations per year of certain invertebrates can be expected. Plants grow larger faster under warm conditions if moisture is adequate.

However, where it is extremely cold or hot, most organisms have had to become specifically adapted for survival. Vertebrates, for example, seek shade when it is hot, or they possess physical characteristics that protect them, such as insulating layers of fur, feathers, or fat. Some animals adapt to temperature extremes through physiological adaptations, such as diapause (a state of arrested development), hibernation (winter dormancy) or *estivation* (summer dormancy). Some plants have evolved differences in leaf size and shape, leaf and bark reflectivity, the presence of hairs and spines, and orientation of leaves toward the sun to enable them to adapt to light and temperature variations.

Plants and invertebrates require a certain amount of heat to develop from one point in their life cycle to the next. This measure of accumulated heat, known as physiological time, can be approximated in units called *degree-days*. Predictive models use degree-day accumulation to time crop and pest management actions, as discussed in Chapter 6.

Light. Light provides the predominant source of energy in an ecosystem. Without light most organisms could not exist; it is vital to the development of most ecosystems. Different organisms react differently to the intensity, quality, and duration of light. Some animals avoid light and are rarely seen unless we search for them at night or use appropriate monitoring tools such as traps. Others are attracted to lights. Plants and animals react to differences in light intensity or wavelength. Some plants, for instance, are able to grow in the shade; others are sun adapted. Exposure to ultraviolet light can injure some microorganisms. These individual adaptations affect the nature and structure of communities in an ecosystem.

Light is also a regulatory factor in an ecosystem. Organisms may adjust their activities according to day length, or *photoperiod*. The photoperiod can trigger physiological processes that are responsible for the growth and flowering of many plants and spore release in some fungi. In birds and mammals, the photoperiod triggers molting, fat deposition, migration and breeding; along with

FIGURE 2-14

*Legumes such as alfalfa have developed a mutualistic relationship with certain bacteria (*Rhizobium *spp.) that provides the alfalfa with more than enough nitrogen for healthy plant growth. These enlarged alfalfa root nodules contain the* Rhizobium *bacteria.*

temperature, it triggers the onset of *diapause* in some insects.

Climate. *Climate* refers to temperature, humidity, precipitation and other atmospheric conditions over a long period of time and can refer to local, regional, or global conditions. In contrast, weather refers to the state of the atmosphere at a given time and place, for instance, over a period of a day or a week. Climatic variations may also produce differences in wind patterns, ocean currents, and rainfall patterns.

Ecosystems are typically associated with specific climatic patterns and can be found only where those climatic conditions occur. For instance, rainforests, deserts, and tundra are to a great extent determined by the climatic patterns in the area.

Gases. Oxygen and other gases are required for the respiration of all living organisms. The concentration of atmospheric gases can limit growth and development of organisms in the ecosystem. For example,

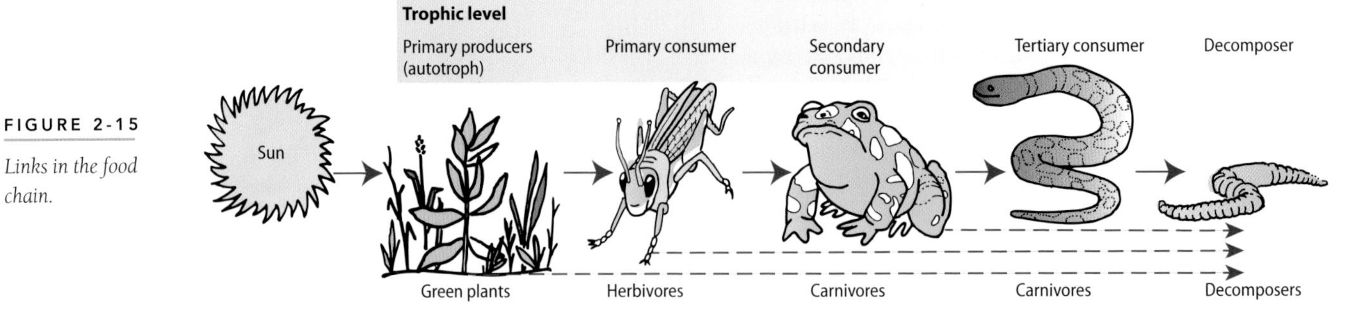

FIGURE 2-15

Links in the food chain.

aerobes are organisms that live only if oxygen is present. As carbon dioxide levels increase deeper in the soil profile, oxygen levels decrease, limiting aerobic activity and resulting in a slowed rate of decomposition. Increases in carbon dioxide can alter competitive interactions. Carbon dioxide is required for the photosynthetic process of green plants, but this gaseous compound is rarely limiting.

Biotic Components

The living organisms found in an ecosystem are called the biotic components. The actions of one species in an ecosystem can affect the population growth and well-being of many other species. Individuals of different species compete with one another for food, light, space, or moisture. One species may depend on another as a food source (e.g., predators or parasites), or two species may mutually aid each other *(mutualism)* (Figure 2-14). Interactions between species can be positive, improving chances for survival and assisting in species adaptation; they may be negative, resulting in the elimination or reduction of the population; or they may have no effect at all.

Trophic Structure

The flow of energy in an ecosystem can be characterized in specific patterns. Energy flows from producers (green plants) to primary consumers that feed on them, to the secondary consumers or predators that in turn consume them, and all the way to the *decomposers* of dead organisms at the bottom of the energy chain. The transfer of energy from one trophic level to the next is called the *trophic structure* of the ecosystem and has an energy cost (loss during each transfer) that typically limits the levels to four or five. The simplest way to analyze the flow of energy through an ecosystem is to construct a food chain (Figure 2-15).

At the base of every food chain is the *autotroph*, or producer (green plants). Plant consumers, or *herbivores*, occupy the second level of the food chain, the primary consumer level. Herbivore consumers, or *carnivores*, occupy the third level of the food chain, the secondary consumer level (carnivores may include predatory or parasitic organisms). Carnivores that eat other carnivores occupy the fourth level of the food chain, the tertiary consumer level. The final level of the food chain consists of decomposers, primarily fungi, bacteria, and invertebrates that break down dead matter into organic substances. Decomposers, in turn, have their own predators and are the basis of a different food chain in the soil.

Many organisms do not fit neatly into a single level of the food chain. The food habits of some consumers can change with the season or during different stages in their life cycle. Some consumers, such as the raccoon, are called *omnivores* because they feed on both plants and animals. Organisms that feed on dead plant or animal matter are called *saprophytes*, but some organisms that commonly feed on living organisms as predators, parasites, or herbivores may also be capable of surviving as saprophytes when other food is scarce; these organisms are called facultative saprophytes.

Food Web. In reality, linear food chains are too simplistic to realistically describe true ecosystems. At every level of a food chain, many organisms may feed on a single organism so that feeding and energy flow form complex *food webs*. Yet all links in the

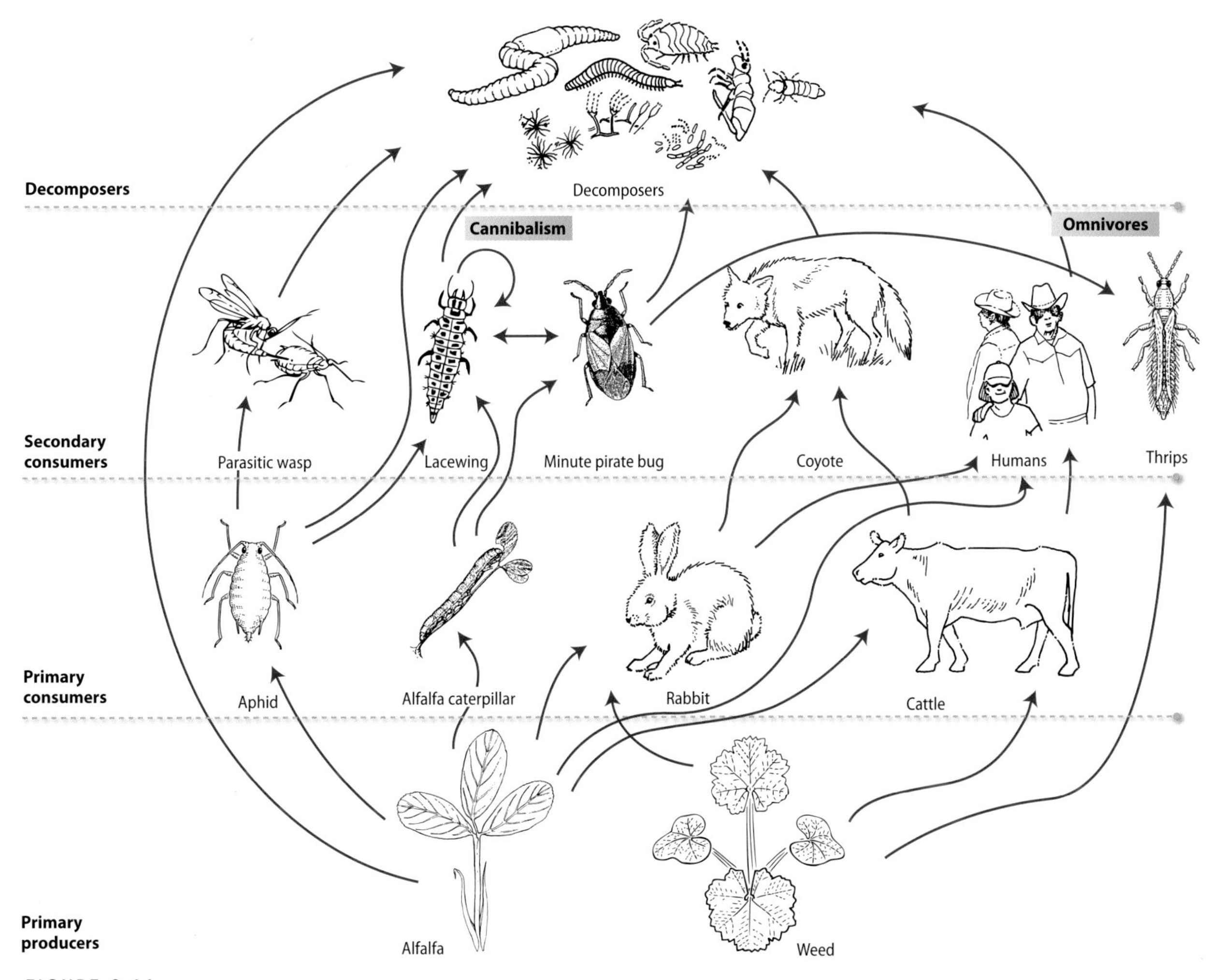

FIGURE 2-16

A schematic diagram of a food web in alfalfa. The web becomes more complex as more populations are introduced into the ecosystem. Each arrow represents a transfer of food or energy from one organism to another. Note that some animals, like omnivores, feed on more than one level and that cannibalism, which is common among some predators, is a kind of secondary consumerism.

food chain lead from producers to primary and secondary consumers. Food webs in highly productive ecosystems are complex and support many species (Figure 2-16). Patterns of complexity in food webs are important to ensure survival. Being part of a complex food web allows consumers to substitute food resources when one source is in short supply.

Competition. Competition results when two organisms vie for the use of the same limited resource, whether it is food, water, space, or light. Competition can occur within the species *(intraspecific)* or between two or more species *(interspecific)*. Intraspecific and interspecific competition are important in determining the number and kinds of organisms that are found in an ecosystem. Intraspecific competition plays an important role in populations that are territorial or self-regulating.

Interspecific competition is most significant in species with similar lifestyles or resource needs. The outcome of interspecific competition is viewed as a driving force in the natural selection of species and is a major factor in species divergence and specialization. Typically, the result of interspecific competition is that individuals of one species reduce the population of another species or force it to occupy a different ecological niche by moving to another space or adapting to a different food or other resource. If one species entirely eliminates another from an area, the phenomenon is referred to as competitive exclusion.

Mutualism. Mutualism occurs when two species develop a positive, reciprocal relationship in which each has a dependency on the other and both populations benefit from the association. Through this relationship, both species strengthen their chances for survival, fitness, or growth. Mutualism is extremely widespread and is beneficial to an ecosystem. The relationship of *mycorrhizae* to plant roots is an example of a mutualistic relationship between fungi and plants. Mycorrhizae are soil-dwelling fungi that can make roots of plants more resistant to infection by certain pathogens (such as *Fusarium*, *Phytophthora*, and *Pythium* spp.), improve plants' ability to absorb nutrients, and may aid in water uptake. Trees such as pines could not establish, survive, and grow without them. In turn, mycorrhizae depend on the nutrients supplied by the trees for their growth and development. Another good example of mutualism is honey bees and the crops they pollinate.

MANAGED ECOSYSTEMS

Every agricultural field, urban landscape, park, managed forest, rangeland, or roadside represents an ecosystem that is managed in some way to benefit people. Some are manmade, simplified ecosystems that are intensely managed. Others, such as forest systems and rangeland, are more long-term ecosystems that have been modified and managed to favor the growth of a few plant species.

Agricultural ecosystems, or *agroecosystems*, are predominantly *monocultures*. In a monoculture, the age and genotype of crop plants are relatively uniform, and species diversity is limited. Complex food webs are simplified because of the lack of diversity at the primary producer level. The autotrophic component consists almost entirely of the crop plant. Weed species are undesirable and eliminated. Also, the physical diversity of the system is much lower than in natural ecosystems.

The uniformity of some monoculture systems can encourage pest outbreaks. Disease epidemics or pest outbreaks can move rapidly in a monoculture system. The establishment of a complex community of organisms, in many monocultures, is often discouraged by management practices such as broadly toxic pesticides or removal of all weeds. However, in some IPM programs, managers try to increase diversity to help manage pests by introducing biological control agents or *intercropping* to provide habitat for natural enemies or as trap crops for pests.

In agroecosystems, inputs and outputs are manipulated by managers and depend largely on supplemented sources of energy and nutrients (Figure 2-17). Productivity is measured in terms of yield. High yields are maintained by inputs such as cultivation, irrigation, fertilization, genetic selection, pest control, and the fuel used to run the machinery that carry out these operations. However, unless carefully managed, excessive irrigation, fertilization, or pest control can drive the cost of producing a crop higher than the yield can return.

Limiting Factors

All organisms have basic requirements to live and thrive in a given situation. Crop plants have certain light, soil moisture, fertility, and climatic requirements to ensure optimal growth. Likewise, the population of other organisms in the ecosystem, such as pests, is regulated partially by light, temperature, and water, as well as food sources, predators, competitors, and other organisms. When one or more of these factors is in short supply or overabundant, the growth and development of the affected organism can be reduced, resulting in stress or death. Any of these factors is called a *limiting factor*.

In managed ecosystems, people may employ practices to provide optimal conditions for plant growth by maintaining limiting factors at appropriate levels. For instance, a limiting factor in the growth of some young plants may be sunlight. Shade produced by larger weeds limits the rate of growth. Addition of other growth requirements, such as water or nutrients, will not increase growth rate until the sunlight deficiency is corrected. Weed control is aimed at ensuring that this factor does not become limiting to the crop plants.

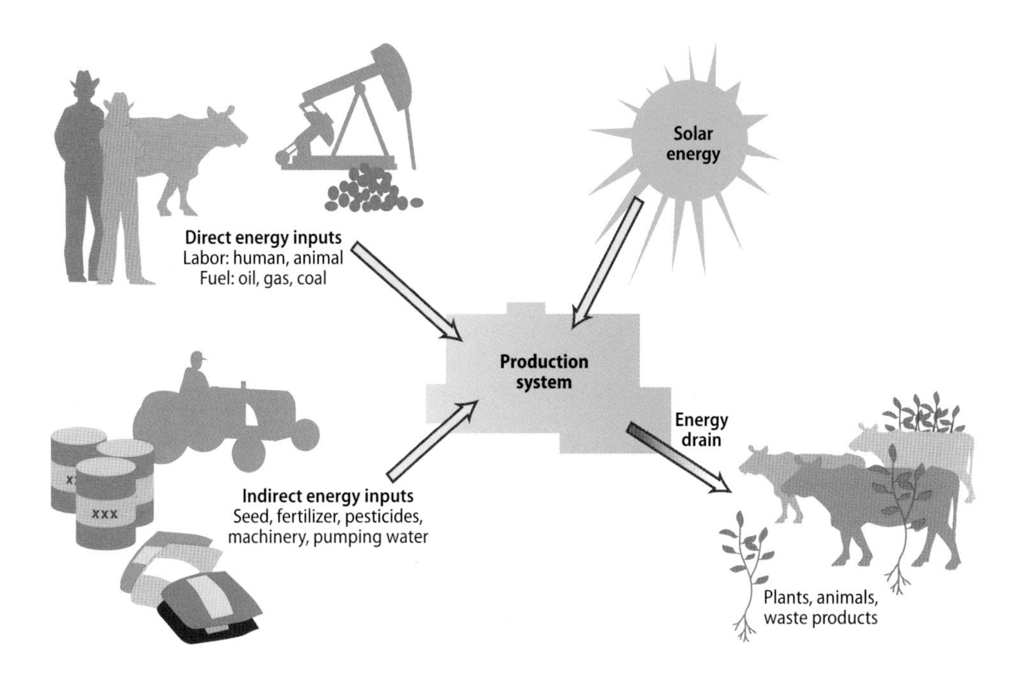

FIGURE 2-17

Energy inputs in an agroecosystem include direct inputs such as labor, fuel, and electricity, as well as indirect inputs such as seed, fertilizers, pesticides, and water. Energy drains are the products or materials, such as plant, animal, or waste products, that are removed from the system.

PEST ECOLOGY

Pest population growth is influenced by the abiotic and biotic components of the ecosystem such as climate, food (host plant), space, other organisms, and population interactions. Diseases, for example, can develop only if there is a susceptible host, a virulent pathogen, and an environment favorable for infection and growth (see the *disease triangle* discussion in Chapter 4, Figure 4-33). Predators, parasites, competitors, and other *antagonists* can be important limiting factors for weeds, insects, and pathogens. Management practices also impact pest and pathogen populations. The spacing of the plants, addition of nutrients, irrigation, planting and harvest date, and growing the crop continually on the same site can contribute to the severity of pest outbreaks. Perennial crops such as fruit trees tend to maintain more complex communities than short-lived annuals such as lettuce, which may remain in the field only a few months.

Equilibrium Population Density

Within an agroecosystem, each species exhibits a typical population level, often called its *characteristic abundance*, as determined by its ecological niche. For instance, whereas it may be typical for an acre of an orchard crop to have thousands of caterpillars, only a few raptors, such as hawks, could be supported. Limiting factors such as food, shelter, or natural enemies enforce this characteristic abundance within certain upper and lower levels, which fluctuate around a mean level called the *equilibrium position*.

The equilibrium position is maintained primarily by constraints set by the physical environment and interactions with other species. The most important factors involved in maintaining a population around its equilibrium position density are density-dependent factors, such as parasitism, competition, or predation, that exert more pressure when populations are high and have a diminished influence at low population density.

r and K Strategists. Species that have high rates of reproduction and rapid growth are known as *r strategists.* These species rapidly colonize new areas and thrive when competition is not severe. Characteristics of r strategists include a high rate of reproduction, rapid dispersal, efficient host finding ability, and small size. Many pest species are r strategists, including aphids, mites, and common weeds such as barnyardgrass, starthistle, and chickweed. Most foliar pathogens are typical r strategists. Agroecosystems typically remain dominated by r strategists as a result of repeated disturbances from management actions such as cultivation, harvesting, planting, and pesticide application.

When competition for resources is high, species with better competitive abilities but slower reproduction and growth rates may be more successful. Species that demonstrate these survival mechanisms are known as *K strategists.* Characteristics of K strategists include larger size, longer life cycles, and higher survival rates of offspring but lower reproductive rates. Some species have characteristics that place them somewhere between typical r and K strategists. Fewer pest species are K strategists; examples include some woody weeds, some root pathogens that slowly kill trees, and large tree-boring beetles. Mature ecosystems with the greatest *biodiversity* are most likely to be dominated by K strategists.

BIODIVERSITY

Biodiversity is the number of different species of plants, animals, and microorganisms in an ecosystem. The term can also refer to other levels of biological diversity, including genetic variability in a species, diversity in communities, and the diversity of communities and ecosystems in the biosphere. Biodiversity generally (but not always) is greater in natural, more mature ecosystems than in younger, managed ones. Maintaining natural biodiversity is important not only to conserve the diversity of species in nature but also to provide a resource for future agricultural crops and animals.

3 The Integrated Pest Management Concept

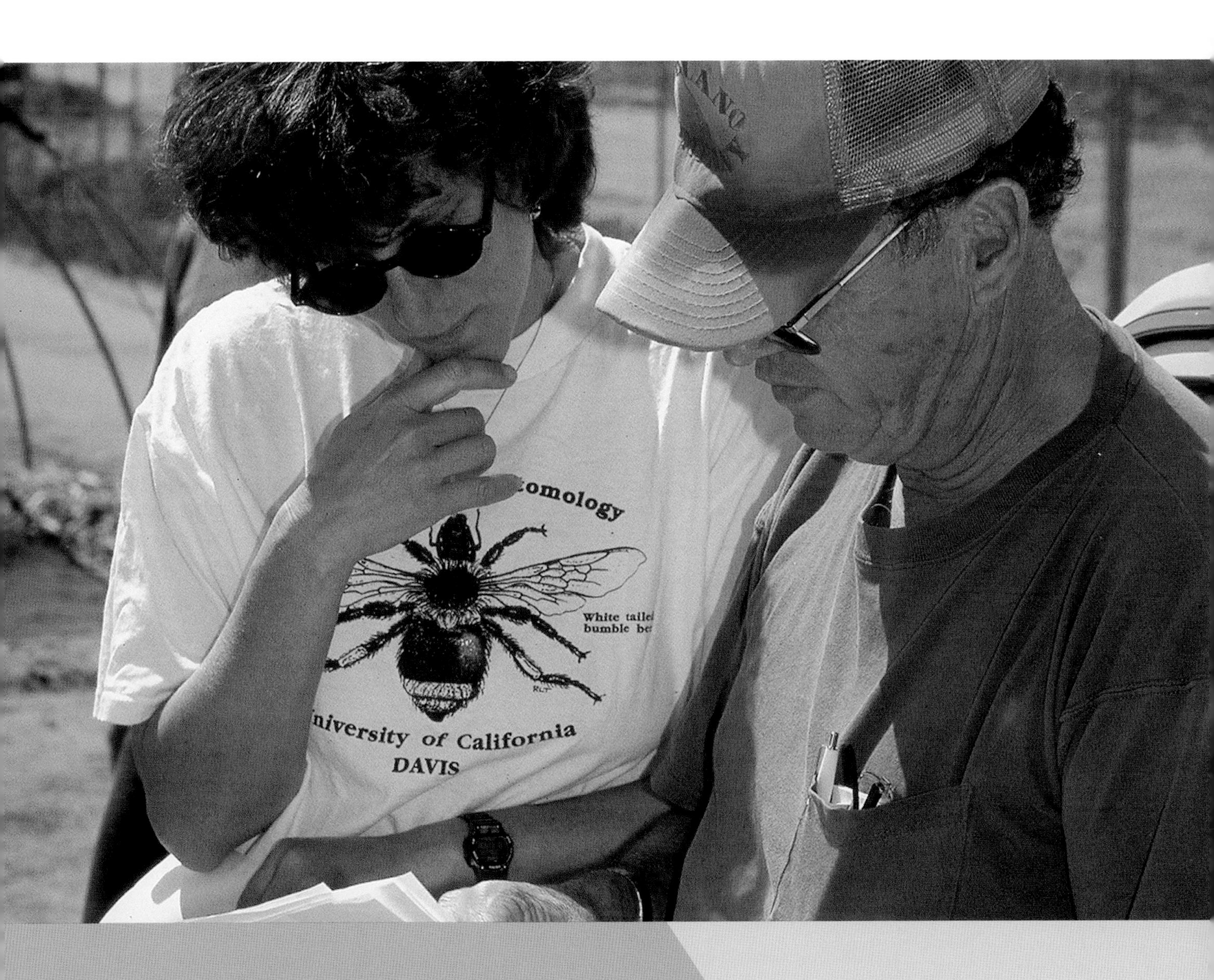

THE PURPOSE OF PEST MANAGEMENT

Pest management includes any activities directed at controlling unwanted organisms or avoiding their damage, and it may involve methods that prevent, suppress, or eradicate pest organisms. A successful pest management program chooses the right tools or combination of control methods to reduce pest losses to acceptable levels in a cost-effective manner that is environmentally safe and protects human health. Management methods may include a variety of cultural, biological, physical, mechanical, and chemical management options.

Economics is a major factor in most pest management decisions. Pest populations can lower crop yield or quality and impact the long-term health of perennial crops such as fruit trees, thereby reducing profits. The cost of pest control activities, including labor, equipment, and material, is a factor in each pest management decision. Human and environmental health hazards are costs that must also be taken into account. Other factors important in pest management decision making are severity of the pest, impact of the damage, effectiveness of the control, and the time delay before an action becomes effective. These considerations are based on past records of pest occurrence and damage, trends in pest development, and the likelihood of future attacks. Successful pest management decisions require information about the pest species—its biological characteristics, distribution pattern and population density, impact on the managed ecosystem, and the cost and likely effect of control.

PEST MANAGEMENT STRATEGIES

Preventive methods discourage damaging pest populations from developing and include planting weed- and disease-free seed and growing varieties of plants that are resistant to diseases or insects. *Cultural controls* such as cultivation to kill weedy plants before they go to seed or choosing planting and harvesting dates unfavorable for the pest are techniques that prevent pest establishment. Other preventive methods might include preplant soil disinfestation (such as fumigation, heat treatment, or *soil solarization*), removal of overwintering sites, and the selection of planting sites that are not already infested with pests that attack the intended crop.

Suppressive pest control methods reduce existing pest populations to tolerable levels. Most pest control actions fit into this category and include the release of biological control agents, mowing, cultivating weeds, or in-season pesticide sprays. The management method chosen does not usually eliminate pests but limits their damage or competition. Two or more suppressive methods are often combined to enhance control.

Eradication strategies, in which no pests can be tolerated, are aimed at totally eliminating the pest from a designated area. Eradication has the appeal of offering a complete solution and in special instances may be a desirable strategy. Newly invading exotic pests posing a health or severe economic threat are the usual targets of eradication programs. Coordination of eradication efforts is usually the responsibility of government agencies. Examples of this approach

FIGURE 3-1

Two targets of eradication efforts in California are the Mediterranean fruit fly, Ceratitis capitata *(left); and Japanese dodder,* Cuscuta japonica *(right; photo by T. Lanini).*

to pest management are efforts directed at eliminating the parasitic plant Japanese dodder from some areas of California and the Mediterranean fruit fly and the oriental fruit fly from southern California (Figure 3-1). For most pests, however, eradication is not feasible and is not compatible with integrated pest management systems.

What Is a Pest?

Pests are organisms that reduce the availability, quality, or value of a human resource. Pests compete with people for food or fiber, interfere with raising crops or livestock, damage property, tarnish ornamental plantings, transmit disease, invade our waterways or natural areas, or are otherwise a nuisance. Any organism can conceivably become a pest under some circumstances. Common groups of pests include weed species; vertebrates such as birds and rodents; invertebrates such as insects, ticks, and mites; mollusks; nematodes; and plant pathogenic microorganisms such as bacteria, *viruses*, fungi, and phytoplasmas.

Pests can be classified as *key pests*, *occasional pests*, or secondary pests. Key pests may cause major damage on a regular basis unless controlled. In walnuts (Table 3-1), for example, codling moth is a key pest because it often requires regular control efforts to prevent economic damage to the crop. Treatments for codling moth can induce outbreaks or suppress populations of other walnut pests. Occasional pests become intolerable only irregularly, often due to climate, environmental influences, or as a result of human activities. Secondary pest problems occur as a result of actions taken to control a key pest. For instance, aphids and soft scale can become secondary pests when pesticides applied to control codling moths kill their natural enemies. Some weed species may become secondary pests. Often these secondary pest weeds are tolerant or resistant to the herbicides applied to control key weeds or were previously suppressed by competition from the key weeds.

THE EVOLUTION OF PEST MANAGEMENT

Since people began cultivating crops, they have managed pests to increase crop yield, quality, and profitability. Early agronomic practices such as crop rotation, burning crop residues, tillage, and hand-removal of pests often were the primary methods of pest control. The first pesticides contained copper, sulfur, lead, organic salts, antimony, and arsenic and included botanical compounds such as nicotine and pyrethrum. Many of these early materials were quite toxic and expensive; their use was limited to applications against a few key pests. Equipment to

TABLE 3-1.

Key, secondary, and occasional insect and mite pests in California walnuts.

Common name	Scientific name	Comments
Key Pests		
codling moth navel orangeworm walnut husk fly	*Cydia pomonella* *Amyelois transitella* *Rhagoletis completa*	These pests must be managed in most orchards every year or economic damage will occur. Pesticide treatments for these pests often cause outbreaks of secondary pests.
Secondary Pests		
walnut aphid dusky-veined aphid frosted scale walnut scale San Jose scale spider mites	*Chromaphis juglandicola* *Callaphis juglandis* *Parthenolecanium pruinosum* *Diaspidiotus (=Quadraspidiotus) juglansregiae* *Diaspidiotus (=Quadraspidiotus) perniciosus* *Tetranychus* spp.	These pests are often well controlled by natural enemies in orchards that do not receive applications of broad spectrum insecticides. They become damaging primarily when sprays applied to manage key pests kill their natural enemies.
Occasional Pests		
Pacific flatheaded borers redhumped caterpillars oystershell scale Italian pear scale	*Chrysobothris mali* *Schizura concinna* *Lepidosaphes ulmi* *Lecanium pruinosum*	These pests may infest some orchards once every few years when environmental conditions favor their development.

safely handle and effectively apply these materials was not well developed. Control effectiveness was often erratic.

Following World War II, the increased availability of *synthetic organic pesticides* and advances in application technology had a profound impact on pest control practices. Pest control materials that were broad spectrum, persistent, effective, cheap, and required little labor were suddenly available. The increasing availability and use of such materials as DDT and related halogenated hydrocarbons, organophosphate, and carbamate materials, as well as phenoxy herbicides (2,4-D), paralleled a growing mechanization of farming practices and a decrease in the number of people involved in farming. Technology improved application equipment; tractors moved more efficiently across fields, making pesticide applications more effective and easier to apply.

Pest control became synonymous with pesticide use. Conventional pesticides pervaded all aspects of pest management. Widespread reliance on the use of pesticides to control pests changed many agronomic practices and permitted the production of crops in areas where they previously could not be grown and during times of the year when high pest populations had formerly made production unprofitable. Optimism emerged that pesticides offered the possibility of an essentially pest-free environment.

However, heavy reliance on a single pest control method began to limit the usefulness of certain pesticides. *Pest resurgence*, *secondary pest outbreaks*, *pesticide resistance*, negative effects on nontarget organisms, and concerns with human exposure and environmental damage renewed interest in environmentally sound pest management strategies. As early as the late 1950s and 1960s, entomologists at the University of California and elsewhere began promoting integrated approaches to counteract *insecticide* resistance and secondary pest outbreaks by incorporating biological control methods with chemical control. They called these programs integrated control.

Integrated pest management (IPM) evolved from the integrated control concept. The premise of IPM was built on the philosophy that natural control should be maximized, enhanced, and relied on whenever possible. Pesticides should be used only when the population of a pest reached a threshold level that caused economically significant damage and where natural controls were not effective.

IPM has expanded and changed since the 1960s. Many new tools have come into use that are more compatible with biological control. IPM concepts, which were first directed primarily at insect pests in alfalfa and cotton, have expanded to other crops, pests, and disciplines, resulting in pest control approaches such as integrated weed management, integrated disease management, and orchard integrated pest management systems as well as IPM programs for nonagricultural situations such as landscapes, structures, and mosquito control. Presently, IPM is more generally referred to as an ecological approach for managing multiple pests with a variety of complementary tactics to reduce pest damage to tolerable levels. Decisions are based on the results of field monitoring combined with research-based biological and ecological information.

Ideal multidisciplinary, multitactic, and multipest IPM programs have been developed only for very few cropping systems. Because of their complexity, IPM programs are typically implemented incrementally, starting simply. In recognition of this incremental development and adoption, IPM scientists have described typical steps or integration levels through which IPM programs may evolve (Figure 3-2). Level I focuses on monitoring and managing a single species or species complex. Level II considers how practices impact multiple pests (e.g., insects, pathogens, and weeds) and focuses on complimentary, biologically based management options. In Level III, multicrop, multiseason, and multitactic considerations are well integrated into the decision-making process.

The University of California's Year-Round IPM Programs (online at www.ipm.ucdavis.edu) give good examples of how monitor-

FIGURE 3-2

Example of the three levels of IPM integration and the range of tactics and strategies employed in each level. This is a cumulative continuum: practices adapted at Level I are generally continued in Level II. Adapted from Benbrook 1996.

	No IPM	Level I IPM (Low level) →	Level II IPM (Medium level) →	Level III IPM (Biointensive level) →
general characteristics of system	☐ proper calibration, operation, and cleaning of spray equipment ☐ scouting for pests ☐ sanitation and good agronomic practices	☐ scouting plus pesticide applications when thresholds indicate need ☐ avoid or delay resistance and secondary pest problems ☐ optimal timing of treatments ☐ some preventive practices (short rotations, resistant varieties, cultivation) ☐ choose selective pesticides	multitactic approaches applied, e.g., ☐ limit or remove pest habitat and augment biodiversity ☐ resistant varieties, cover crops, and longer rotations ☐ enhance beneficials, use soil amendments ☐ insect and disease forecasting models ☐ use biorational pesticides	reliance on preventive measures to limit pests and enhance beneficials ☐ multiple steps to enhance plant health and soil quality ☐ focus on conservation of beneficials and habitat ☐ microbial biocontrol of root pathogens ☐ release of beneficials ☐ consider long-term cropping patterns and multicrop interactions
monitoring practices	scouting for detection of dominant pests; visual inspections	trapping for major pests, visual inspection for minor pests and beneficials	scouting and trapping for major pests and beneficials, monitoring crop phenology, weather monitoring	scouting to determine beneficial release timing
treatment thresholds	treatment based on calendar, crop phenology, or pest detection, not numerical thresholds	pesticides applied according to thresholds associated with monitoring program; treatments timed to minimize impacts on beneficials and nontargets	treatment thresholds adjusted to account for activities of beneficials	development of thresholds for releasing beneficials and biopesticides

ing and management recommendations for multiple pests of major crops can be integrated in an IPM program. Year-Round IPM Programs guide users through a full year or growing season of monitoring pests, taking preventive measures, making management decisions, and planning for the following season. A sample checklist for a Year-Round IPM Program, for peaches, is shown in Figure 3-3.

Since IPM's inception, research into alternatives and educational efforts to develop and implement IPM has been ongoing. Concerted efforts have been made to deliver new ideas and technologies to the field. Research has continued, and integrated pest management has matured to encompass an interdisciplinary approach to pest management.

Components of an IPM Program

Pest management systems are complex; they differ by crop, production system, locale, and philosophy of the practitioners. They are dynamic and in a constant state of flux due to the market value of the crop or other managed resource, adaptation of pests to environmental conditions, and availability of control tools including pesticides, *host plant resistance*, cultural controls, and invasions by new pests. Each field, orchard, or landscape possesses unique communities and is subject to varying environmental pressures; no two locales should be managed in quite the same way. An IPM program that matches pest management practices and methods with the unique nature of the management unit can be designed by following the five major components common to all IPM programs:

- pest identification
- field monitoring and population assessment
- control action guidelines
- preventing pest problems
- integrating biological, chemical, cultural, and physical/mechanical management tools

Pest Identification. Because most pest management tools, including pesticides, are effective only against certain pest species, an IPM program requires that you know which pests are present or are likely to appear. Pests must be accurately identified and their potential impact on the system properly evaluated. Correct identification is essential for selecting pest management options. Two species can be morphologically similar but biologically different (Figure 3-4). Closely related species may include other pests, beneficials, or nonpests in a single crop or landscape. Pest identification is discussed in Chapter 4.

Field Monitoring. Field monitoring means regularly going into the field, orchard, or landscape and systematically checking for pests or damage symptoms. For many crops and pests, special techniques and sampling procedures have been developed to improve the accuracy, information value, and efficiency of monitoring activities (Figure 3-5). Field monitoring provides information on daily or seasonal conditions, such as the status of pests, crop, weather, or soil factors. This information is used to predict and evaluate potential pests. Because conditions vary, individual fields or landscapes should be monitored separately. Regularly check the pest species present; the maturity and health of the crop, plant, or commodity protected; the weather; the plant environment, including soil conditions; and, when appropriate, the population levels of pest and *beneficial organisms*. In some cases, careful observations of plant health or pest presence is all that is involved in field monitoring, but in other cases field monitoring requires quantitative sampling of pest numbers or damaged plant parts. Chapter 6 discusses field monitoring procedures in more detail.

Keep written records of monitoring results, weather, and management activities (Figure 3-6). Records can be hand-written, but tools such as hand-held electronic devices, global positioning systems (GPS), and geographic information systems (GIS) are increasingly useful for collecting, organizing, and communicating pest management information. During the season, these records indicate whether pests or *natural*

Peach Year-Round IPM Program
Annual Checklist

www.ipm.ucdavis.edu

Supplement to UC IPM Pest Management Guidelines: Peach

These practices are recommended for a monitoring-based IPM program that enhances pest control and reduces environmental quality problems related to pesticide use.

Water quality becomes impaired when pesticides and sediments move off-site and into water. Air quality becomes impaired when volatile organic compounds (VOCs) move into the atmosphere. Each time a pesticide application is considered, review the Pesticide Application Checklist at the bottom of this form for information on how to minimize water quality problems.

This program covers the major pests of peach. Details on carrying out each practice and information on additional pests can be found in the guidelines. Track your progress through the year with the annual checklist form. All photo identification pages and examples of monitoring forms can be found online at: http://www.ipm.ucdavis.edu/PMG/C602/m602yiformsphotos.html

Done	**Dormant/delayed-dormant season activities** Mitigate pesticide usage to minimize air and water contamination.**
	Apply fungicide treatments as needed according to PMGs: • Peach leaf curl • Shot hole disease
	Manage orchard floor vegetation: • Survey weeds and keep records (*example form available online*). • Manage weeds with pre- or postemergent herbicides or nonchemically in organic orchards. • If ground cover present, manage according to system needs. If frost is indicated, mow before bloom.
	Make an oil treatment for scales and European red and brown mite eggs. • If you saw increasing damage from scales last year, take a dormant shoot sample to see if an insect growth regulator should be added to the oil treatment.
	Treat peach twig borer and obliquebanded leafroller with environmentally sound material or delay treatment until bloom.
	Other pests you may see: • Fruittree leafroller egg masses • Armillaria root rot • Voles • Pocket gophers • Stink bugs • Tree borers

	reen tip to petal fall) contamination.**
	Valley, February 20 in Sacramento Valley) bruary 20) d Sacramento valleys) *able online*).
	e dispensers in orchard after first moth is
	plant bugs, as well as known invasive moth).
	When rainy conditions promote disease, time fungicide treatment according to PMGs: • Brown rot at 20 to 40% bloom and full bloom. • Jacket rot treatment at full bloom. • Powdery mildew treatment at petal fall. • Scab, if orchard has a history of this disease.

	Monitor for diseases:
	• Rust
	  o Monitor twig cankers beginning late March.
	  o Treat with fungicide if needed according to PMG.
	• Shot hole
	  o Fruiting structures in leaf lesions as long as weather is wet.
	  o Manage if needed according to PMG.

	Observe the orchard for vertebrates and manage as necessary: • Gophers • Ground squirrels
	If orchard floor vegetation present, manage as needed.
	Other pests you may see: • Armillaria root rot (oak root fungus) • Bacterial canker • Phytophthora crown and root rot

FIGURE 3-3

Example Year-Round IPM Program checklist for peaches integrates management of pathogen, invertebrate, and weed pests.
From UC IPM website, www.ipm.ucdavis.edu.

Peach Year-Round IPM Program

Page 3 of 4

Done	**Fruit development period activities (petal fall to harvest)** Mitigate pesticide usage to minimize air and water contamination.**
	Put up pheromone traps for: • Peach twig borer (March 20 in San Joaquin Valley, April 1 in Sacramento Valley) • Obliquebanded leafroller (April 15 in San Joaquin Valley and Sacramento Valley)
	Monitor shoot strikes for damage from oriental fruit moth and peach twig borer, especially in mating disruption orchards. • Keep records (*example form available online*). • Manage if needed according to PMGs.
	If wet weather persists, continue to monitor for rust: • Manage if needed according to PMG.
	If orchard has a history of scab: • Treat 3 weeks after full bloom. • Treat again 2 weeks later if scab was severe the previous year.
	Make fertilizer applications at appropriate intervals.
	Where ground covers are present, take sweep samples for pests, beginning from early April to early June for: • Plant bugs (*Lygus* and *Calocoris*). • Katydids. • Stink bugs. Manage if needed according to PMG.
	Sample fruit damage every other week after color break.
	Monitor powdery mildew and treat if needed according to PMG.
	If ground cover present: • Survey weeds. • Complete a late-spring weed survey form (*example form available online*). • Mow, spray, or cultivate ground cover as needed.
	Monitor spider mites from May through August: • For best evaluation, conduct two 5-minute searches and keep records on a monitoring form (*example form available online*). • Manage if needed according to PMG.
	Select leaf samples in July to analyze for nutrients. Pay particular attention to nitrogen, potassium, and some of the micronutrients such as zinc and boron. • Take 60 to 80 mid-shoot leaves from moderately vigorous fruiting shoots.
	If rain is predicted during the last 4 weeks before harvest, treat for ripe fruit rot.
	Other pests you may see: • Armillaria root rot • Bacterial canker • Phytophthora root and crown rot • Peach silver mite • Black peach aphid • Scab • Verticillium wilt • Tree borers

ter contamination.**
ted.
ter contamination.**
or:
oth populations are building.
rot inoculum and prevent shoot death.

s in an IPM program, review and complete this
ronmental and efficacy problems.
gement Guidelines for the target pest
UC IPM WaterTox database. (For more
OX/simplewatertox.html.)
ormation, see *Pesticide Choice,* UC ANR
u/pdf/8161.pdf.).
spectrum of activity, and pesticide resistance).
treatment when resistance risk is high.
cides on target.
e areas (for example, waterways or riparian
for pesticide handling, storage, and disposal guidelines.

- Check and follow restricted entry intervals (REI) and preharvest intervals (PHI).

After an application:

- Record application date, product used, rate, and location of application.
- Follow up to confirm that treatment was effective.

Consider water management practices that reduce pesticide movement off-site. (For more information, see UC ANR Publication 8214, *Reducing Runoff from Irrigated Lands: Causes and Management of Runoff from Surface Irrigation in Orchards,* http://anrcatalog.ucdavis.edu/pdf/8214.pdf.)

- Limit irrigation to amount required using soil moisture and evapotranspiration (ET) monitoring. (For more information see UC ANR Publication 8212, *Understanding Your Orchards Water Requirements,* http://anrcatalog.ucdavis.edu/pdf/8212.pdf.)
- Install an irrigation recirculation or storage and reuse system.
- Consider the use of cover crops.
- Consider vegetative filter strips or ditches. (For more information, see *Vegetative Filter Strips,* UC ANR Publication 8195, http://anrcatalog.ucdavis.edu/pdf/8195.pdf.)
- Install sediment traps.
- Use polyacrylamide (PAM) tablets in furrow irrigation or sprinkler irrigation systems to improve soil infiltration and prevent off-site movement of sediments.
- Redesign inlets and outlets into tailwater ditches to reduce erosion. (For more information, see UC ANR Publication 8225, *Reducing Runoff from Irrigated Lands: Tailwater Return Systems,* http://anrcatalog.ucdavis.edu/pdf/8225.pdf.)

Consider orchard floor management practices that improve soil structure and reduce erosion. (For more information, see UC ANR Publication 8202, *Orchard Floor Management Practices to Reduce Erosion and Protect Water Quality,* http://anrcatalog.ucdavis.edu/pdf/8202.pdf.)

Consider practices that reduce air quality problems.

- When possible, choose pesticides that are not in an emulsifiable concentrate (EC) formulation, which release volatile organic compounds (VOCs); this is especially important from May to October. VOCs react with sunlight to form ozone, a major air pollutant.

FIGURE 3-3 cont.

Example Year-Round IPM Program checklist for peaches integrates management of pathogen, invertebrate, and weed pests.
From UC IPM website, www.ipm.ucdavis.edu.

FIGURE 3-4

Proper identification is essential for making pest management decisions. The striped cucumber beetle, Acalymma trivittatum *(**bottom**), a common pest found in corn, can be mistaken for the similarly shaped aphid-feeding lady beetle,* Hippodamia sinuata *(top).*

FIGURE 3-6

Keeping written records of monitoring activities, such as the results of a weed survey, can help PCAs determine whether future control actions will be needed and can provide insight on the effectiveness of past management activities.

FIGURE 3-5

For many insect pests, pheromone traps are used to improve sampling accuracy. When combined with other tools, such as degree-days and sunset temperatures, potential pests can be more accurately evaluated and management activities better timed.

enemy populations are increasing or decreasing. They also aid in forecasting possible pest outbreaks. Pest population assessments, together with records of control measures, cultural practices, and weather conditions, help determine whether future control actions will be needed. Simple tables and graphs of data help define patterns; maps can identify where infestations are occurring and whether their extent is changing over time. Over the years, these records provide valuable historical data for long-term population management and understanding how factors such as new pest management practices or development of pesticide resistance impact pest populations. Monitoring programs are discussed in Chapter 6.

Control Action Guidelines. Control action guidelines help decide whether management actions, including pesticide applications, are needed to avoid eventual loss from pest damage. They are useful only when combined with careful field monitoring and accurate pest identification. Guidelines for insects and mites are generally numerical thresholds based on certain sampling techniques. They are intended to reflect the population level that will cause economic damage. Guidelines for other pests may be based on the history of a field or region, the stage of crop development, weather conditions, and other observations. Many of these guidelines do not involve numerical thresholds.

Tolerable Injury Levels. The concept that a low pest population density or a certain amount of pest damage can be tolerated is fundamental to integrated pest management. The tolerable pest level depends on the pest, crop, and stage of the crop. In agricultural crops, the levels are generally based on what is perceived as economically unacceptable damage. A processor may set the *tolerable injury level*, for example, by stipulating that a maximum of 1% worm damage will be allowed for fruit. In other cases, the level may be based on the perception of the amount of damage the consumer will tolerate. Sometimes the tolerable level is very low because even very low populations of a pest can rapidly expand to intolerable levels. This is true of many foliar pathogens and even insect pests such as codling moth.

The tolerable injury level has been defined by some as the pest density (or damage level) where the cost of pest control is less than the cost of the damage that the pest infestation causes. This concept has resulted in the widely used term *economic injury level*. However, the economic factors involved in the development of many of these injury levels are often crude. For instance, such calculations generally do not take into account price fluctuations for crop value or the cost of control measures employed.

For landscape pests, pest managers use the concept of *aesthetic injury level*, which is the level of pest damage or pest populations the general public will tolerate. In mature landscapes, many pests usually don't cause economic damage. Pest control actions are usually undertaken because the presence of the pest or the damage it causes is aesthetically displeasing. Aesthetic tolerances vary among people and by the prominence of the damaged plant. For instance, a certain level of damage may be tolerable in a background plant but not tolerable on the same species planted at the entrance of a building. Some people find the presence of any insect or weed intolerable while others are not bothered by them.

Defining an aesthetic injury level on which people can agree is difficult and subjective. Management decisions are based on the attitudes of the PCA, the public, or the client, as well as potential pest problems. An effective education program can often increase the public's tolerance for pest presence, thereby increasing aesthetic tolerance levels. Decision making is further complicated in landscapes because of the diversity of plant species, their susceptibility to pests, and the public's differing perceptions of intolerable damage and pesticide applications.

Although the injury level concept was originally developed for use in insect pest management, its application in other pest disciplines, such as nematology, pathology, and weed science, is being actively researched. The inherent differences between weeds, pathogens, and insects, however, complicate the development of injury levels. For example, the development of injury levels for weeds must take into account the diversity of weed species in a single field as well as the longevity of weed seeds and vegetative propagules. Injury levels for plant disease management are complicated by the need for preventive management practices and the difficulty in assessing pest impact. For other pest species, the concept of a tolerable injury level is difficult because damage is often not visible until an economic loss has occurred. For example, the tolerable injury level for pocket gophers in some situations is zero because once a population is established enough to be detected, economic injury is likely. Similarly, many structural or household pests such as termites, cockroaches, bedbugs, rats, and mice are considered at intolerable levels if detected in a building.

Treatment Thresholds. The problem with the tolerable injury level concept is that once pests have reached that level, it is usually too late to control them before pest populations reach unacceptable levels. Therefore, the critical issue in IPM programs is to define control *action* or *treatment thresholds* (sometimes also called *economic thresholds*)

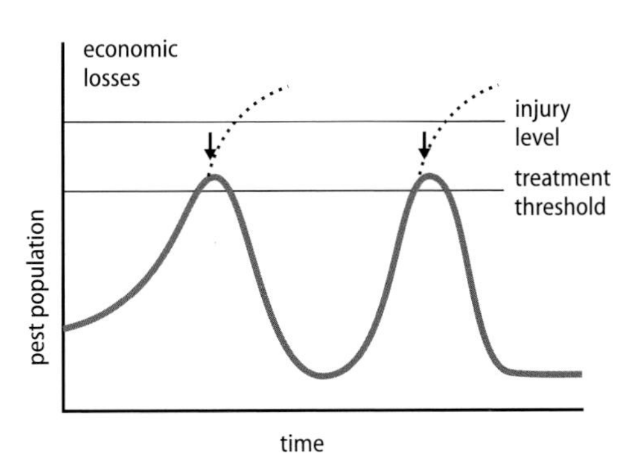

FIGURE 3-7

This graph illustrates an insect population increasing to the treatment threshold. Management action against the pest at two different points in time illustrates how the population declines, keeping the pest population below the injury level and avoiding economic losses. The dotted lines above the injury level illustrate the pest population if no management action is taken, resulting in economic losses.

that specify the population density at which control measures must be applied to prevent crop loss or damage from going beyond acceptable levels. Figure 3-7 illustrates injury levels and treatment thresholds, as well as the effect of management actions for a typical insect pest.

For many pests, treatment must be applied well before unacceptable levels are reached so population levels of the pest are not considered in the treatment decision. This is particularly true of pathogens, which can rapidly develop into epidemics if environmental conditions are favorable. For many pathogens, *fungicides* are applied preventively when weather conducive to disease development is expected. In some cases, triggers indicating that a treatment is needed are based on moisture levels, temperature-moisture combinations, or simply on the occurrence of rain. The difficulty of using treatment thresholds with pathogens is increased because many fungicides only prevent infection and do not eradicate infestations that have already begun, so the potential for lowering infection is not as great as with insect population control using insecticides.

Similarly, because most fields have an abundance of weed seeds, treatment thresholds for weeds are generally based on the conduciveness of field and environmental conditions for the development of certain weed species rather than on population levels of weed species. However, for very serious weeds such as certain *perennials*, treatment thresholds may be appropriate to initiate specific management actions. Treatment thresholds based on aesthetic preferences may also be appropriate for landscape situations.

Treatment thresholds depend not only on the growth and damage potential of the pest, but also on the *efficacy* of the control procedures themselves. For instance, an augmentative release of a natural enemy may be recommended at a lower population threshold than an insecticide for the same pest because of the time required to establish biological control. Treatment thresholds have been developed for some pests through university research based on field trials and analysis of yields and quality under different pest population levels and management regimes. In other cases, a PCA improvises treatment thresholds to assist in decision making. These can be revised annually based on the success of the programs. Both types of thresholds may be valuable to the pest manager.

Control Action Timing. Applying pesticides and other management practices at optimal times for effectiveness is another fundamental part of an integrated pest management program. For some pest species, treatment thresholds indicate whether the population warrants control. However, they do not always indicate when treatment is most effective. For example, the need for scale treatment is indicated by dieback during the growing season. However, insecticides applied at this time are often not effective. The most effective treatment timing for many scale species is the *dormant season* or in spring when crawlers are first apparent on double-sided sticky tape traps.

Because treatment thresholds for many pathogen and weed pests are quite low, guidelines that help time control practices according to crop development or environmental conditions are central to management of these pests. For plant pathogens,

FIGURE 3-8

For treatment thresholds to be reliable, the same sampling technique and procedure must be followed each time a sample is taken. Here a sweep net sample for lygus bug is being taken in cotton. Sweep net samples can vary greatly if the method is not standardized.

disease development modules that rely on weather inputs are often used to determine the best time for treatment.

Relationship of Monitoring to Control Action Guidelines. Most treatment thresholds and *control action guidelines* require accurate monitoring of pests, pest damage, or environmental factors using quantitative sampling methods. For example, treatment thresholds for insect pests may be expressed as the number of individuals per sweep (Figure 3-8) or a certain number of individuals per leaf. To obtain a reliable estimate of the pest population requires that a specified number of *samples* be taken at a given number of sites chosen randomly (or in some other predefined manner) throughout the field or orchard. It is important that careful sampling procedures be followed when comparing population estimates with treatment thresholds. A consistent sample unit such as a leaf, fruit, or length of row is required for consistent results. It is also important that the same procedures be followed each time the field is resampled so that results are comparable between dates. Likewise, when weather data are used to time management practices for insect or pathogen control, it is essential to confirm that the data are from standardized weather monitoring equipment that were collected following recommended protocols and also from a location that is representative of the crop management site.

TABLE 3-2.

Ways to prevent pest problems in the landscape.

- ☐ Select an appropriate site for the plant.
- ☐ Prepare and/or treat soil properly before planting.
- ☐ Choose resistant cultivars, rootstocks, or plants.
- ☐ Choose appropriate irrigation and soil management strategies.
- ☐ Time cultural activities to discourage pest development.
- ☐ Keep plants healthy using best management practices.
- ☐ Use landscape fabrics, mulches, and weed-free amendments.
- ☐ Plant vigorous, fresh grass seed at the correct seeding rate.
- ☐ Clean cultivating and mowing equipment between sites.
- ☐ Mow turf at correct height.
- ☐ Prevent excessive foot traffic and soil compaction.

Preventing Pest Problems. Often, preventive measures can be taken to provide for optimal crop production or resource management with a minimum of pest problems. Examples of ways to prevent pests and their damage in the landscape are listed in Table 3-2. Using practices that prevent problems is basic to IPM. Although many of these measures are not directly the responsibility of the pest manager, the pest manager should be aware of the availability of preventive measures, as well as when and where they have been used. Often a PCA can suggest that some of these actions be taken to reduce or eliminate pest pressure.

Host resistance is a preventive pest management tool that takes advantage of the genetic attributes of certain plant cultivars and allows the plants to resist or tolerate pest attack. Host resistance is one of the most successful and ecologically sound pest management techniques and is used widely, especially in the management of plant pathogens, nematode pests, and to a more limited extent, *arthropod* pests. For example, there are tomato cultivars that are resistant to root knot nematodes, *Verticillium*, *Fusarium*, tomato spotted wilt, and tobacco mosaic virus. Cultivars resistant or tolerant to downy mildew of lettuce and crucifers, celery Fusarium yellows, white mold of beans, cucumber mosaic virus of cucumber, root rots of peas, and tobacco mosaic virus of pepper are commonly used in California. Likewise, there are many examples of resistant ornamental plants, including roses resistant to powdery mildew and black spot, India hawthorn resistant to entomosporium leaf spot, and crape myrtle resistant to powdery mildew.

Successful examples of resistance or *tolerance* to insect pests include the use of phylloxera-resistant grape *rootstock*, apple rootstocks resistant to woolly apple aphid, and varieties of alfalfa resistant to blue alfalfa aphid, pea aphid, and spotted alfalfa aphid. There are many more examples, and much more research is being devoted to this aspect of pest management. New developments in host plant resistance should be followed closely. When new resistant cultivars are introduced, pest managers should carefully monitor for host plant susceptibility to other pest species.

Several characteristics make host plant resistance an excellent tool in an integrated pest management program, including its specificity and minimal disruption of the environment. Host resistance can be a long-lasting control measure, especially if properly used and if more than one gene is responsible for resistance. For the grower, host resistance is usually cost effective. It requires little specialized equipment and is compatible with other methods of crop and pest management. Sometimes, however, seed of resistant cultivars may be much more expensive than seed of susceptible plants.

Many cultural practices are used to prevent or eliminate damaging pest populations prior to planting. Crop rotation and *fallowing* can be used to manage certain nematodes, weeds, and plant pathogens that cannot invade fields rapidly from neighboring areas or survive long periods of adverse conditions. Crop rotation and fallowing interrupt the pest's life cycle, thereby reducing or eliminating future infestations. Planting crops that are competitive with specific weeds in the rotation can eliminate harmful weed populations from future crops. For instance, weeds in strawberries can be partially controlled by rotation with cereal grains or various legumes. In grain crops, weeds are manageable by rotation with crops such as alfalfa, dry beans, potato, tomato, safflower, or sugar beets. These rotations can also manage nematodes and certain diseases (see Chapter 5, Table 5-10). In some cases, a change in planting and harvesting dates can also be made to avoid pests.

FIGURE 3-9

Grassy weeds growing around tree trunks provide protection for meadow voles and retention of water that promotes crown rot infections. A control practice that removes weeds also reduces meadow vole infestations and decreases the chance for crown rot.

Site selection can also be critical. For example, susceptible crops should not be planted in fields having a history of severe soilborne diseases, nematodes, or weeds that are particularly troublesome to that crop. In addition to the site, PCAs must be aware of the surrounding habitat, especially when controlling vertebrate pest populations. Controlling vertebrate pest populations in neighboring areas, when possible, will reduce the chance of reinvasion.

Another technique used to prevent pests and their damage is *habitat modification*.

Pests become problems in managed systems only when they are provided with the requisites they need for survival (e.g., food, shelter, alternate hosts, and proper environmental conditions). Habitat modification intentionally changes the managed ecosystem to limit availability of one or more of these requirements, making the environment less suitable for pest populations (Figure 3-9). Alternatively, habitat modification may be used to improve survival or effectiveness of a pest's competitors or natural enemies.

Integrating Management Tools. PCAs have a variety of pest control tools at their disposal. Some of these, like crop rotation, host resistance, and habitat modification, are used in the preventive mode. Other tools, such as certain pesticides, cultivating, or mowing weeds, are used primarily to curtail pest populations that are approaching damaging levels. Many tools can be used both to prevent damage and to control populations.

Most pest control tools do not eliminate all pest individuals, only a percentage of the population. Many are effective against one stage but ineffective against another stage. Some biologically based or less toxic pesticides may control only part of a pest population, but this may be all that is needed to keep a pest suppressed just enough to allow other mortality factors like natural controls to reduce the population to a tolerable level. Some pest management tools may affect several different pests. Weed control, for example, may also result in fewer vertebrate pests. The pest manager must always keep in mind the effect of any management practices on other pest organisms. Most management tools fit into one of four major categories: biological, cultural, mechanical and physical, and chemical. These categories are described here and discussed in greater detail in Chapter 5.

FIGURE 3-10

A parasitic wasp, Aphidius smithi, *lays its eggs inside a pea aphid. Naturally occurring biological control agents such as these insect parasites can be very important in the management of many pests.*

Biological Control. Biological control is any activity of one species that reduces the adverse effect of other species. Living natural enemies are the agents of biological control. For insects, the most important biological control agents are insect parasites *(parasitoids)* (Figure 3-10), pathogens, and predators that kill pests directly. For plant pathogens, weeds, and nematode pests, competitors and antagonistic organisms play a larger role. Herbivores are also important biological control agents of some weeds. Biological control integrates well with other management options in IPM and has the advantage of being relatively safe for human health and the environment.

Biological control agents are present in almost all ecosystems, even on the most intensely managed farms. Many organisms are not considered pests simply because naturally occurring biological competition in fields, orchards, and landscapes keep them in check most of the time. These natural enemies are often at very low levels and may not be easily observed; however, without them, many pests would increase to destructive levels and become pests. A basic tenet of integrated pest management is to take advantage of free, naturally occurring biological control whenever possible. In IPM, efforts are made to restore, enhance, or mimic the biological control that occurs in natural situations.

Cultural Control. Cultural controls are the modification of normal crop or landscape management practices to decrease pest establishment, reproduction, dispersal, and survival. Cultural practices include some of the oldest pest management tactics used,

FIGURE 3-11

Changing irrigation practices is a cultural control that can reduce pests. Water allowed to pond around trunks creates conditions conducive to root and crown diseases.

often exploiting weak links in the pest's life cycle or behavior. They generally require a good knowledge of crop and pest biology, ecology, and phenology to be used most effectively. Cultural controls can be especially effective for the control of some vertebrate pests. A reduction in favored food resources or removal of cover used as shelter by vertebrate species such as meadow voles can decrease the overall pest population.

Many cultural practices influence pest populations. Reduction in damage caused by insects, mites, weeds, and other pests can be achieved through the selection of plant varieties, timing of planting and harvest, crop rotation, the use of trap crops, and irrigation management (Figure 3-11). *Sanitation*, host-free periods, and intercropping are a few of the strategies that can be combined with other management tactics in an integrated pest management program for optimal pest control. Even when cultural practices cannot be modified to control pests, it is important to understand their potential impact on future problems.

Mechanical and Physical Controls. Mechanical and physical controls are measures specifically taken to kill the pest directly or to indirectly make the environment unsuitable for pest entry, dispersal, survival, or reproduction. Weak links in the pest's life cycle or specific behavioral patterns are often targeted.

Examples of physical controls include steam pasteurization of the soil, soil solarization, and cold storage of stored products. Pest barriers such as screens or sticky substances are also used as a physical control.

Cultivation for weed control is an important mechanical weed management practice. Cultivation is sometimes also used in the management of insects or pathogens, for example, burying plant litter that may harbor overwintering insects or pathogen *inoculum*. Mechanical traps for vertebrates or cone traps for flies or wasps are also examples of mechanical controls. Suction devices for insect control, such as bug vacuums, are another example.

Chemical Control. Pesticides can be important tools in integrated pest management programs (Figure 3-12). For many pests, they may be the only tool used to control damaging populations. The challenge in an integrated pest management program is to use them effectively and efficiently with minimal impact on nontarget organisms and the environment. Decisions about whether a pesticide application is needed, what material to use, and how and when to apply it are

FIGURE 3-12

Pesticide applications are one tool in an IPM program. They provide more effective control when used in combination with other management options.

central to the success of an IPM program. The decisions regarding the necessity of a pesticide application are some of the most important decisions PCAs make. The more attention PCAs put into deciding whether treatment is needed, the more likely that unnecessary treatments can be avoided, thus saving the costs of pesticide applications and lowering potential risks to the environment and human health associated with pesticide use and the potential for buildup of pesticide resistance.

In an IPM program, pesticides are generally used in combination with other management options, such as cultural practices, to achieve more effective long-term control than can be achieved using either approach alone. For management of insects and mites, natural enemies may also be important controlling factors, so it is essential that the PCA choose pesticides that are least disruptive of biological control.

Whenever a pesticide is used, be aware of its chemical class and *mode of action*, its impact on natural enemies, its potential hazards to human health and the environment, its safety precautions and label restrictions, and what to do in case of an emergency. In addition, from a pest management perspective, PCAs should know the identity of the target pest, the beneficial organisms present (especially natural enemies and *pollinators* that might be affected by the application), and alternative methods or materials that may be available for control. Various application methods are available for different pesticide products, ranging from aerial or ground foliar sprays to soil applications of granular formulations, fumigation, and chemigation. Application methods have a significant impact on effectiveness and health and environmental impacts.

The Costs and Benefits of Controls—Alternatives and Pesticide Use. Fundamental to the success of an integrated approach to pest management is that the cost of control should not exceed the economic return or increased value of the plant host due to the management activity. In agricultural operations, if the cost of controlling the pest exceeds the resulting economic return from marketing the crop, the control action is not warranted. However, in reality, assessing the economic benefit of implementing a management practice is often difficult. For instance, it is difficult to calculate actual returns in advance when the price the grower will get from a crop is not known until harvest or when the costs of a practice, such as installation of a permanent drip irrigation system that reduces weeds, may be expensive the first year but reap rewards for many years to come. In urban landscapes and some forestry systems, value is often determined by aesthetic qualities, not crop yield, so the cost-benefit assessment is more subjective; the pest manager, therefore, must determine the value of maintaining the desired aesthetic quality versus the economic and environmental costs of achieving it.

Although they may be more expensive initially, the use of certain types of practices may bring benefits that more than pay for the investment. For instance, produce to be marketed as "certified organic" is usually priced higher to account for its more expensive inputs (including a 3-year transition period), lower yields, and perceived increased value in comparison with conventionally grown produce. Some processors may disallow the use of certain pesticides of concern or be willing to pay growers more for produce grown without use of these materials. Elimination of pesticide applications may reduce risks and the need for worker training or personal protective equipment. In areas where the risk of human hazard is high, such as schools, urban parks, or other populated areas, the increased costs of more expensive alternatives may be well worth reducing the potential for problems associated with pesticide use.

There are significant economic and environmental advantages in using an IPM strategy. When management systems are based on sound knowledge of these complex biological systems, unnecessary pesticide treatments can be eliminated. By taking a holistic, ecologically based approach, management practices are implemented with the long-term goal of keeping multiple problems below tolerable levels rather than focusing on short-term control of a single pest. This strategy can result in long-term cost savings as well. When treatments are deemed necessary, selective materials are chosen to reduce chances of secondary pests and development of pesticide resistance, which can increase management costs.

If the pest population was reduced or eliminated, pest damage averted, and host quality did not suffer or was enhanced, the pest management decision can most likely be judged effective. But the cost-benefit evaluation of a pest management decision extends beyond the immediate impact on the pest population. To truly evaluate pest management decisions, their impact on the total ecosystem must be assessed. An evaluation needs to include the impact of the decision in terms of production costs, profits, and the risk of employing the recommended pest management strategy—especially if the decision involves a specific change in current management practices. Also, environmental impacts and health impacts within and beyond the treated area must be considered.

Working within an Ecosystem

PCAs are called upon to make pest management decisions on field crops, livestock areas, orchards, aquatic systems, rights-of-way, urban landscapes, and forests. Each of these areas is an ecosystem managed by people for economic, safety, or aesthetic reasons, typically requiring some intervention to be maintained. Every ecosystem is uniquely influenced by geographic, climatic, edaphic (soil-related), and socioeconomic factors, as well as by the crops, weeds, arthropods, and microorganisms involved.

To devise appropriate integrated pest management strategies, pest managers must consider the type of crop or resource being managed; the methods used to grow the crop or manage the resource; the input of labor, capital, and resources compared with the resulting output; and the ultimate plan for the resource (i.e., crop sale to cannery, urban landscaping, timber production, etc.). Pest managers must also keep in mind the environmental constraints of the managed site and the risk of human exposure, along with the specific pest populations that may arise.

The Physical Environment. The physical environment sets limits on the functioning of the ecosystem and every species in it. Available heat limits the rate of growth as well as species survival. Water limits the growth of plants and has a major impact on outbreaks of disease. Soil type influences water-holding capacity as well as nutrition. The intensity of sunlight can be a limiting factor in photosynthesis.

The pest manager can manipulate some attributes of the physical environment with such practices as irrigation and the addition of soil amendments. However, other physical conditions cannot be easily modified and must be taken into account in the overall management system. For example, rainfall patterns and sudden changes in temperature can trigger changes in the ecosystem, such as disease outbreak. While rainfall, in some situations, is a predictable occurrence, the timing, duration, intensity, and the resulting impact on pest populations can be unpredictable. A good monitoring and sampling program will help the PCA cope with these unpredictable occurrences.

Water. All crop and landscape plants need proper amounts of water to achieve optimal growth. Too much or too little water can directly damage plants and cause changes that make the plant more susceptible to pests or pest damage. For instance, overwatering is a major contributor to the development of root and crown diseases such as those caused by *Phytophthora* or *Pythium*. On the other hand, drought stress and insufficient water can increase shoot and branch dieback, providing an attractive environment for other pathogens such as Cytospora canker or Botryosphaeria canker on many ornamental trees. Tools such as tensiometers and evapotranspiration monitoring are available to assist in determining a plant's need for water.

The type of irrigation can also contribute to pest infestations. Overhead irrigation systems increase humidity and leaf wetness, which allow bacteria and fungal spores of certain pathogens to spread and infect. The pest manager must use knowledge about the impact of water on the crop and the pests in forecasting problems and recommending management strategies. Ideally, the PCA would be consulted in the design or operation of an irrigation program.

Soil. Soil is the major support medium for plant growth. It affects and is affected by plants, weeds, insects, pathogens, and other organisms, as well as the practices used to manage the ecosystem. Soil is the anchoring medium for plants and weeds; it supplies nutrients and water for both. It is a reservoir containing many organisms, including microorganisms and invertebrates that aid in the regulation of plant nutrient uptake, organic matter decomposition, soil fertility, and water retention.

Soils have individual properties and, in many ecosystems, more than one soil type is likely to be encountered. Soils are defined by such characteristics as texture, depth, *pH*, organic matter, slope, cation exchange capacity, and water-holding capacity. These features interact with one another and with the abiotic and biotic components in the ecosystem to form a complex and dynamic system.

Soil characteristics can have a significant influence on weeds, soil pathogens, insects, and nematodes, as well as on the feasibility of management practices. Soils high in clay or silt become extremely cloddy and compacted when tilled in wet conditions. Coarse-textured soils that are low in organic matter have low buffering or adsorptive capacity, which can increase herbicide activity. Soils too high in organic matter, however, bind many soil-applied herbicides tightly, and inadequate herbicide activity occurs. Plants grown in soils with high levels of nitrogen may be more prone to pests such as fire blight in pome fruit or increased aphid populations.

Light. Some activities of plants, invertebrates, and disease organisms are closely regulated by light. Light is essential for photosynthesis. Substantial shading or short days reduce the growth of plants; average daily temperature is also usually lower when days are short. Day length changes often trigger important developmental changes in plants and animals, such as changes in *dormancy* and diapause, reproductive cycles, and flowering responses. Moonlight is associated with the mating flights of some moths. Some weed seeds require exposure to light to germinate, and all plants need some light to grow vigorously and mature.

Heat. All organisms require a minimum level of heat for growth and survival. For plants and cold-blooded organisms, ambient temperatures affect development rates, with plants and insects growing faster when temperatures are higher. Very high temperatures and very low temperatures slow or stop development and can kill. Organisms that are successful in extreme environments have adopted strategies to avoid severe heat and cold, including migration, diapause, or dormancy.

Weather. Weather directly affects the activities of most organisms. For some, wind and rain can enhance dispersal, and temperature and humidity affect reproduction and development. Early frosts or severe storms can kill many individuals and contribute to the regulation of population density. Conditions of high humidity or rain can provide an environment conducive to the development of plant diseases or insect pathogens. Additionally, rain leads to more weed growth, providing shelter and food resources for other pests. The pest manager must be aware of the impact of how weather patterns affect specific pest species in managed ecosystems.

Crop and Host Biology. The ultimate focus of a management program is to achieve a healthy plant, whether it's a crop plant or an ornamental plant. For that reason, it is important to know how all elements in the managed ecosystem, including the physical environment, cultural practices, and other organisms such as pests, affect plant growth.

Plants deficient in any of the elements required for optimal growth display symptoms that often resemble those of plant disease or other damage; they compete poorly with weeds and are less able to tolerate insect and

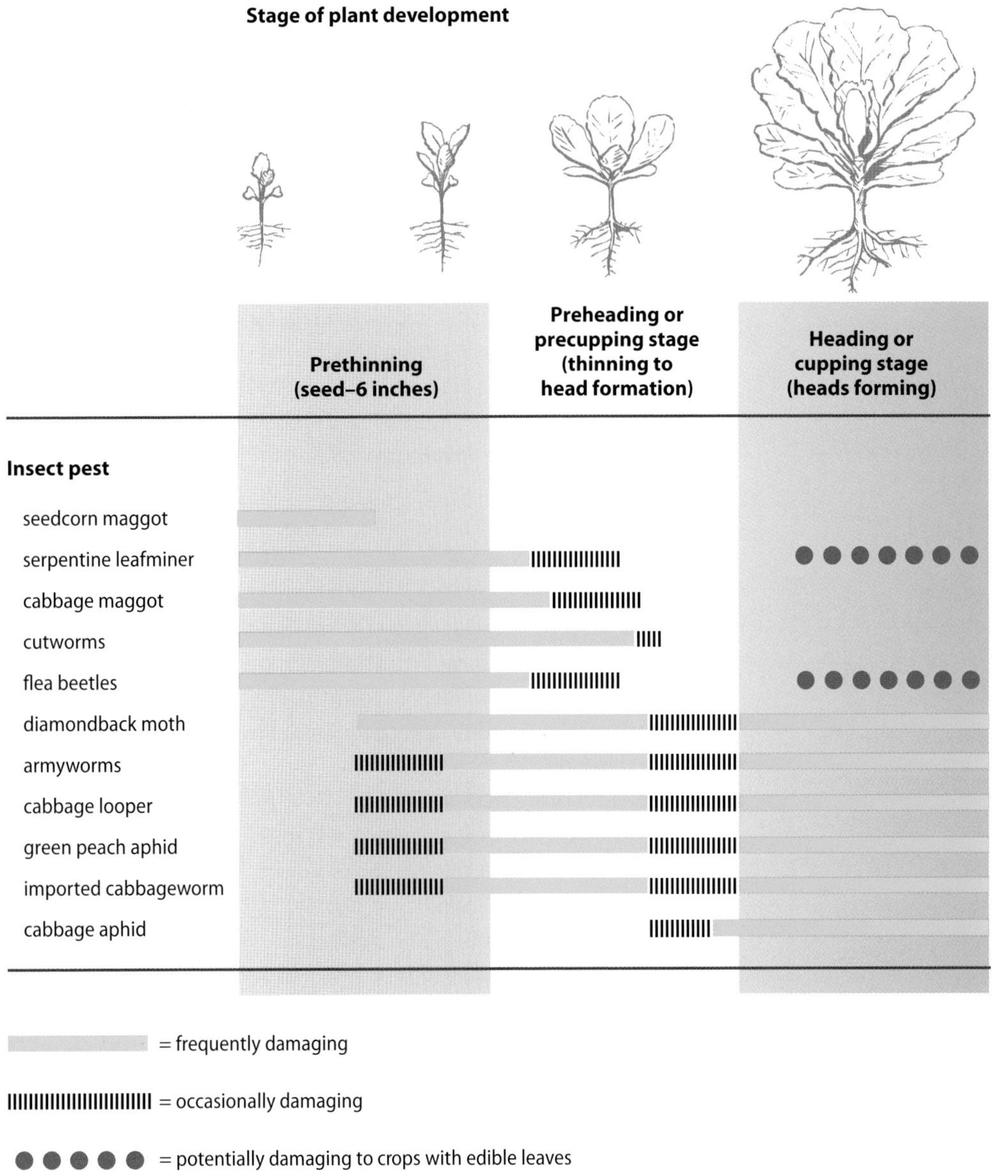

Stage of plant development

Prethinning (seed–6 inches)

Preheading or precupping stage (thinning to head formation)

Heading or cupping stage (heads forming)

Insect pest

seedcorn maggot
serpentine leafminer
cabbage maggot
cutworms
flea beetles
diamondback moth
armyworms
cabbage looper
green peach aphid
imported cabbageworm
cabbage aphid

= frequently damaging

= occasionally damaging

= potentially damaging to crops with edible leaves

FIGURE 3-13

Most pest species are of primary concern during specific stages of plant development. Shown here are the insect pests most likely to cause damage at various stages of development of cole crop plants.

disease damage. The objective of management is to provide the best conditions for growth and to minimize any negative effects of cultural practices, harvesting, pest damage, and other stresses.

The pest manager must be aware of the susceptibilities of the host during the life stages important in crop production. Annuals normally go through three developmental stages: germination and seedling development, *vegetative growth*, and reproductive growth. The germinating seedling is susceptible to a variety of soil pathogens, vertebrates, and invertebrates. After the plant enters its vegetative stage and has produced 4 to 6 *true leaves*, a different pest complex is likely to become important. As the plant continues its vegetative growth, it becomes more tolerant to certain types of damage. In fruit crops, pests that attack flowers or fruits cause serious damage, even at low levels. As these pests are often hard to spot and control, vigilant monitoring is required at this time. When it is the leafy parts of the crop that are sold, tolerance for defoliators is much lower than for crops that produce fruit. Figure 3-13 shows the susceptibility of a cabbage crop to insect pests at various growth stages.

Management of fruit and nut trees must be directed toward long-term maintenance of overall plant health and vigor. In the first year, stress caused by disease, nematodes, weed competition, or insufficient irrigation can hinder root development, which is critical for tree establishment. In the second and third years, young trees are most susceptible to shoot- or twig-boring insects and weed competition, which can significantly reduce plant growth and encourage lethal plant diseases, such as bacterial canker, and vertebrate pests. During the bearing years, the most serious pests are those that injure the marketable fruit or nut directly or interfere with the tree's water- and nutrient-conducting systems, reducing the long-term vigor of the tree.

In the landscape, trees and shrubs live for many years. For deciduous species, seasonal growth begins in the spring with root and shoot elongation. Buds swell and expand into foliage and flowers. As the dormant season approaches, food produced by leaves is increasingly diverted from growth and maintenance to storage in roots, trunks, and limbs. During dormancy, the plant slowly consumes the stored food by converting it to energy until conditions again become favorable and the seasonal growth cycle is repeated.

Pest Biology. Knowledge of the biology of the pest allows for better timing of management actions and selection of the best control options. In a managed system, the PCA must know the potential for damage by the specific insect, pathogen, weed, or vertebrate pest and the options available to manage it. Plants are susceptible to most pests only at certain times. Knowing the vulnerable stages of the plant and the pest stages that cause damage sets the parameters for effective monitoring and treatment.

Understanding why pest species are able to exploit the managed system can also help in determining appropriate management actions. The practices that enhance the growth and development of the plant or crop are often advantageous for the pest species as well. Many insects and pathogens thrive best when high rates of nitrogen application and irrigation stimulate lush new plant growth. The weeds that compete most successfully with the host for light, water, and nutrients are often those that emerge before the crop with the first rain or irrigation.

Certain characteristics are common to many successful pest species. Generally speaking, arthropod, plant, and pathogen species that become pests possess strong dispersal abilities, and their populations increase rapidly (they are r-strategists, as discussed in Chapter 2). Additionally, arthropod and pathogen pest species often have strong host-finding abilities or a life cycle closely tied to that of the host. Population-regulating mechanisms such as biological control agents, competitors, or limits to food supply, may be absent or reduced when pests become a problem. Pests have, in most cases, adapted to the host's environment and are able to exploit these conditions to the disadvantage of the host.

Environmental conditions in the crop can be manipulated to the disadvantage of the pest species. Fallowing disrupts the life cycle of certain disease organisms and nematode species. Crop rotation can be effective against many soil pathogens, nematodes, vertebrates, and certain weed species. Removal of basal leaves from grape vines makes the environment less favorable for Botrytis bunch rot fungi and leafhoppers. Overhead sprinklers can reduce mites but may increase some diseases. Knowledge of crop and pest biology allows the PCA to time pest management strategies to the pest's most vulnerable stage.

The Influence of Control Strategies on the Environment. The impacts of control strategies are not limited to the borders of the managed ecosystem. Pest management actions can affect the quality of air, water, and soil, as well as the diversity of species at sites well beyond the treatment area. Any evaluation of a control action must assess the effect on adjacent crops,

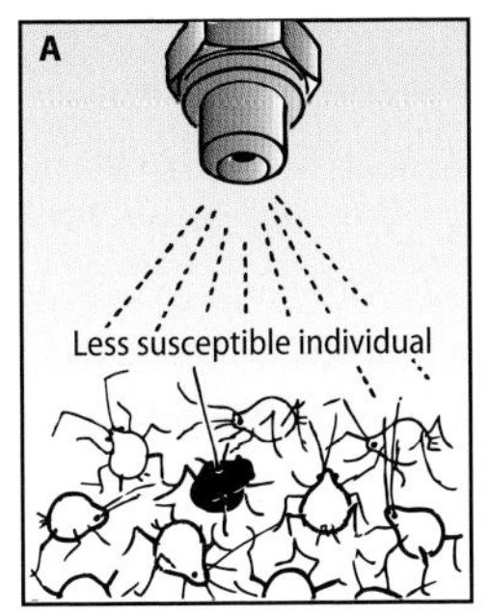

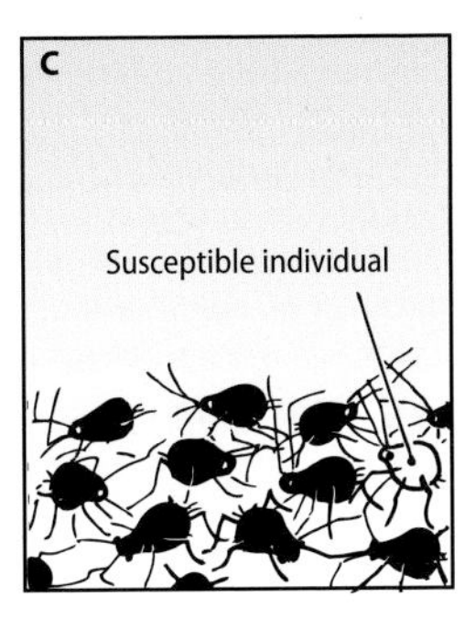

FIGURE 3-14

Pest populations develop resistance to pesticides through genetic selection. (A) Certain individuals in a pest population are less susceptible to a pesticide spray than other individuals. (B) These less-susceptible pests are more likely to survive an application and to produce less-susceptible progeny. (C) After repeated applications, the pest population consists primarily of resistant or less-susceptible individuals, and applying the same material or other chemicals with the same mode of action is no longer effective.

nontarget species such as honey bees and natural enemies, and the peripheral effects of soil and water contamination.

WHY USE IPM?

In the past, pest control often relied on a single pest control method, such as repeated pesticide applications. Sole reliance on single pesticides led to pest resurgence, secondary pest outbreaks, resistance to pesticides, and environmental contamination. Integrated pest management combines a variety of cultural, biological, mechanical, and chemical controls to optimize effectiveness and reduce the problems that can result from overreliance on one tactic. Even more importantly, by incorporating information developed from research and field monitoring, IPM allows more reliable decisions to be made, thus improving effectiveness of control actions.

Pesticide Resistance

Repeated application of pesticides has led to the development of populations of pest species that are resistant to pesticides that once controlled them. Once resistance begins to develop, higher rates of the same pesticide are necessary to achieve the same amount of control. Eventually, little or no control is achieved despite repeated applications at maximum rates.

Resistance develops through natural selection when some individuals in a population have genetic characteristics that allow them to tolerate or overcome the toxicity of a pesticide (Figure 3-14). These individuals are more likely to survive an application, and when they reproduce, the genes that have allowed them to survive the pesticide application are passed to succeeding generations. If applications of the same pesticide continue, over time the pest population will consist primarily of resistant individuals. Biological factors influence the development of resistance. Short-lived, rapidly developing pests with many offspring, limited mobility, and many generations per year, such as aphids or mites, are most likely to develop resistance rapidly.

Pesticide resistance has become a serious problem, limiting the number of effective pesticides available in many cropping systems. More than 500 species of insects and mites are known to be resistant to at least one pesticide. Resistance has been documented in 100 weed species, 100 species of plant pathogens, and in nematodes and rodents. See the section "Resistance Management Strategies" in Chapter 5 for tips on avoiding pesticide resistance.

Pest Resurgence

Reliance on pesticides as the primary control option has also led to pest resurgence. Pest resurgence occurs when the pest and its complex of natural enemies or its competitors are killed by a pesticide application (Figure 3-15). Any segment of the natural enemy population that survives the application often starves since the controlled pest population is too low to provide an adequate food source. Although the pest population initially declines from pesticide mortality, without the restraining effect of natural enemies, the pest population is able to rebound, often to higher levels than were present before the pesticide application.

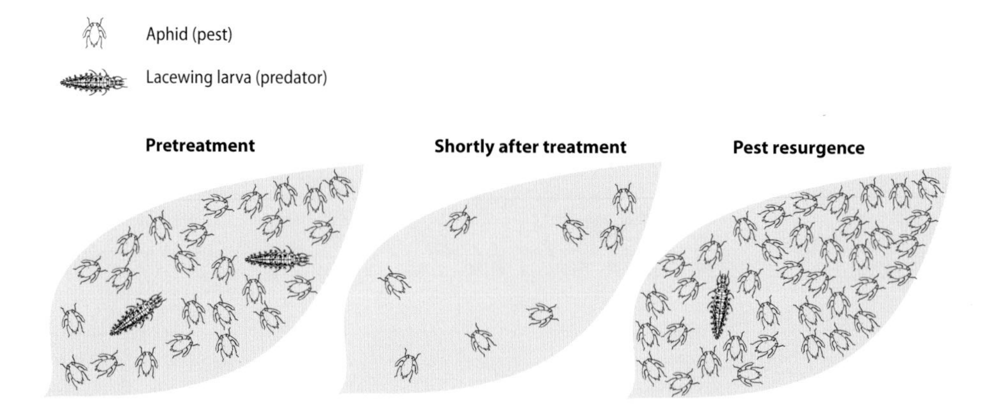

FIGURE 3-15

Target pest resurgence can result when natural enemies are destroyed. In comparison with their effect on pest species, pesticides often kill a higher proportion of the populations of predators and parasites. The immediate effect of spraying is not only a reduction in the number of pests, but also an even greater reduction of the natural enemies. The resulting unfavorable ratio of pests to natural enemies permits a rapid increase, or resurgence, of the pest population.

Secondary Pest Outbreak

Secondary pest outbreaks are also primarily the result of the destruction of the natural enemies that had been keeping the secondary pest under effective biological control (Figure 3-16). These outbreaks occur when a species that was previously innocuous or a minor pest erupts following the application of a pesticide meant to control another pest, usually a key pest. For instance, spider mites and leafminers frequently become secondary pests when their natural enemies are killed following the application of certain broad-spectrum insecticides targeting lepidopteran or other key pests.

Species Displacement

In some situations, extensive use of a single herbicide has created a phenomenon known as weed species displacement, a problem that is similar to secondary pest outbreaks. Species displacement occurs when herbicide applications to control one weed species or species complex eliminate the weed of primary concern but leave a few other weedy plants uncontrolled. If the same herbicide is used repeatedly, secondary weed species that are either resistant or tolerant to the herbicide being applied are able to thrive in the treated area. This new, thriving weed species may turn out to be more competitive and harder to control than the original weed species.

Pollinators

Bees and other insects that pollinate crops may be poisoned by pesticide applications. Honey bees are important pollinators for many crops and are themselves an important agricultural commodity. Numerous wild bee species, including leaf-cutting bees, alkali bees, and bumblebees, are also important for some crops. Pesticides are especially harmful to bees if applied during the bloom period. While insecticides are most commonly associated with bee poisonings, some herbicides and fungicides harm bees, too. Herbicides applied to bee forage plants can be potentially harmful to hives as well. Information on the hazards of pesticides to honey bees can be found in the UC IPM Pest Management Guidelines for various crops at www.ipm.ucdavis.edu.

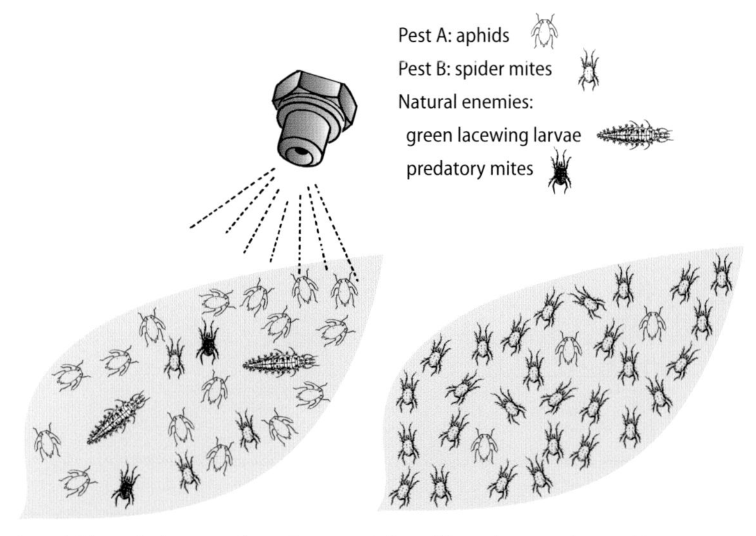

FIGURE 3-16

Secondary outbreaks of insects and mites are often caused by the destruction of natural enemies. For example, spider mites are often present on plants at low densities, but they become excessively abundant and cause damage when pesticides applied against other species kill the natural enemies of these formerly innocuous species. Here a pesticide applied to kill aphids (pest A) not only killed aphids, but also killed predaceous green lacewing larvae and predatory mites, leading to a secondary pest outbreak of spider mites (pest B).

Environmental and Health Risks

The impact of pesticides on human health and the environment continues to be of concern. People come into contact with pesticides in many ways. Applicators and fieldworkers are at highest risk, but the general public may be exposed to pesticide residues in food, water, or surfaces in their homes and gardens. There are also concerns regarding pesticide movement into groundwater, *surface water*, and air, and about their *persistence* in the soil. Many communities are extremely sensitive to the idea that they or their children might be exposed to *any* pesticide. IPM provides a framework for ensuring that applications are limited to situations in which they are needed and for choosing the safest materials for the situation. See Chapter 8 for more information on health and environmental problems associated with pesticide use.

More Reliable Control

IPM focuses on managing pest populations through a combination of tactics. Although the relationships between pests and the managed ecosystem are complex, reliance on multiple complementary tactics increases the effectiveness of pest management. Many biological techniques kill only a portion of the pest population; their effectiveness increases when combined with cultural controls, mechanical controls, or selective pesticide applications. Incorporation of multiple tactics in pest management requires more knowledge about the cropping system and the pest complex.

A PCA must make pest management decisions that are appropriate to the ecosystem and the site being managed. An IPM approach allows flexibility by relying on up-to-date field information. The decision to treat is justified by knowledge of pest biology, established decision-making guidelines, and the results of regular field monitoring. All available pest management options, including the decision that no action is needed, are considered for their impact on the ecosystem and on production economics.

IPM is site specific: it varies by ecosystem, plant, crop, cropping system, and location. IPM is dynamic: the ability of pests to adapt to adverse environments and materials, host plant resistance, and cultural and biological controls constantly alter pest pressure and the pest management options available. IPM is knowledge intensive: a solid foundation of plant and pest biology and their relationship to the environment is needed to combine control tactics into a systematic approach for optimal pest control. IPM is realistic: it addresses society's concerns regarding health and the environment, the economic issues involved in managing ecosystems, and a manager's needs for reliable pest control. A successful IPM program must be economically viable.

4 Understanding Pests

Photo by *David Rosen*

Not all organisms found in crops, forests, or landscapes are pests. Many organisms are beneficial; others are of no consequence to the area being managed. However, a small number of organisms interfere with the availability, quality, or value of a managed resource. These organisms are called pests. For many pest species, a few individuals or light damage caused by their activities can be tolerated. But when organisms have an intolerable negative effect such as interfering with crop or animal production, damaging ornamental plantings, transmitting plant or animal disease, or are generally a nuisance, some management action must be taken.

Pest status is achieved when the population or activity of an organism conflicts with human needs or values. Successful pests often have strong competitive abilities and the capacity to reproduce rapidly over a short time span or under special conditions. They have the ability to adapt to uncertain and variable environments and have strong dispersal and host-finding capabilities.

PEST IDENTIFICATION

Misidentification of pests can contribute to the failure of pest control measures. Many pests look similar, and some can easily be confused with beneficial or innocuous organisms (Figure 4-1).Frequently, damage symptoms are incorrectly associated with an organism that happens to be present at the time the symptoms are observed when, in fact, that organism is not causing the problem. The pest that actually caused the damage may have left the site or may be hard to detect, such as rodents that damage a tree

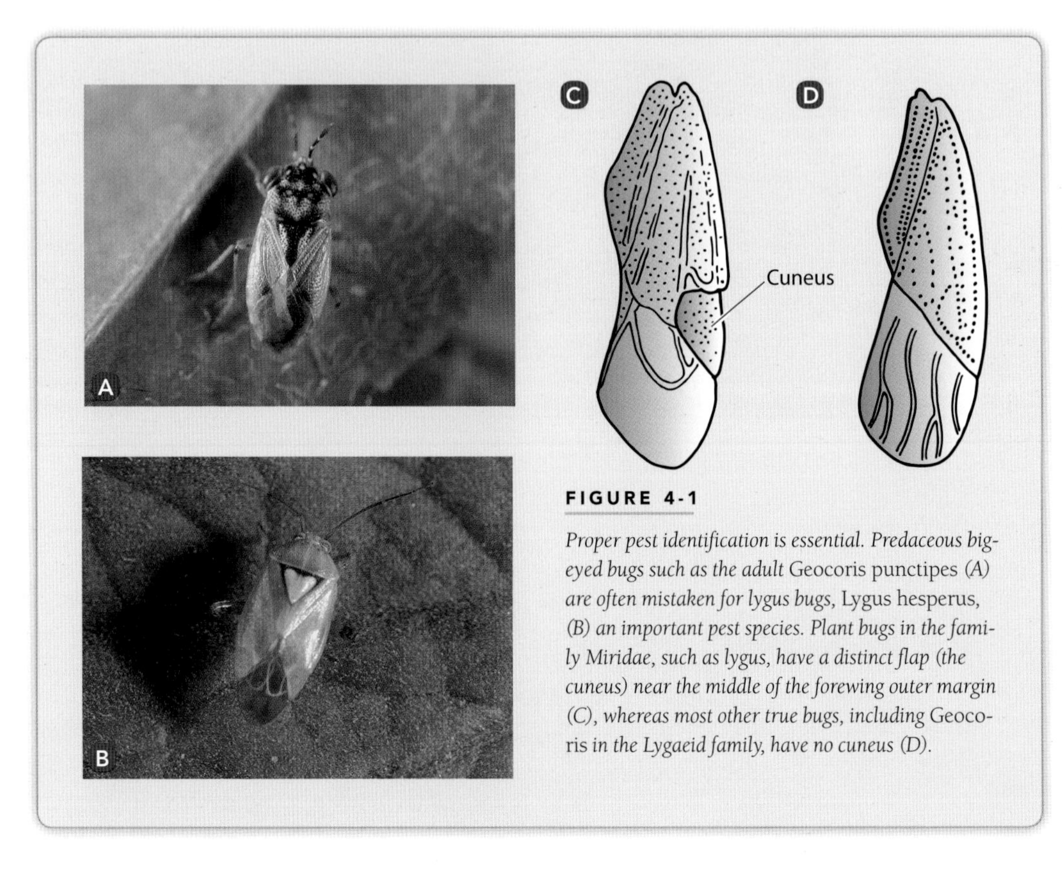

FIGURE 4-1

Proper pest identification is essential. Predaceous big-eyed bugs such as the adult Geocoris punctipes *(A) are often mistaken for lygus bugs,* Lygus hesperus, *(B) an important pest species. Plant bugs in the family Miridae, such as lygus, have a distinct flap (the cuneus) near the middle of the forewing outer margin (C), whereas most other true bugs, including* Geocoris *in the Lygaeid family, have no cuneus (D).*

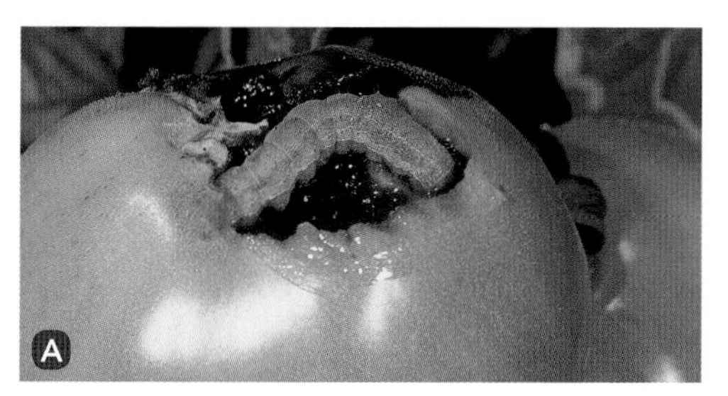

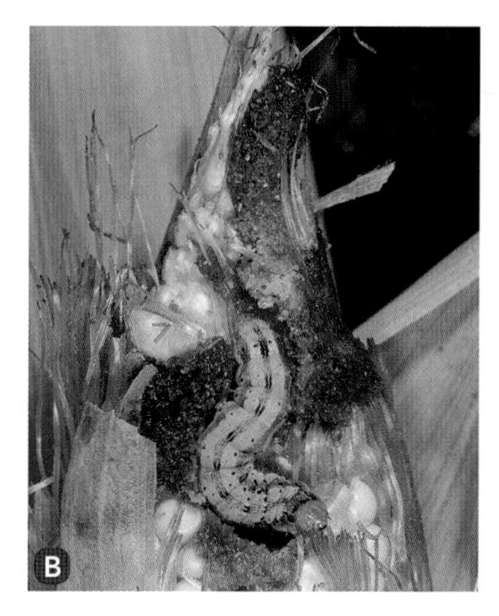

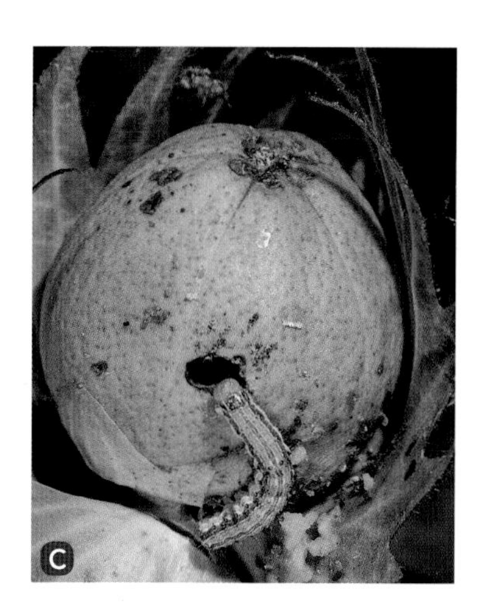

FIGURE 4-2

Common names are often not a dependable way to identify pests. Some pests may have more than one common name. For example, the pest Helicoverpa zea *is known as the tomato fruitworm (A), corn earworm (B), or cotton bollworm (C), depending on which crop it is infesting.*

trunk and then leave or a plant pathogen confined within the root system and thus unable to be seen above ground on leaves or stems. Damage symptoms can also be caused by factors other than pests such as over- or underwatering, toxins in the soil or water, air pollutants, cold, heat, hail, wind, or genetic disorders.

Be prepared to identify all common pest and beneficial species significant to the crop or ecosystem being managed. Associating damage with a specific species can be complex because the simultaneous presence of multiple pests complicates the situation. *Identification keys*, consultation with identification experts, and the use of laboratory analysis are tools to help make the correct diagnosis. Once the pest is identified, seek out information on its life cycle, growth requirements, and other characteristics to help determine control strategies. Check the "Resources" section at the end of this book for additional sources of identification information. Also, be aware that some pests, such as pathogens and nematodes, can be diagnosed reliably only by experienced professionals; do not hesitate to seek their advice.

Names of Pests and Other Organisms

The same pest occurring in different parts of the country or in different crops may be called different names by people in the field. A good example is *Helicoverpa zea*, which is known as the corn earworm when it attacks corn, the tomato fruitworm when it attacks tomatoes, and the cotton bollworm when it attacks cotton (Figure 4-2). Multiple common names such as these can cause a great deal of confusion when trying to identify a pest or discuss it with an authority.

To overcome this type of confusion, scientists use a unique, two-word Latin name for each animal, plant, and microorganism species. In the case of the corn earworm–tomato fruitworm–cotton bollworm, the scientific name is *Helicoverpa zea*. This *scientific name* provides the surest identification for the pest species because the same scientific name is used throughout the world. The first word, *Helicoverpa*, is the *genus*, or generic, name, and it is capitalized. The second word, *zea*, is the specific name, or *specific epithet*, and it is not capitalized. The genus and the specific epithet combined form the species name. Both words are italicized or underlined and are in Latin so scientists can understand what plant or animal others are referring to, regardless of language. After its first use in text, the genus name is often abbreviated, for instance, as *H. zea*. Sometimes scientific names are followed by a nonitalicized name indicating who first described it, for example, *Helicoverpa zea* (Bodie). Parentheses indicate that the genus name has been changed since the first discovery. When several species in the same genus are discussed together, species may be abbreviated as spp. (*Helicoverpa* spp.). When referring to only one unnamed species, sp. is used (*Helicoverpa* sp.). Where a subspecies is used, as with *Bacillus thuringiensis* subspecies *kurstaki*, the abbreviation is ssp.

Scientific names are part of a hierarchical organization or classification system that reflects their relationship to other organisms. The highest division among organisms is the *kingdom*. Most plants and animals are easily classified into two groups: the plant kingdom and the animal kingdom. It is more difficult, however, to classify microorganisms because of their large numbers and diversity. Several smaller, less well recognized kingdoms include the fungi, protozoa, procaryotae, and viruses.

Each kingdom is divided into six subcategories or taxonomic groups: phylum, class, order, family, genus, and species. Organisms are separated into each group according to unique characteristics that set them apart from other organisms. For example, within the animal kingdom, in the phylum Arthropoda all organisms have jointed appendages and an external skeleton, while the animals in the phylum Chordata have a backbone, a spinal nerve cord, and an internal skeleton. The convergent lady beetle example below shows the classification of this species in the various taxonomic groups:

Kingdom: Animalia (animals)
- Phylum: Arthropoda (arthropods)
  - Class: Insecta (insects)
    - Order: Coleoptera (beetles)
      - Family: Coccinellidae (lady beetles)
        - Genus: *Hippodamia*
          - Specific epithet: *convergens*

The species as defined by the genus and the specific epithet is the basic unit of classification. A species is unique from all other organisms even though there may be genetic variations among individuals within a species in traits such as color, size, or the ability to attack a specific crop cultivar. Generally, individuals of the same species can interbreed with each other but cannot normally interbreed with different species. When identifying pests in the field, identify them to the species level with proper scientific names.

It is also important for PCAs to be aware of *common names* of pest organisms. These are often the names that growers or other clients use and are the pest names normally referred to on pesticide labels. As indicated earlier, many pest species may be called by several different common names, causing confusion. In addition, common names do not provide any information about the relationship of one organism to another; knowing that species are closely (or distantly) related can be useful in making pest management decisions. For example, the common names of johnsongrass, sudangrass, and grain sorghum provide no clues as to the relationships of these three plants, but their scientific names, *Sorghum halepense*, *Sorghum sudanense*, and *Sorghum bicolor*, immediately indicate that they are closely related species in the genus *Sorghum*. Good sources of common and scientific names for pest species can be found in the "Resources" section.

Identification Keys

Identification keys provide a systematic way to identify living organisms. Many keys are written by taxonomists primarily for other taxonomists and include all the known species in a family, order, or other taxonomic group in a large area such as North America. These keys are difficult to use because they include many closely related species that are very similar in appearance and may require dissection to positively identify. In the field, these species are often not likely to be found together; each may be found only in certain areas or habitats and perhaps feeding on different plant hosts. More simple identification keys have been developed for specific crops or situations, and these are much easier for field practitioners to use because only the species known to be associated with that habitat are included.

Most keys are dichotomous, which means that they are based on a series of sequentially paired statements that the user must choose between. An example of a dichotomous key is shown in Figure 4-3. To use a *dichotomous key*, select the statement that best fits the pest being identified and proceed to the next pair of statements. Continue working through the key in this manner until the pest's identity has been revealed. Well-illustrated identification keys with photographs or drawings are easiest to use. Illustrated keys provide a visual description of the pest to facilitate faster identification; some printed keys are also compact and durable enough to be used

Key to Orangeworm Larvae in California Citrus
(partial key; from ***IPM for Citrus,*** ANR Publication 3303)

To use this key, start at the first pair of descriptions and choose the one that applies to the specimen to be identified; continue to the number indicated. Always read both descriptions and compare them to the specimen; if there are pictures or drawings provided, compare them to the specimen as well. Continue down the key until the description accurately matches the specimen.

1. Scent glands present on top of prothorax, sometimes retracted in slit.
   Black Anise Swallowtail

   Scent glands absent.
   See 2.

2. Tufts of hair on top of prothoracic and abdominal segments 1 to 4 and 8.
   Western Tussock Moth

   Tufts of hair absent on prothoracic and abdominal segments.
   See 3.

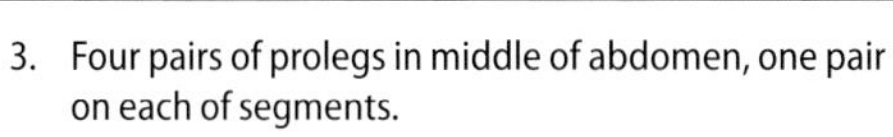

3. Four pairs of prolegs in middle of abdomen, one pair on each of segments.
   See 4.

   Only one pair or two pairs of prolegs in middle of abdomen.
   Citrus Looper

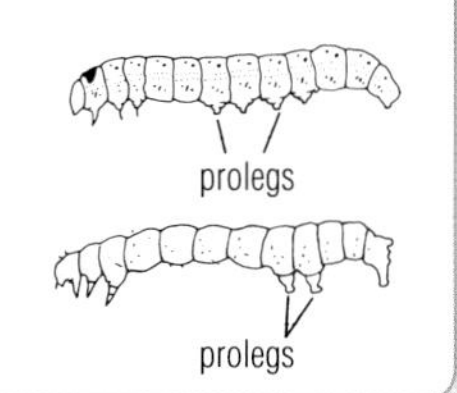

FIGURE 4-3

Example of a dichotomous key. Users choose between a series of paired statements, as if following the branches of a tree, until an identification is made.

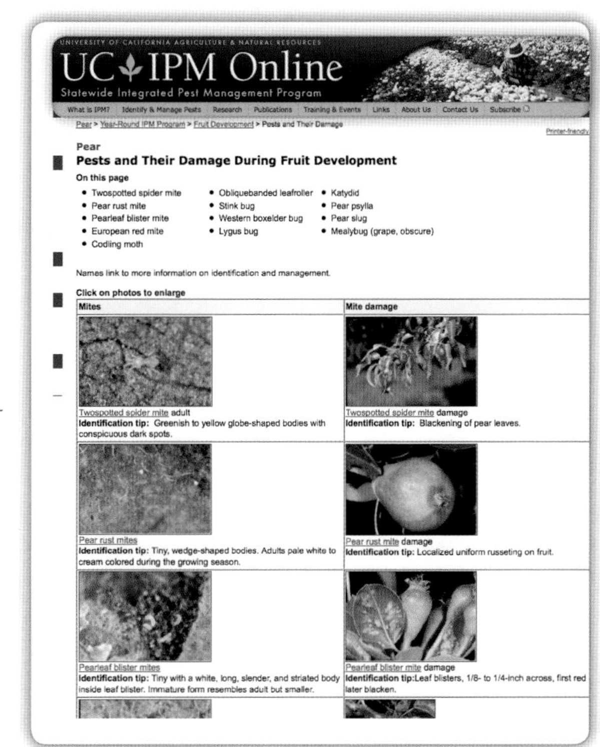

FIGURE 4-4

Online pest identification guides such as this one on pests damaging pears from the UC IPM Management Guidelines: Pear *on the IPM website (www.ipm.ucdavis.edu) will help you distinguish damage in specific crops.*

in field conditions. Although many keys for insect and weed pests rely on physical features of the pest organism itself, some keys are based on observation of symptoms or *signs*. These keys are especially helpful in identifying vertebrates and pathogens because the causal agent is often gone or difficult to see.

A great variety of well-illustrated online keys are also available. Many provide relatively quick and easy-to-use identifications. Some are dichotomous, and others are multi-access keys. Multi-access keys use a database of identification characters that allow users to start with one or more characters and then narrow down the choices until the pest or weed identity is confirmed. New keys are being developed regularly. In California, check the University of California Statewide IPM Program website, www.ipm.ucdavis.edu, for online keys to many pests. In other states, check local Cooperative Extension IPM websites.

In addition to keys, many Cooperative Extension IPM programs and other organizations make available a variety of printed and online resources to help PCAs, growers, and others identify pests and understand their biology and management. An example from the University of California is shown in Figure 4-4. Find out what is available locally for the crops or situations you work in.

Identification Experts

Although simple keys and photo identification guides that show pests and damage symptoms are a great asset in the field, sometimes accurate identification can be made only by trained experts, in some cases using special techniques and equipment. For instance, some microbial pathogens require special laboratory testing for identification, including incubation at specific temperatures to induce growth, use of selective nutrient media, and examination under electron microscopes. Useful molecular tests, such as enzyme-linked immunosorbent assay *(ELISA)* or polymerase chain reaction (PCR), are described in the section "Laboratory Tests" in Chapter 6.

It is essential that pest managers seek outside expertise when confronted with a pest species that has never been seen before and is not included in publications on local pests. Have these pests identified by local authorities so you can be certain what pests you are dealing with and to be sure they are not new invading species.

Identification experts may be associated with public agencies or private laboratories. County agricultural commissioners and their staff are helpful resources for pest identification and in California are associated with the California Department of Food and Agriculture (CDFA), which maintains a pest identification laboratory. Similar services are available in other states. Most county Cooperative Extension offices have people with expertise in pest identification and are able to refer unfamiliar specimens to other university experts. Private laboratories are the appropriate choice for routine specimens or for identifying large quantities of material. Choose a lab that is well recognized for its expertise in the type of pest you need identified.

When outside expertise is sought, it is often necessary to submit samples of the pest to be identified. Most labs have specific procedures to follow; check with them for their preferences. Samples must be as fresh as possible for accurate identification. Detailed notes on field conditions and location, extent of damage or infestation, population intensity, and all applicable conditions should be included with the sample. For insects and mites, collect all life stages present. Weed samples should include the entire plant, including the roots and flowers, if present. Pressed plant samples are acceptable if weed specimens cannot be maintained in good condition. Pathogen samples should include the diseased portion, the interface area between the diseased portion and the healthy plant tissue, and healthy tissue. Nematode samples should be a composite of soil samples, including roots, from the field (Figure 4-5). See Chapter 6 for more complete details on sampling.

Most laboratories process nematode samples using extraction methods that detect only the worm-shaped invasive juveniles and males of cyst nematodes and the worm-shaped stages of root knot, needle, and other nematode species. When getting nematode samples analyzed, be sure the laboratory notes the extraction method used. Also, laboratories vary in the service they offer. Some laboratories interpret lab findings; others do not. If the laboratory does not provide this service, ask them where you can go for help in interpreting results.

Other Identifying Characteristics

Damage symptoms are useful clues for the identification of many pest organisms. The type of damage, such as chewed, stippled, or discolored leaves, can lead to the identification of groups of organisms that cause those symptoms. Bored holes, frass, or sap on tree trunks may indicate boring insects. Growth abnormalities, deformations, or color changes to host tissues can provide clues that indicate symptoms of certain pathogens.

In the case of vertebrate pests, tracks, burrows, mounds, or holes that could be den entrances can indicate the presence of particular species. Rodents often leave identifying gnaw marks on tree trunks or other objects or dig unique burrows in the ground; often tracks or fecal pellets can be found to assist in identification.

Damage symptoms, however, must be considered with care because many pests and abiotic factors cause similar symptoms. It is always best to identify the causal organism (i.e., the pathogen) to prevent misdiagnosis. Furthermore, if the pest is no longer present, there may be no need for control.

How to Identify a New Pest Situation

Occasionally new pests emerge and their impacts on managed systems have to be accurately assessed. The first concern of the pest manager is how the new pest will affect the managed ecosystem—does it need to be controlled immediately? If the answer is positive or uncertain, contact the county Cooperative Extension office or agricultural commissioner's office for help. Local Cooperative Extension offices also have expertise in identification and help in devising appropriate management strategies for new pests.

SAMPLING

1. Take random samples, but sample in a consistent manner. Mark off area to be sampled into grids. Randomly select areas to be sampled.
2. Keep samples from suspected diseased areas separate from samples taken in healthy areas.
3. Dig up and include roots and soil of diseased and healthy plants. Take root samples from below the level of surface feeder roots because temperature and moisture fluctuations at the surface will affect the nematode species living there.
4. Include partially rotted or decayed roots.
5. Include aboveground plant parts if it is suspected that nematodes are infesting these areas. Keep aboveground parts separate from root and soil samples.

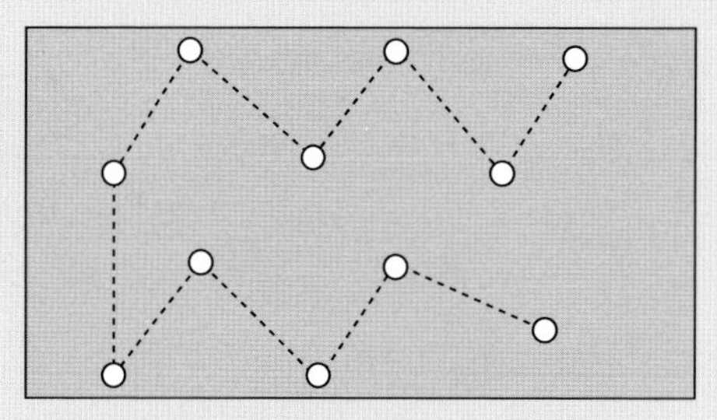

Recommended sampling pattern for collecting soil samples from a sampling block in a fallow field.

PREPARATION

1. Place samples in clean plastic bags and keep them out of the sunlight. Keep samples in an insulated ice chest or refrigerator until they can be shipped. Do not freeze.
2. Pack samples in a well-insulated carton or disposable foam ice chest. Be sure container is sturdy to prevent damage to its contents.

LABELING

Attach a label to the outside of each sample bag. Include the following information on the label:

1. Name and address of grower or property owner.
2. Crop or plant type, including variety.
3. Location of field or property (names of nearby crossroads).
4. Portion of planted area that sample represents. Include a map if necessary.
5. Brief description of the crop history (include previous crops).
6. Observations by you and the owner or operator of previous problems and of when present problem was first detected.
7. Dates samples were taken.

SHIPPING

1. Contact the person or laboratory who will receive samples to determine the best method of shipping and to inform them that samples will be arriving.
2. Mark the package clearly and request that the shipper keep it in a cool location.
3. Ship packages early in the week so they will arrive before a weekend.

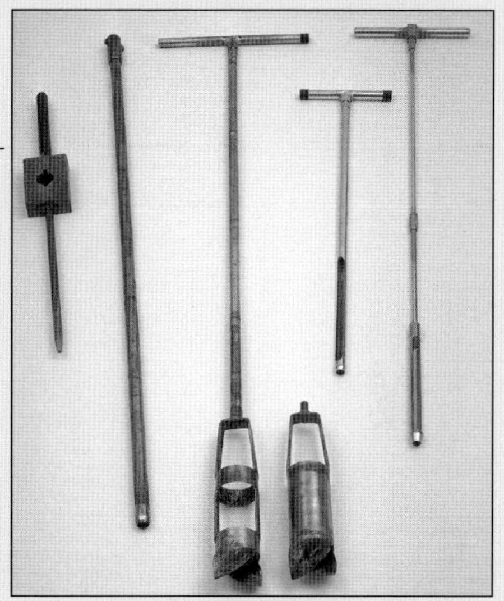

Tools for taking soil samples. The Veihmeyer tube (left) has a slotted hammer for driving the tube into the soil and removing it. Soil augers (center) have a variety of bits for different soil samples. Oakfield soil tubes (right) usually are easiest to use, especially for samples down to 2 feet (60 cm).

FIGURE 4-5

General guidelines for sampling and sending plants for identification of plant-infesting nematodes. See the crop UC IPM Pest Management Guidelines at www.ipm.ucdavis.edu for specific guidelines for individual crops.

A variety of factors can lead to the development of a new pest. Sometimes organisms that have been present in the area but have not been previously identified as problems become pests. Changes in management practices or plant density, the introduction of a crop or horticultural species, or a new or different location for a planting can contribute to the development of new pest situations. Environmental changes, such as the introduction of irrigation or unusual weather, can also be a contributing factor. In other cases, a new exotic pest species is introduced from abroad.

Exotic and Invasive Pests. Exotic pests occasionally arrive from other states or countries. In some cases, these exotic pests can pose severe threats to agriculture, natural areas or landscapes, homes, and people. Exotic pests may arrive on shipments of produce, wood, or nursery plants or in packing materials from other areas, or they may hitchhike with travelers in luggage or on vehicles. They usually arrive without the natural enemies that keep them in check in their native range. Some of the most serious invasives are those that vector plant disease, such as the Asian citrus psyllid that can vector the deadly huanglongbing (HLB) disease of citrus. Many exotic invasive pests are of major concern in California. The glassy-winged sharpshooter (an insect) (Figure 4-6) and purple loosestrife (a weed) are two invasive species that are established in some

TABLE 4-1.

Distinguishing characteristics among the six arthropod classes that include major pest and beneficial species.

Class	Size	Distinguishing characteristics
Arachnida (spiders, mites)	40,000 species 9 orders	No antennae, 4 pairs legs, 2 main body parts, wingless, all habitats. Includes pseudoscorpions, sun spiders, jumping spiders, mites, ticks.
Chilopoda (centipedes)	3,000 species 4 orders	1 pair antennae, body long and segmented (15–181 segments) with 1 pair legs per body segment, first pair legs contain poison glands, 1 pair jaws, trachea, wingless, terrestrial, nocturnal, mostly predaceous.
Crustacea (crabs)	26,000 species 30 orders	2 pairs antennae, 1 pair jaws, 2 pairs maxillae, 5 or more pairs jointed legs, calcareous exoskeleton, gills, biramous (paired) appendages, wingless, mostly aquatic. Includes shrimps, water fleas, barnacles, lobsters, crayfish, pillbugs, sowbugs.
Diplopoda (millipedes)	8,000 species 7 orders	1 pair antennae; body long, cylindrical, and segmented (9 to more than 100 segments), with 2 pairs legs per body segment; wingless; prefers moist areas; terrestrial; decomposers.
Insecta (insects)	2 million+ species 31 orders	1 pair antennae, usually 3 pairs legs and 2 pairs wings (as adults), mouthparts variable, chitinous exoskeleton, trachea, all habitats. Includes beetles, butterflies, moths, flies, roaches, lice, fleas, termites, bees, wasps.
Symphyla (symphylans)	< 100 species 1 order	1 pair antennae, long and segmented (15–22 segments), 10–12 pairs legs, wingless; prefers humus soil, decaying wood, and other moist conditions.

areas and threaten to invade other areas. Newer exotic species of concern include Diaprepes root weevil, light brown apple moth (LBAM), and various aquatic weeds. Some of the worst invasive plants in California, saltcedar and yellow starthistle, have caused substantial changes to California's wildlands. Insect-carried diseases such as West Nile virus threaten public health and also affect horses and native birds.

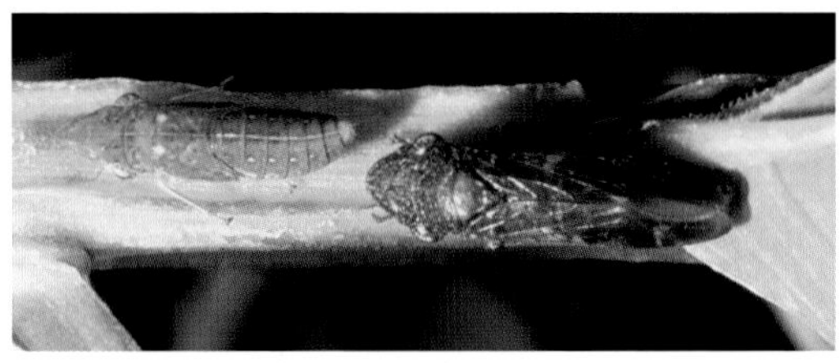

FIGURE 4-6

Be on the lookout for new invasive pests such as this glassy-winged sharpshooter (nymph on left, adult right), which invaded southern California in the 1990s but is still not established in northern California. It carries a bacteria that can cause serious diseases in many crops and ornamentals, including Pierce's disease of grapes.

If you think you have found an exotic pest, contact the county agricultural commissioner's office or Cooperative Extension office for accurate identification and information. The CDFA and the U.S. Department of Agriculture (USDA) Animal and Plant Health Inspection Service will assess what actions, including eradication, may need to be taken to address exotic pests.

INVERTEBRATES

Invertebrates are animals without backbones. They include nematodes and segmented or true worms, snails and slugs, and the arthropods, which include insects, spiders, mites, crustaceans, and their relatives. Many invertebrates are classified as pests. Some transmit pathogens to people, animals, or plants. A large number of invertebrates feed on plants. Some are considered nuisance or aesthetic pests. Others are predaceous and beneficial.

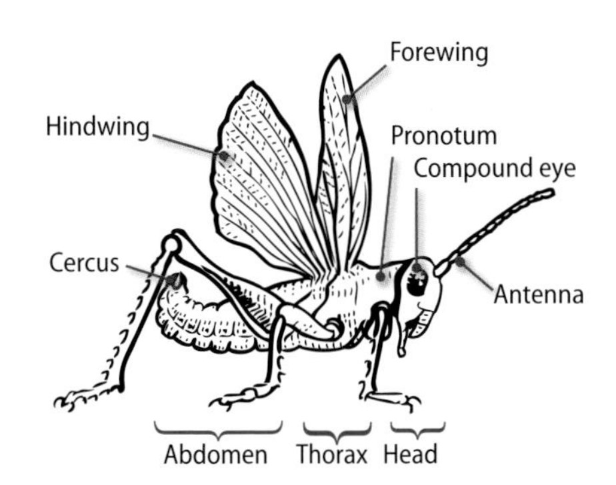

FIGURE 4-7

The body parts of this grasshopper represent the general structures that can be seen on most adult insects. However, some adult insects are wingless or have only one pair of wings.

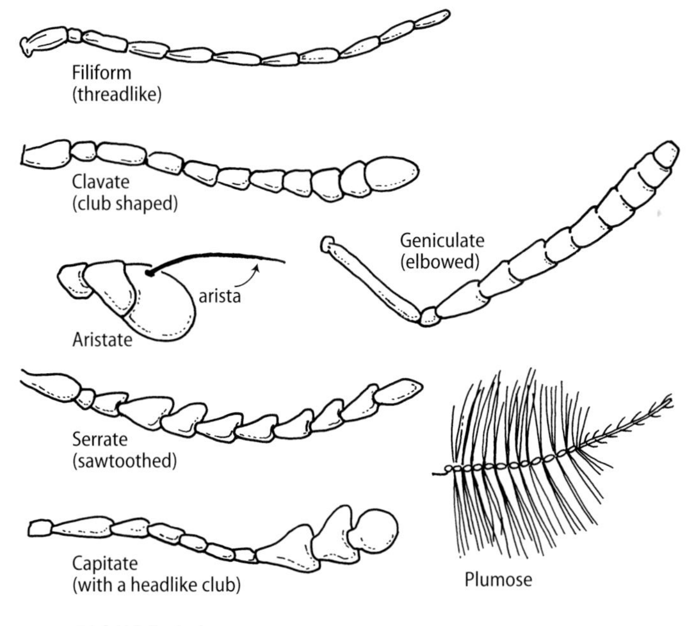

FIGURE 4-8

Various types of insect antennae. Antennal structure can be an important character in identifying some insects.

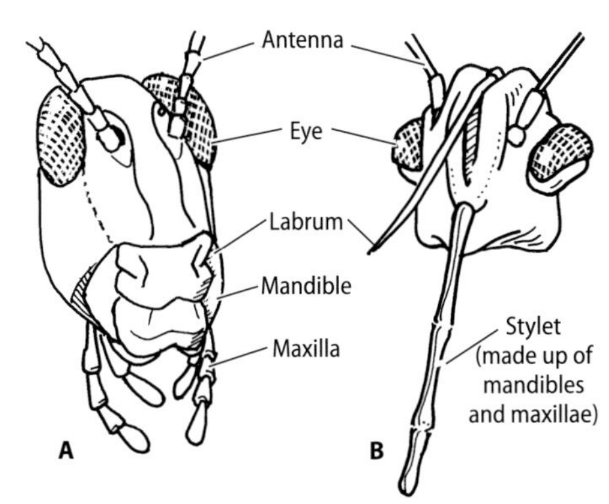

FIGURE 4-9

Knowing the type of mouthparts an insect has helps identify the order to which it belongs and the types of damage it causes. The two major types of mouthparts among pest insects are (A) chewing and (B) piercing-sucking.

Arthropods

Arthropods, phylum Arthropoda, are organisms with an external skeleton and jointed body parts. The arthropod group includes six classes with members that are significant pests (Table 4-1) as well as several minor classes. Insects are by far the most abundant and diversified class of arthropods, with 31 different orders. Spiders, ticks, and mites belong to the next-largest class, the arachnids. Other major arthropod classes are the crustaceans (including sowbugs and pillbugs), the centipedes, and the millipedes. Symphyla is a minor class of arthropods, but it includes the symphylans, which can be of some economic importance as pests. A first step in identifying arthropods is to look for the body structures that are unique to different classes, as described in Table 4-1.

Insects. The adult insect has three distinct body parts—the head, *thorax*, and abdomen (Figure 4-7). The thorax is divided into three segments—the prothorax (front segment), the mesothorax, and the metathorax—with one pair of jointed legs attached to each segment. Insects are the only invertebrates that have wings, but not all adult insects have wings. If present, two pairs of wings are normal, although in a few groups such as true flies, one pair of wings has been greatly reduced. Wing features such as venation, membranes, and pigmentation are important identifying features. Tails, stingers, or forceps at the tip of the abdomen can also be important in identification.

All insects have one pair of antennae on the front of the head. Insect antennae vary greatly in size, shape, and form (Figure 4-8). Insects often have two kinds of eyes, simple and compound. For most adult insects, three simple eyes (ocelli) are located on the front upper part of the head; compound eyes are situated toward the top and to each side of the head.

Mouthparts are useful in identifying insect pests. There are two main types of insect mouthparts: chewing and sucking (Figure 4-9). Sucking mouthparts vary in appearance. The mandibles are either lacking or form part of the proboscis. In many families, such as in mosquitoes and aphids, there are

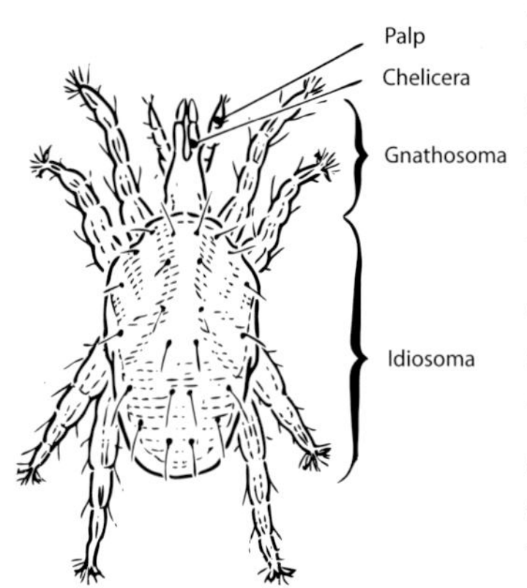

FIGURE 4-10

The body arrangement of mites and ticks is different from that of insects or spiders. Adult mites and ticks generally have four pairs of legs; immature forms usually have three pairs.

2 to 6 piercing structures called stylets that take up liquids from a plant or animal and inject saliva into the host to keep the flow going. The coiled proboscis of a butterfly, on the other hand, does not pierce but is adapted for siphoning nectar and other liquids. Similarly, the mouthparts of certain flies are specialized to sponge up liquids. Thrips have primitive rasping-sucking mouthparts somewhat intermediate between chewing and sucking types. The principal characteristics of the major insect orders and the damage symptoms with which they are associated are found in Table 4-2.

Mites. Mites differ from insects in that they have two body parts and no antennae (Figure 4-10). A small head, called a gnathosoma, has a pair of chelicerae and palps (the organs of food acquisition) attached at the end. The abdomen and thorax are combined and referred to as the idiosoma. Mites usually have four pairs of legs that are attached to the abdomen; some immature forms and a few adults only have three pairs.

Mites are very small and as a group very diverse in form and habits. Only an expert can positively identify most mites to the species level. However, you can learn to identify major groups of mites, which is helpful because mites within a family often have similar habits and management options. Features used in the field as identification clues include host plant; mite size, shape, and color; the presence and pattern of spots or spines; the number of legs; and the timing, location and grouping of egg laying. The eggs also provide identifying characteristics such as color, number, and presence or absence of spines.

Field identification of mites to the family level is helpful because mites within a family have similar life cycles and management options. The most common pest groups are the spider mites and red mites (family Tetranychidae) and eriophyid mites (family Eriophyidae). The best-known predatory mite species are found in the family Phytoseiidae.

Spiders. Spiders are very common invertebrates; all are predaceous. While beneficial in most cases, spiders are considered a pest when venomous species such as black widows inhabit residential areas. As with ticks and mites, spiders lack antennae and have two main body parts—a strongly constricted abdomen and the *cephalothorax*, which is the head and the thorax combined (Figure 4-11). Four pairs of legs, a variable size and configuration of (usually 8) eyes, a pair of pedipalps, and the mouthparts attach to the cephalothorax. The mouthparts include a pair of fangs, or chelicerae. Several pairs of spinnerets, the web-spinning organs, are located on the terminal end of the abdomen. Common spiders can be identified to family by body shape, web type, hunting or other behavior, and the arrangement and relative size of their eyes.

Other Arthropods. Other arthropod classes have varying arrangements of body parts and legs. For instance, centipedes (Chilopoda) are elongate, flattened, and wormlike. They have a distinct head and definite jointed legs; 15 or more pairs of legs are arranged in 1 pair per body segment. Centipedes have a single pair of antennae but unlike insects do not have a thorax or wings (Figure 4-12). On the first segment behind the head, there are a pair of poison claws that are used to paralyze prey.

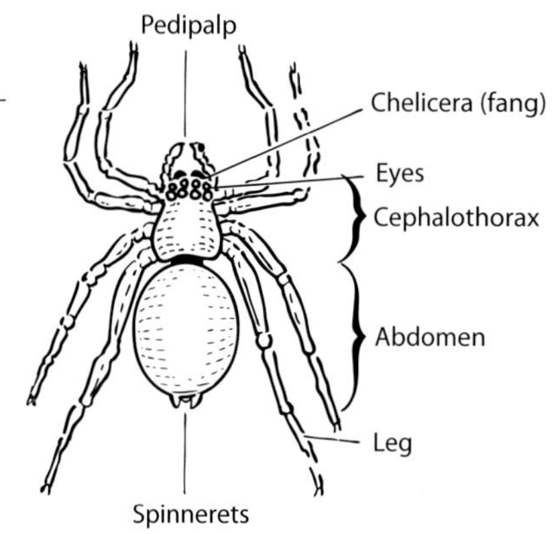

FIGURE 4-11

Spiders have two main body regions, the cephalothorax and the abdomen. They have four pairs of legs, a pair of pedipalps, and a pair of chelicerae that terminate in fangs used to inject venom. At the end of the abdomen are clusters of spinnerets, part of the spider's web-producing mechanism.

FIGURE 4-12

Centipedes have many body segments, most of which give rise to a pair of legs. Centipedes have a pair of poisonous fangs that arise from the head.

TABLE 4-2.

Typical characteristics and damage symptoms of the principal insect orders.

	Order	Mouthparts	Antennae	Wings	Metamorphosis	Comments
	Collembola (springtails)	chewing	4 to 6 segments	none	none	Not true insects. Primarily decomposers; found in woodland soils, decaying vegetables, mushroom houses.
	Coleoptera (beetles)	chewing	variable	2 pairs	complete	Depending on species, may damage fruit, leaves, roots, and burrow into wood, etc.; many beneficial species.
	Dermaptera (earwigs)	chewing	long	2 pairs	gradual	Some species attack growing shoots of plants, seedlings; irregular holes or chewed edges in leaves of older plants; shallow holes on stone fruit; also feed on a variety of dead and living organisms, such as insects and mites.
	Dictyoptera (cockroaches, mantids)	chewing	filiform, long, multisegmented	2 pairs, some wingless	gradual	Pest species of cockroaches infest stored food and can chew fabric and paper products; some carry disease. Mantids are predators.
	Diptera (flies, mosquitoes, gnats)	sucking, rasping, chewing	variable	1 functional pair, some wingless	complete	Some species blood-sucking and can transmit disease to livestock and humans; the larvae of some species mine leaves, tunnel in fruit, roots, and bore into stems; some are predaceous or parasitic on various insects.
	Ephemeroptera (mayflies)	vestigial in adults, chewing in nymphs	short	2 pairs	gradual	Feed on dead or living aquatic vegetation; sometimes considered a nuisance.
	Hemiptera: suborder Heteroptera (true bugs)	piercing-sucking	variable	2 pairs	gradual	Feed on juices of plants producing piercing damage on leaves, corky tissue on fruit; some prey on other insects.
	Hemiptera: suborder Homoptera (aphids, scales, mealybugs, whiteflies)	piercing-sucking	variable	2 pairs, some wingless	gradual	Pierce plant tissues and feed on the sap of plants; cause leaf and fruit deformities, loss of plant vigor, stunted growth, and plant dieback; most species excrete honeydew, which supports growth of black sooty mold fungus; some species vectors of disease-causing pathogens.

	Order	Mouthparts	Antennae	Wings	Metamorphosis	Comments
	Hymenoptera (bees, wasps, ants, sawflies, horntails)	chewing, lapping, sucking	variable	2 pairs	complete	Some species venomous and can inflict painful stings; some species bore or chew wood structures; some are considered nuisance pests around homes and outdoor areas. Bees and other species important pollinators. Parasitic wasps important in control of certain insect species.
	Isoptera (termites)	chewing	variable	2 pairs	gradual	Attack dead hardwood, chewing into wood structures and creating tunnels or galleries.
	Lepidoptera (butterflies, moths, skippers)	reduced in adult, chewing in larva	variable	2 pairs	complete	Larvae of pest species chew fruits, vegetables, nuts, grains, cotton, forage crops, and fabrics; contaminate stored foods.
	Mecoptera (scorpionflies)	chewing	filiform, long	2 pairs, some wingless	complete	Little or no economic importance; some species predaceous on other insects.
	Neuroptera (lacewings, dobsonflies)	chewing	variable	2 pairs	complete	Mostly predaceous and beneficial, preying on aphids, ants, and other small insects.
	Odonata (dragonflies, damselflies)	chewing	short	2 pairs	gradual	Mostly beneficial; adults feed on horseflies and mosquitoes.
	Orthoptera (grasshoppers, crickets, katydids)	chewing	variable	2 pairs, some wingless	gradual	Chewing damage to leaves; can cause defoliation.
	Plecoptera (stoneflies)	chewing	filiform, long	2 pairs	gradual	Important to humans as fish food.
	Psocoptera (barklice, booklice)	chewing	long, multisegmented	2 pairs, some wingless	simple	Chewing damage on paper, starch, grains.
	Thysanura (silverfish and firebrats)	chewing	long	none	none	Pests in the home feeding on cereals, books, glued paper, and clothing starch.

	Order	Mouthparts	Antennae	Wings	Metamorphosis	Comments
	Siphonaptera (fleas)	piercing	short	none	complete	Bite humans and animals, may cause allergic reaction; some species transmit pathogens of disease; some species intermediate host of tapeworms.
	Thysanoptera (thrips)	intermediate rasping-piercing-sucking	short to medium	2 pairs, fringed, some wingless	intermediate between gradual and complete	Puncture plant cells and suck up the fluid; deform fruit, flowers, and other plant parts; some species beneficial, predatory on aphids and mites.
	Tricoptera (caddisflies)	vestigial	long, filiform	2 pairs	complete	Important to humans as fish food.

Drawn from Borror and White 1970.

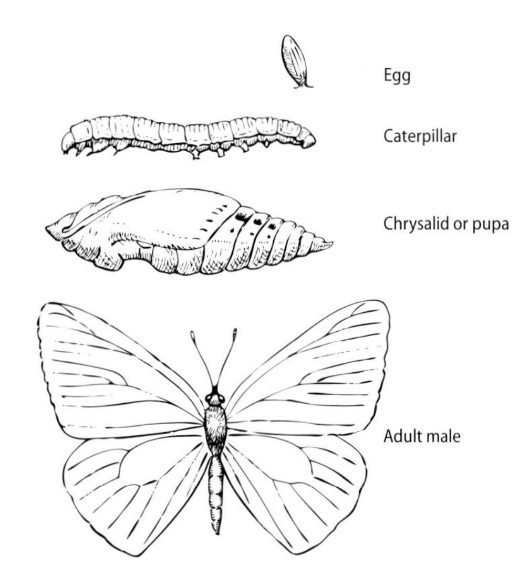

FIGURE 4-13

Stages of development of a typical moth or butterfly, showing complete metamorphosis. Most moths and butterflies go through four or five molts during the caterpillar stage.

Millipedes (Diplopoda) resemble centipedes, but millipedes have many more legs than centipedes, with 2 pairs of short legs on each body segment, and they do not have poisonous appendages. Millipedes are wormlike and cylindrical with 1 pair of short, segmented antennae.

Symphylans (Symphyla) are slender, whitish, and similar to centipedes. Adults have 10 to 12 pairs of legs and a pair of antennae.

Pillbugs and sowbugs (Isopoda) are soil-dwelling crustaceans with a hard, shell-like covering that is made up of a series of segmented plates. They have 7 pairs of legs and are dark gray or brown but may be almost purple after molting.

Life Cycle and Growth Requirements. Most arthropods hatch from eggs into immatures that increase in size by molting or shedding their outer body covering (exoskeleton) and growing a new, larger one. Often they modify their shape with each successive molt, a process known as *metamorphosis*. The period between one molt and the next is known as an instar. Immature arthropods pass through several instars before becoming adults.

Some species of insects undergo major morphological or structural changes between the immature stages and adulthood. This transformation occurs within a nonfeeding pupal stage; these insects are said to have complete metamorphosis (Figure 4-13). Immatures in these groups are called larvae and, in many cases, have different feeding habits from adults, so that only either the larval or adult stage causes damage. Examples of insects with complete

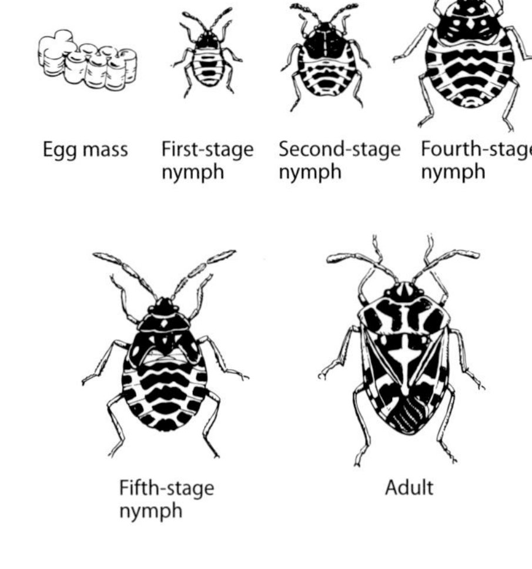

FIGURE 4-14

Development of the harlequin bug, showing incomplete metamorphosis and wing development.

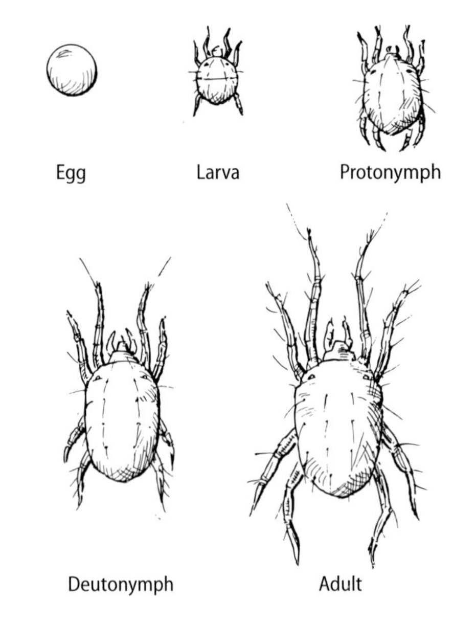

FIGURE 4-15

Development of a typical plant-feeding spider mite. The stage that hatches out of the egg is called a larva. Remaining immatures are called nymphs.

metamorphosis include flies, wasps, moths, butterflies, and beetles.

Other insects, such as grasshoppers, aphids, and true bugs, go through gradual or incomplete metamorphosis and do not have a pupal stage (Figure 4-14). Their immatures are called nymphs; they differ from adults primarily in their size and absence of wings. Adults and immatures of these insects have the same food habits.

The development of mites, spiders, and other arthropods is similar to the gradual metamorphosis of insects. For instance, mites hatch from eggs and pass through several immature stages before becoming adults; the stage that hatches out of the egg is normally called the larva (Figure 4-15). Later-stage nymphs look similar to adult mites. Immature mites and spiders have feeding habits similar to adults of the same species.

Arthropods have one to many generations per year, depending on species, location, and environmental conditions. In hot weather, species such as spider mites or aphids reproduce rapidly, and in warmer areas such as some parts of California, they can reproduce all year. On the other hand, other arthropods, such as some borers, take more than a year to mature. The number of generations and the length of time for each generation to develop are important in determining monitoring activities and control methods.

Eggs. The eggs of arthropods vary in size, shape, color, attachment, arrangement, and where they are laid (Figure 4-16). Eggs are usually laid in a location where the young will have an immediate food source upon hatching. Learn to identify the eggs and their typical location for pest and beneficial species. Eggs can be used to detect a pest's presence or forecast a pest's destructive stage. For instance, egg traps are used to monitor the egg-laying activities of the navel orangeworm, *Amyelois transitella.* Leaf samples are used to monitor the egg laying of the tomato fruitworm, *Helicoverpa zea*, on processing tomatoes.

Immatures. Learning to identify immature forms of arthropods is essential. Immature stages are often as damaging or more damaging than the adult arthropods of the same species. Furthermore, control actions are often directed at immatures, either because they do the primary damage or because they are easier to control than adults. Many identification keys are aimed at identifying adults and do not feature characters that are useful in distinguishing larvae or nymphs. This is particularly true of insects that undergo complete metamorphosis, such as beetles, moths, and flies; their immatures look completely different from adults. Adult keys are also difficult to use for immature insects and mites with gradual metamorphosis; these immatures do not feature the wings, pigmentation, or hairs or bristles required to identify adults. Look for keys, photographs, and other identification aids that help with identifying immatures. Figure 4-17 shows how to distinguish larvae of moths, beetles, and sawflies.

Habits. The life cycles and activities, or habits, of each insect or mite vary according

FIGURE 4-16

Examples of insect eggs. Pest species include (A) corn earworm egg; (B) cabbage looper egg; (C) imported cabbageworm egg; (D) the barrel-shaped eggs of the harlequin bug; (E) the overlapping fish-scale-like eggs of the omnivorous leafroller; (F) beet armyworm egg mass with some of the cottony covering removed; (G) European red mite eggs with the slender spike rising from the center. Natural enemy species include (H) praying mantis egg case; (I) two tachinid fly eggs on the head of an amorbia caterpillar; (J) a syrphid fly egg; (K) convergent lady beetle eggs; (L) eggs of spined assassin bug.

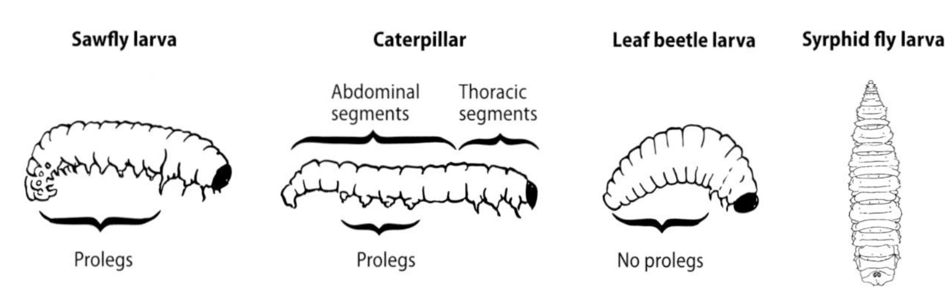

FIGURE 4-17

Larvae (caterpillars) of moths and butterflies can be distinguished from the larvae of beetles and sawflies by the number and arrangement of their appendages. All these insect larvae have three pairs of true legs, one pair on each thoracic segment. Sawfly larvae also have appendages (called prolegs) on six or more of their abdominal segments. Caterpillars have prolegs on two to five abdominal segments but never on their first two abdominal segments. Beetle larvae have no prolegs. Fly larvae, or maggots, such as the syrphid fly larva shown here, have no true legs.

to species. Understanding these characteristics can provide information for identification as well as monitoring and control strategies. Almost no generalizations can be made about the habits of arthropods because as a group they occupy almost every habitat and niche. However, knowing the specific niche of the pest in question can help in planning an efficient detection, sampling, and management program.

Successful Characteristics. Many insects and mites that have become serious pests possess characteristics that allow them to thrive under the conditions in managed crops or landscapes. Many have evolved with the crop plants or animals they infest; the life cycles of some species that attack plants are so tailored to the host that infestations can be avoided by altering the planting or harvest date. In addition, many insect and mite species have developed the capacity to search for specific hosts through chemical cues. Genetic variation and selection allow insects and mites to further adapt to the cultural, ecological, and environmental conditions that favor crop growth and development. One example of this adaptability has been the development of resistance to insecticides. Under the selection pressure of the insecticide, resistant individuals multiply rapidly in the absence of competition, and after several generations their genotype may dominate the population.

The success of many insect and mite pests is due in large part to their high reproductive capacity. A few individuals of some pest species can produce an outbreak in only a few days under favorable conditions. Mobility is another factor. Legs enable insects to walk, jump, or swim for short distances; wings allow for movement over great distances. Various stages of insects and mites also move by air currents or by attaching themselves to people, animals, plant materials being moved, as well as to devices such as field implements.

The ability to go into diapause enables some insects to survive extreme conditions such as heat, cold, or drought. Diapause is a dormant state that occurs in the life of some insects in which growth, reproduction, differentiation, and metamorphosis cease and do not resume unless specific environmental cues are triggered. Winter dormancy is sometimes referred to as *hibernation*, and dormancy during high temperatures is called *aestivation*. Diapause can occur in different species during egg, larva, pupa, or adult stages; it can last from several months to a year or more.

Dispersal and Movement. Knowledge of the methods of pest dispersal and movement and their relationship to pest management practices, the ecosystem, and environmental factors can be used to predict pest infestations. As mentioned earlier, a successful characteristic of insects is their ability to move. For some species, dispersal is primarily through flight or walking. Other methods that insects and mites use to disperse to suitable hosts include moving with air and water. Migration can occur on strong dominant winds, seasonal fluctuations, or microclimatic disturbances. The direction of the wind pattern may correlate to infestation patterns. Irrigation water carries some species across fields; rainfall helps move others.

Some species hitchhike their way to new locations. The border inspection stations of California have successfully kept the gypsy moth, *Lymantria dispar*, out of California by recognizing the moth's propensity for laying its egg masses on objects including cars, outdoor furniture and freight trains. Inspectors vigorously inspect outdoor material shipped from the eastern parts of the United States and in this manner have prevented the gypsy moth from becoming established in California. Infested plants, soil, or farm machinery can transport arthropod species from field to field or into new areas.

Knowledge of the dispersal and movement patterns of insects and mites can be useful in planning monitoring. Most insect and mite infestations are localized. When a population of concern is noted, sampling in concentric patterns from the *hot spot* can determine the intensity of the buildup and in which direction the population is moving.

Types of Damage. Arthropods cause damage in many ways. On plant crops or ornamentals, they may chew twigs, leaves, or fruit; suck sap; bore into branches, trunks, or fruit; chew or suck roots; lay eggs in plant tissues; or disseminate disease organisms. Some insects are pests of structures or stored foods. Other arthropods, such as flies, cockroaches, and ants, are nuisance pests or indicators of unsanitary conditions.

Damage symptoms can be useful for identifying pests (Figure 4-18). Chewing damage

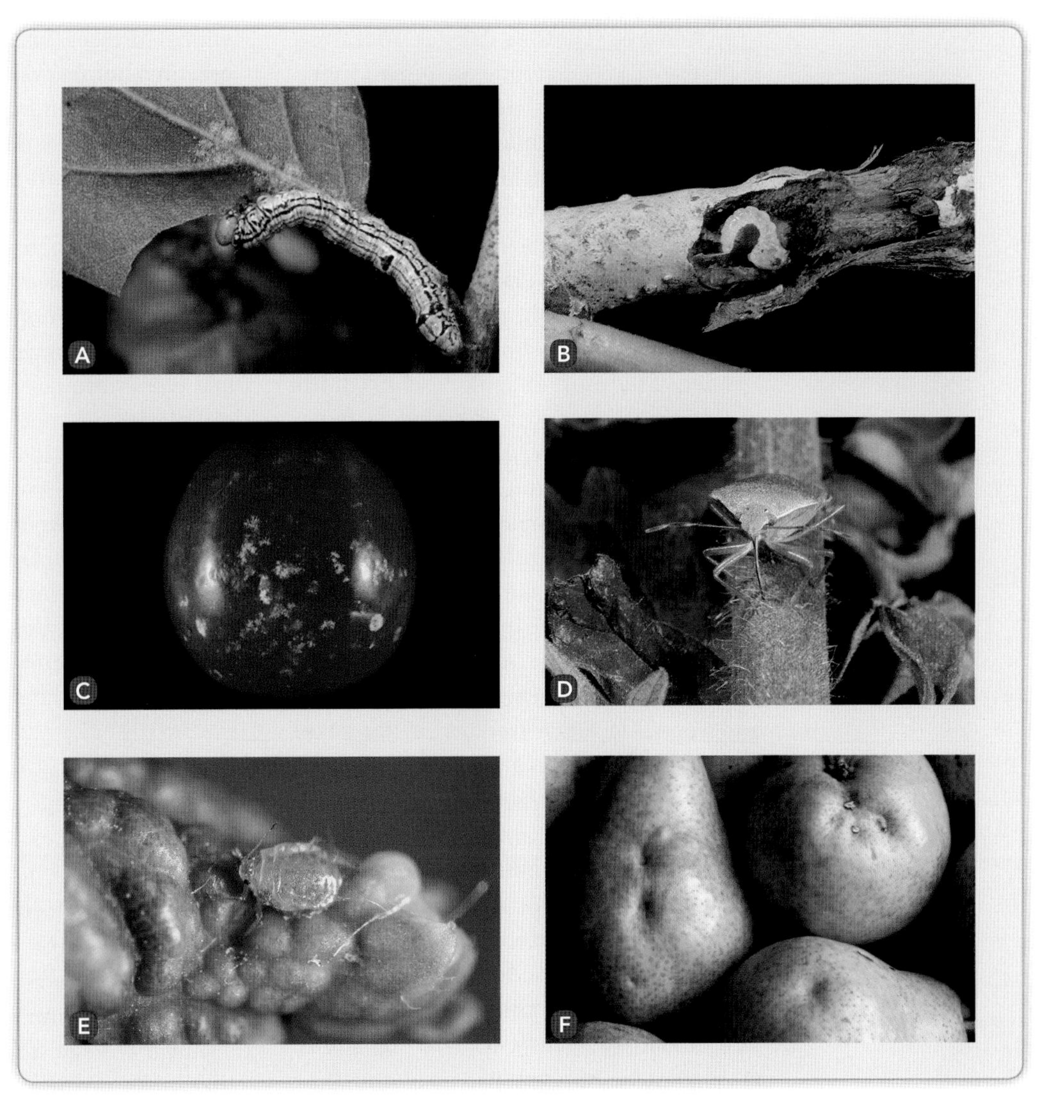

FIGURE 4-18

The type of damage found can often be used to identify the insect pest. Damage typical of insects with chewing mouthparts includes (A) oak leaves chewed all the way through by the California oakworm and (B) cracked bark and internal boring into trunks and limbs caused by borers, such as the Pacific flatheaded borer. Damage typical of insects with sucking mouthparts includes (C) stippling damage on tomatoes made by the Southern green stinkbug (D), leaf curling and distortion caused by the leaf curl plum aphid (E), and puckering on pears caused by consperse stick bug (F).

guides the investigator to insect orders with chewing mouthparts, such as beetles, the larvae of butterflies or moths, orthopterans such as grasshoppers, and earwigs. For instance, in lettuce, ragged holes in leaves, holes in the heads, and heads and leaves contaminated with frass are types of damage associated with various caterpillars, including the cabbage looper (*Trichoplusia ni*) the alfalfa looper (*Autographa californica*), the imported cabbageworm (*Pieris rapae*), or other leaf-feeding caterpillars. To confirm the culprit, search the plants for caterpillars, eggs, pupae, or moths.

Stippling or yellowing damage on leaves, fruit, or twigs; leaf curling; deformed fruit; wilting; or the general decline of the whole plant are damage symptoms associated with mites and piercing-sucking insects such as aphids, scale insects, and true bugs. These symptoms could be the result of plant pathogens as well, so once damage is evident, further monitoring will be needed to establish the identity of the pest organism.

Some of the most difficult pests to control are internal feeders such as borers, leafminers, *gall* insects, or the larval stages of insects that develop inside fruits, nuts, or seeds. Bark beetles, an example of boring insects, are considered among the most destructive insect pests of California forests. Pitch tubes and sawdustlike frass are indications of bark beetle damage. Adult bark beetles lay eggs in tunnels etched between the bark and wood. The larvae feed on the inner bark and phloem and form branching tunnels or galleries, causing further damage to the tree. Leafminers lay their eggs between the upper and lower epidermis of a leaf. Off-color patches, sinuous trails, or holes in leaves indicate damage from leafminers.

Insects whose larvae feed in fruits or nuts are among the most economically damaging pests. Examples include the codling moth, apple maggot, and the cotton boll weevil. This group typically lays its eggs on or just under the flesh, damaging fruit with stings and deep entries. Larvae hatch and bore farther into the fruit, often tunneling into its core. Small piles of frass on the side of the fruit indicate an infestation.

Root feeders damage roots, limiting the plant's ability to take up water and nutrients. Wilting and declining plants are typical symptoms of root feeder damage. Grape phylloxera, *Daktulosphaira vitifoliae*, is an example of an insect pest that damages the host by feeding on the root system. Grape phylloxera feed either on growing rootlets, which then swell and turn yellowish, or on mature hardened roots, where the swellings are often hard to see. Necrotic spots (areas of dead tissue) develop at the feeding sites on the roots.

Some insects are vectors of plant and animal diseases. Insects transmit pathogens in a number of ways: by carrying the pathogen on their bodies, by creating an entry for a disease through egg laying or feeding, or by carrying the pathogen in their bodies and transmitting it through feeding. For example, beet curly top virus is vectored by the beet leafhopper, *Circulifer tenellus*. The leafhopper overwinters on a wide range of annual and perennial weeds and readily acquires the virus when it feeds on infected plants. Beet curly top virus can cause significant losses in tomatoes, beans, and cucurbits. Some of the most serious diseases to affect people, such as epidemic typhus, bubonic plague, and yellow fever, are transmitted through the bites of the human body louse, the oriental rat flea, and the mosquito, respectively.

Mollusks

Snails and slugs are mollusks and belong to the phylum Mollusca, class Gastropoda. Mollusks have no bones, are soft bodied, have a true body cavity, and have well-developed circulatory, nervous, and digestive systems. Snails and slugs are related to clams, scallops, abalone, octopi, and squid.

Identification. Snails and slugs have a similar biology and structure, except that snails have a conspicuous spiral shell (Figure 4-19). The body is divided into four sections: the head with a mouth, sensory tentacles, and simple eyes; the foot with its creeping sole; the mantle, a fold of skin over the back forming the respiratory cavity; and the visceral mass, which contains the main internal organs of the body. In snails, the visceral mass lies within the shell; in slugs, it is found under the mantle. Snails and slugs glide along on their muscular "foot." This muscle constantly secretes mucus, which later dries to form the silvery slime trail that is a clue to their presence. The most common pest snail found in California is the brown garden snail, (*Cornu aspersum*, formerly *Helix aspersa*). Common pestiferous slugs are the gray garden slug (*Deroceras reticulatus*), the banded slug (*Lehmannia poirieri*), and the greenhouse slug (*Milax gagates*). Distinguishing slug species requires a good key and a 10X hand lens. Features used in slug keys include the size, shape, and location of the mantle; position of the breathing pore; color; skin texture; the presence or absence of a keel on the back or a caudal mucous pore; and the color of the mucous. Generally, all slug species are managed similarly, so species identification is not as critical as for some other pests.

Life Cycle and Growth Requirements. Snails and slugs lay from 10 to 200 eggs, depending on the species, beneath the soil. Adult brown garden snails lay about 80 spherical, pearly white eggs at a time into a hole in the topsoil. They can lay eggs up to six times per year. Slugs lay eggs in small, sticky packets. Snails and slugs are hermaphroditic (have both male and female organs).

During cold weather, snails and slugs hibernate in the topsoil. During hot, dry periods, snails estivate by sealing themselves off with a parchmentlike membrane and attaching themselves to tree trunks, fences, or walls.

Habits. Snails and slugs are commonly found in damp areas, soil litter, and foliage of plants. They are most active at night and on cloudy days. On sunny days, they hide from the heat and sun under boards, stones, debris, organic *mulches*, weedy areas around trunks, leafy branches growing close to the ground, and in dense ground covers such as ivy.

Types of Damage. Snails and slugs feed on a variety of living plants as well as on decaying plant matter. Some snails are predaceous. On plants, they chew irregular holes with smooth edges in leaves and can clip succulent plant parts. They can also chew fruit and young plant bark. Because they prefer succulent foliage, they are primarily pests of seedlings, herbaceous plants, flowers, and ripening fruit, such as strawberries and tomatoes that are close to the ground. However, they also feed on foliage and fruit of some trees; avocado and citrus are especially susceptible to damage (Figure 4-20).

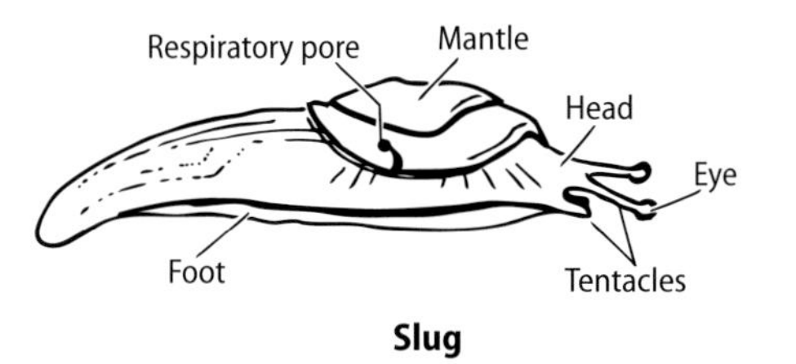

FIGURE 4-19

The bodies of snails and slugs are very similar. The main body parts consist of the head, on which the eye and tentacles are located; the foot; the mantle, which is exposed on the slug and protected by the shell on the snail; the respiratory cavity or pore; and the visceral mass, which contains the main internal organs of the body.

FIGURE 4-20

Snails and slugs are primarily pests of seedlings and herbaceous plants, but they also feed on young woody plant bark and fruit. The brown garden snail is shown here on citrus fruit.

TABLE 4-3.

Important pest nematodes.

Common name	Scientific name
Plant Pests	
citrus nematode	*Tylenchulus semipenetrans*
cyst nematodes	*Heterodera* spp.
dagger nematodes	*Xiphinema* spp.
foliar nematodes	*Aphelenchoides* spp.
lesion nematodes	*Pratylenchus* spp.
potato rot nematode	*Ditylenchus destructor*
rice root nematodes	*Hirschmanniella* spp.
ring nematodes	*Criconemoides* spp.
root knot nematodes	*Meloidogyne* spp.
seed-gall nematodes	*Anguina* spp.
spiral nematodes	*Rotylenchus* spp., *Helicotylenchus* spp.
stem nematode	*Ditylenchus dipsaci*
stubby root nematodes	*Trichodorus* spp., *Paratrichodorus* spp.
stunt nematodes	*Tylenchorhynchus* spp.
Animal Pests	
canine heartworm	*Dirofilaria immitis*
filariasis nematode	*Wucheria bancrofti*
hookworm	*Ancylostoma duodenale*
pig lungworm	*Metastrongylus apri*
pinworm	*Enterobius vermicularis*

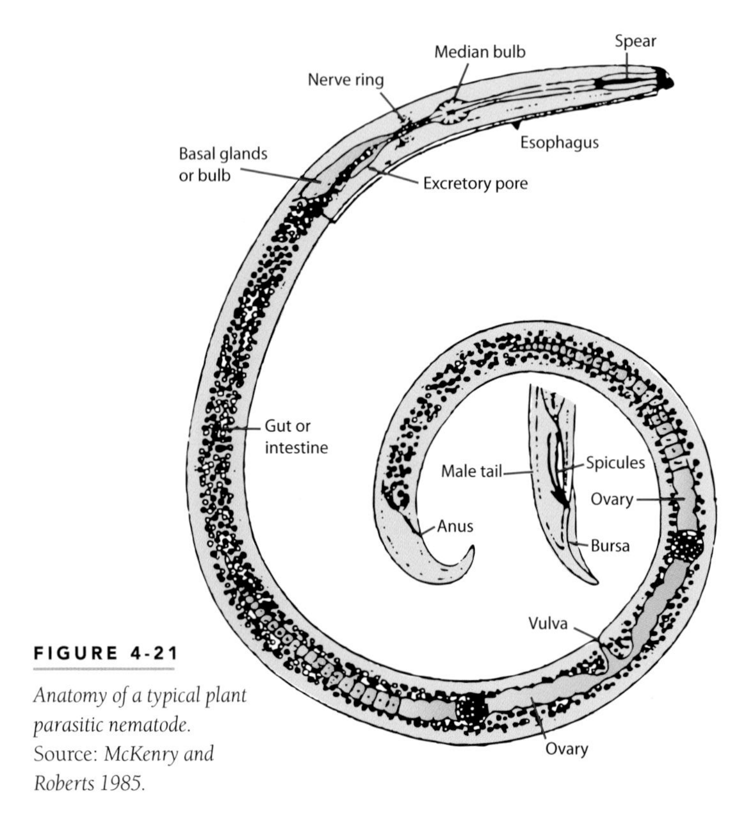

FIGURE 4-21

Anatomy of a typical plant parasitic nematode.
Source: *McKenry and Roberts 1985.*

Nematodes

Nematodes belong to the phylum Nemata (or Nematoda), the third-largest phylum in the animal kingdom. Nematodes are unsegmented roundworms that can be found in soil, water, plant tissues, or animals; they feed on a diversity of organisms including plants, insects, and animals. The nematodes of most concern in pest management are plant-parasitic species that attack crops and species that cause disease in domestic animals (Table 4-3). Most nematodes, however, are not pests; some are beneficial, breaking down organic matter or attacking insects, weeds, plant-pathogenic organisms, or other nematodes.

Plant-parasitic nematodes live on or in a plant host, deriving energy from the host in the form of nutrient substances. Close association with the host influences nematode behavior, reproduction, and dispersal. Nematodes occur in practically all soils, are common in high mountain soils, and can occur to depths of at least 17 feet (5 m) in some agricultural soils.

Identification. Nematodes are microscopic roundworms that are bilaterally symmetrical, unsegmented, generally transparent, and colorless. Most are slender, range in size from 0.25 to 12.0 mm (about 0.01 to 0.5 inch) in length, and cannot be seen without a microscope. A typical plant-parasitic nematode is pictured in Figure 4-21. Nematodes are difficult to identify because of their morphological complexity and microscopic size. Identification of plant-parasitic nematodes requires samples of surrounding soil, roots, and other affected plant parts. Send samples to a laboratory for identification.

Nematodes may be spindle shaped, pear shaped, lemon shaped, or sac shaped. The mouth opening is located on the anterior end, which is usually blunt and rounded. All plant-parasitic nematodes have a stylet that can be protruded to penetrate plant cells for feeding. The opening of the excretory system is located in the anterior portion of the body. Nematodes are covered with a multilayered cuticle that has various surface markings.

Because of their size, field identification of nematodes requires knowledge of specific

FIGURE 4-22

Root-feeding nematodes impair plants' ability to take up water and nutrients, resulting in a decrease in aboveground growth. The small grapevines shown above are suffering from a root knot nematode infestation.

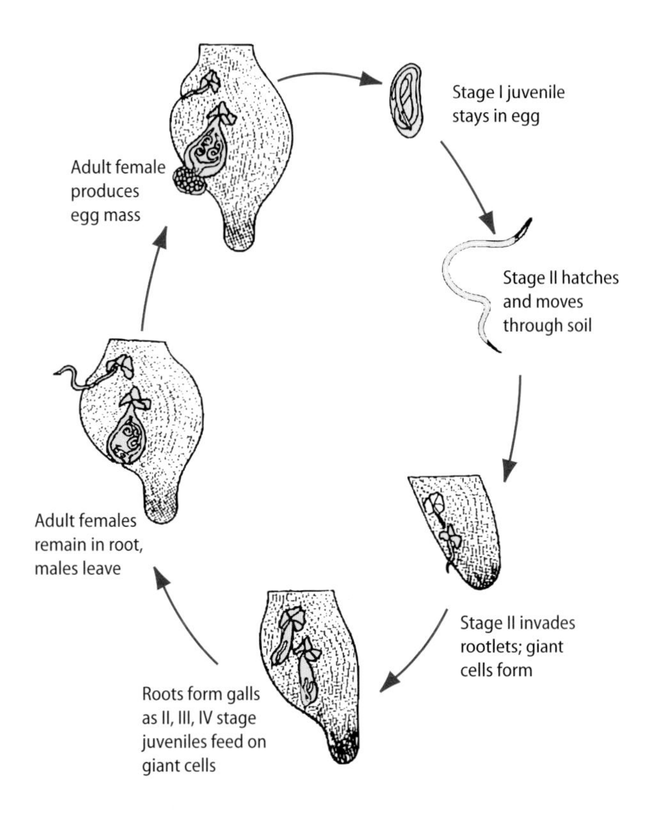

FIGURE 4-23

Root knot nematodes spend most of their life cycle in galls on roots. Second-stage juveniles invade new sites near root tips, and the host produces a gall in response to the nematodes' feeding.

pest and crop associations. Nematodes attack specific crops, and their presence can be determined by symptoms visible above the ground (such as stunting, *chlorosis*, or leaf drop) and symptoms existing below the ground on the roots (galls, lesions, or stunting). Nematode infestations should be suspected whenever a general decline of a particular plant species is observed (Figure 4-22). Delayed plant maturity might also indicate nematode damage. If no other causes for the decline of the plant are obvious, examine the roots for signs of nematodes. Signs include root galls or stubby, stunted, or proliferating roots. Darkened roots with lesions or plants with fewer roots than normal can also indicate a nematode infestation. Again, accurate identification is important; collect samples and send them to a laboratory for analysis. Contact the county Cooperative Extension office to locate a testing laboratory.

Life Cycle and Growth Requirements. The nematode life cycle consists of six distinct stages: an egg stage, four juvenile stages, and an adult stage (Figure 4-23). The nematode molts between each juvenile and adult stage. There is no metamorphosis in nematodes; the young resemble the adults but are smaller. Differences between males and females become observable at the third juvenile stage.

Some nematodes lay their eggs singly in the soil; others lay them in the root tissue. Some species encase their eggs in a gelatinous material. The female cyst nematodes, *Heterodera* spp., produce a brown protective case that keeps the eggs viable for several years. The number of eggs laid varies among species. Some lay from 2 to 10 per day over an extended period; the total laid can vary from 100 to 500 for each female.

In some nematode species, the first molt occurs in the egg; a second-stage juvenile emerges. The juvenile molts twice more while passing through the third and fourth juvenile stages to become a fully developed adult. The length of the life cycle varies with species and environmental conditions and is directly related to temperature, moisture, and availability of food from the host.

Habits. Plant-parasitic nematodes are active only when in moist habitats. They can be classified into three groups based on their life habits.

- *Migratory ectoparasitic species* feed on root surfaces without becoming attached and are free living throughout their life.
- *Migratory endoparasitic species* move about freely but enter the plant during certain life stages and feed from within.
- *Sedentary endoparasitic species* do not move about once they are within host tissue or other organ.

Juveniles and adults of migratory *ectoparasitic* species feed externally on the roots of the host and move about freely during all life cycle stages; eggs are laid in the soil. Examples of migratory ectoparasitic nematodes include the ring (*Criconemoides* spp.), pin (*Paratylenchus* spp.), and stubby root (*Trichodorus* spp.) nematodes. Other ectoparasitic nematodes, such as the sheath nematode (*Hemicycliophora arenaria*), are immobile after attachment to the root. Ectoparasites are found in the soil around the root zone, where they feed and browse before moving to a new feeding site. Preferred feeding sites include the root tip, the region of root elongation and root cell division behind the root tip, and sites of lateral root junction.

Migratory *endoparasite* juveniles and adults such as lesion (*Pratylenchus* spp.) and stem and bulb nematodes (*Ditylenchus* spp.) are mobile during all life stages but may be found within or outside the roots. There is no distinct infective stage because adults and juveniles can enter into and come out of roots. Preferred feeding sites for migratory endoparasites include the junction of the lateral roots and the zone of elongation behind the root tips; most feeding and movement occur in the cortical region. Overwintering occurs in the roots or along the root zone in the soil.

Infection from sedentary endoparasites occurs when an infective-stage juvenile enters a root, becomes sessile, and remains at that site for the rest of its life cycle. Eggs are typically laid in a matrix that is contained in the root tissue but may be exposed on the outer root surface. Adult males remain vermiform; reproduction usually occurs outside the root tissue. Examples of sedentary endoparasites include the root knot (*Meloidogyne* spp.), cyst (*Heterodera* spp.), and citrus (*Tylenchulus semipenetrans*) nematodes.

Some species, such as the foliar nematodes *Aphelenchoides fragariae* and *A. ritzemabosi*, can be endoparasitic or ectoparasitic, depending on the host. These species are aboveground parasites attacking the leaves, stem, or bud tissues of various plants.

Successful Characteristics. Generally, wherever plants are cultivated, some species of plant-parasitic nematodes will be present. Nematodes have evolved successful survival mechanisms that enable them to withstand adverse environmental conditions such as desiccation, brief periods without food, and various soil reactions. Most nematodes are encouraged by the agronomic practices of irrigated, intensive cropping situations.

Dispersal and Movement. Nematodes move very slowly through soil by themselves; their movement is aided when soil pores are lined with a thin film of water. Nematodes spread easily by the movement of infested plants, farm equipment, soil, irrigation water, and animals. Movement over long distances primarily occurs from the transport of farm produce, nursery plants, seed, *tubers*, or bulbs.

Types of Damage. Plants infected with root nematodes can exhibit damage symptoms on upper portions of the plant as well as on the root. Symptoms such as stunting, chlorosis, wilting, curling and twisting of leaves and stems, delayed or uneven maturation of crops, and fruit drop can be associated with nematodes and with many other pests as well. Nematodes inject saliva into the plant, which induces much of the distortion of plants. Root symptoms produced by various nematodes include galls (Figure 4-24), swelling, stubby roots, lesions, and stunting. These symptoms are indicative of nematode presence but are not by themselves diagnostic. An expert should confirm nematode presence. Some nema-

FIGURE 4-24

Galling on roots, shown here, is symptomatic of root knot nematode presence but is not diagnostic. For accurate identification, collect samples and send to a laboratory for analysis.

FIGURE 4-25

A Beldings ground squirrel is one of the many vertebrate pests that negatively impact many crops and landscapes.

todes develop an association with pathogens to develop a nematode-disease complex. For instance, Fusarium wilt is more severe in cotton when nematodes are present. Nematode vectors transmit several plant viruses, including tomato ring spot and grapevine fanleaf.

VERTEBRATES

Vertebrates, which belong to the phylum Chordata, are animals with internal skeletons and backbones. They include fish, amphibians, reptiles, birds, and mammals. A vertebrate pest is any native, introduced, domestic, or feral vertebrate species that has a short-term or long-term adverse effect on human health and well-being; destroys food, fiber, or natural resources; or is a general nuisance. By this definition, any number of animals can be categorized as pests: common examples of vertebrate pests include commensal rats and mice, voles, bats, skunks, muskrats, possums, rabbits, beaver, ground squirrels (Figure 4-25), moles, pocket gophers, coyote, deer, horned larks, crows, and starlings.

Identification

Identification of vertebrate pest species can be relatively easy if the species doing the damage is observed, but in most situations, this does not occur: damage is usually observed when the culprit is nowhere in sight. Fortunately, vertebrate pests typically leave behind signs such as tracks, tooth marks, droppings, dens, burrows, and trails. These signs coupled with a familiarity of the habits of vertebrates can aid in identifying the species.

Identifying the pest that is responsible for the damage can be difficult at times because different animals can attack the same parts of the plant and cause similar types of damage. For example, gophers and moles usually damage turf; root damage on fruit trees can be the result of gophers or voles. Trunk damage can be attributed to voles, ground squirrels, or rabbits; birds, rats, ground squirrels, and voles can feed on fruits (Figure 4-26) or leafy crops. Although typical signs can be associated with vertebrate pest damage on many crops, the signs alone are not conclusive. Refer to the materials listed in the “Resources” section and consult pest identification experts to confirm identification of pest species.

FIGURE 4-26

It is often difficult to positively associate damage caused by vertebrates with the species doing the damage. The strawberry at left shows a general picking away at the fruit, damage typically caused by birds. The center strawberry has seeds picked off the strawberry surface, typically caused by goldfinches. The fruit on the right shows tooth marks consistent with ground squirrel feeding.

Once identification of the damaging pest has been confirmed, a management program can be designed. Management tools depend on pest species, time of year, relative density of the pest, and the location and host plants being damaged. Since vertebrate control materials can be very toxic to nonpest wildlife species and pets, extra precautions must be taken to prevent problems when they are used, especially where endangered species may occur. An integrated pest management program for vertebrate pests may incorporate many tools, including habitat modification, trapping, poison baiting, fumigation, exclusionary devices, repellents, and frightening devices. Most tools are more effective when populations are low, so regular monitoring is critical. More information on these methods and tools is presented in Chapter 5.

WEEDS

Weeds are plants that interfere with the growing of crops or ornamental plants; endanger livestock; affect the health of people; interfere with the safety or use of roads, utilities, and waterways; or are visual or physical nuisances. Grasses, sedges, broadleaf herbaceous plants, shrubs, and even trees are considered weeds if they interfere with human activities, result in economic loss, or are otherwise undesirable.

Identification

All plants are members of the kingdom Plantae. Most weed species are flowering plants, which are further subdivided into two main groups—*monocots* and *dicots*. Monocots produce only a single grasslike leaf in the seedling. Leaves typically have parallel veins that run the length of their axis and flower parts in threes or multiples of three. Monocot weed species can be found in two major families: grasses (Poaceae) and sedges (Cyperaceae). Sedges can easily be distinguished from grasses. The stems of sedges are triangular in cross section, and their leaves are arranged in threes at the base instead of twos as in the case of grasses. Important identifying features for grasses are presence or absence of ligules or auricles in the collar region, where the leaf sheath joins the leaf blade, and flower structures such as awns and glumes in mature plants. Figure 4-27 illustrates some of the structures used to identify grasses.

Dicots, commonly called *broadleaves*, are plants whose seedlings produce two *cotyledons*. Leaves of dicots usually have netlike veins and flower parts usually in fours, fives, or multiples thereof. Dicots commonly have a main taproot. Table 4-4 lists some common dicot weed families and some of their identifying characteristics.

TABLE 4-4.

Identifying characteristics of some common dicot weed families.

	Family name	Common name	Main feature
	Amaranthaceae	pigweed	dense clusters of small dry flowers; one-seeded, tiny and inflated fruit
	Asteraceae	sunflower; (thistle)	flowers in dense heads; includes many thistles; fruit an achene, often with hairs or bristles
	Brassicaceae	mustard	4 petals that narrow to a claw below the wide part; petals spread in a cross-shape form; fruit a silique, often elongated
	Chenopodiaceae	goosefoot	nonshowy flower with no petals; tiny, dark seed
	Convolvulaceae	morningglory	trumpet-shaped flowers, leaves lobed at base, vine-like growth
	Fabaceae	legume	long or coiled pod fruit; butterfly-shaped flower; compound leaves
	Malvaceae	mallow	roundish, palmately lobed leaf; fruit
	Solanaceae	nightshade	star-shaped, 5-petalled flower; fruit is often a berry

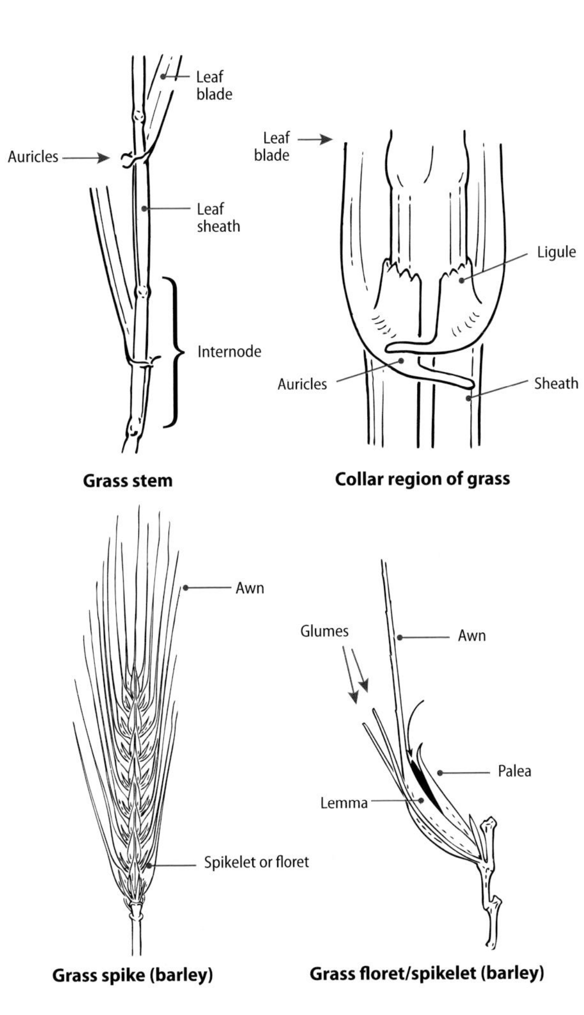

FIGURE 4-27

Plant structures useful in identifying grasses.

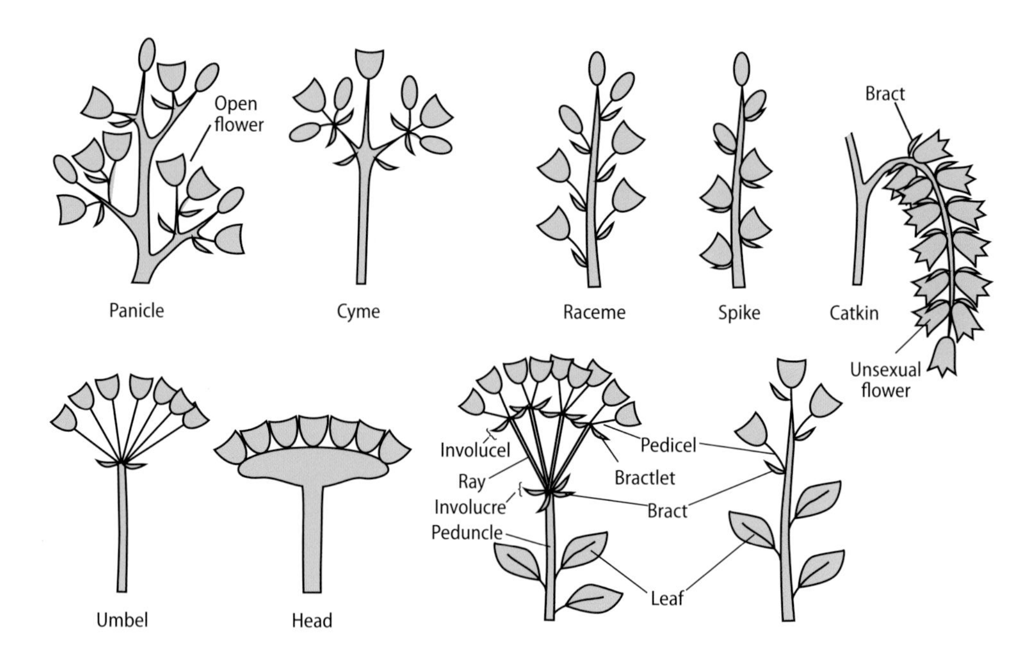

FIGURE 4-28

Some of the various inflorescences used to identify plants.

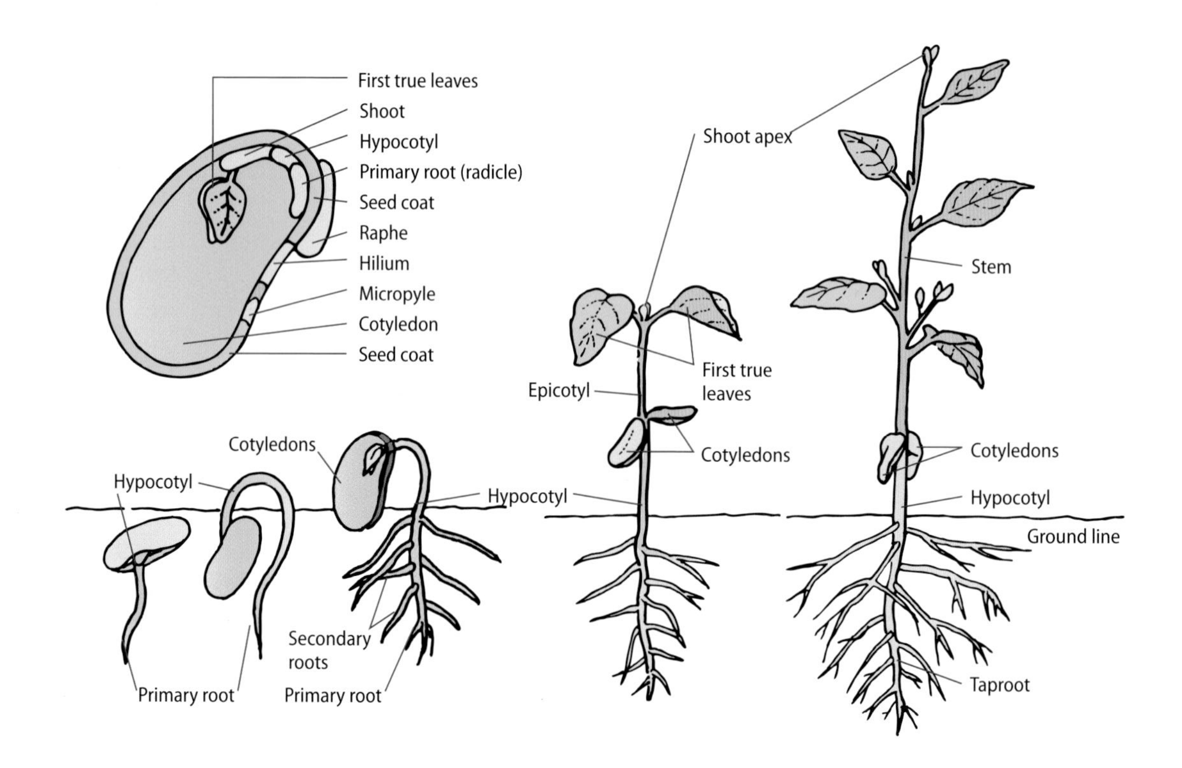

FIGURE 4-29

Seed, germination, and emergence of a bean plant, a typical dicot.

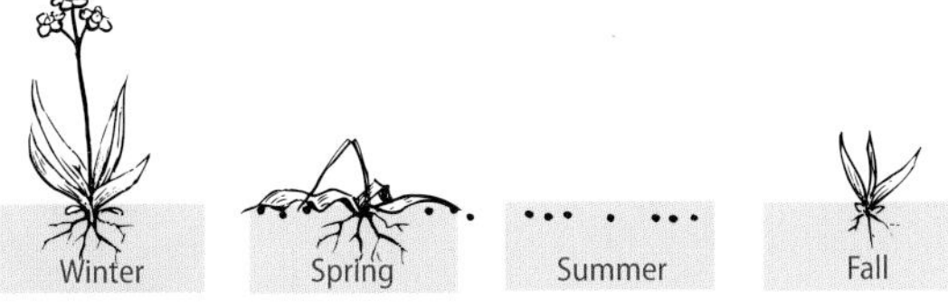

Winter annual
Winter annuals germinate in the fall, mature in the winter, and die in early summer. The seeds remain dormant until the fall.

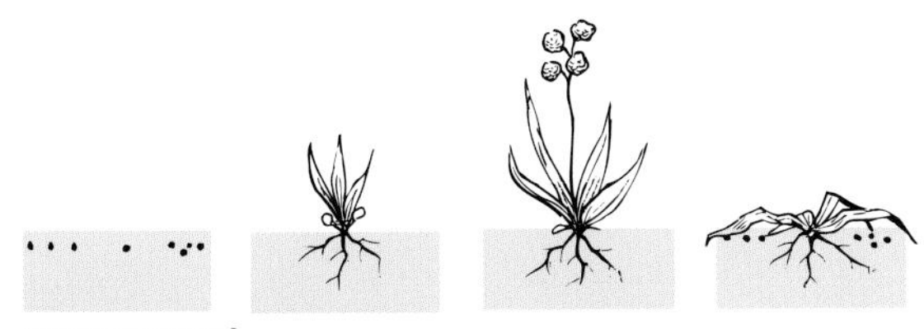

Summer annual
Summer annuals germinate in the spring, mature in the summer, and die in the fall. The seeds remain dormant until the spring.

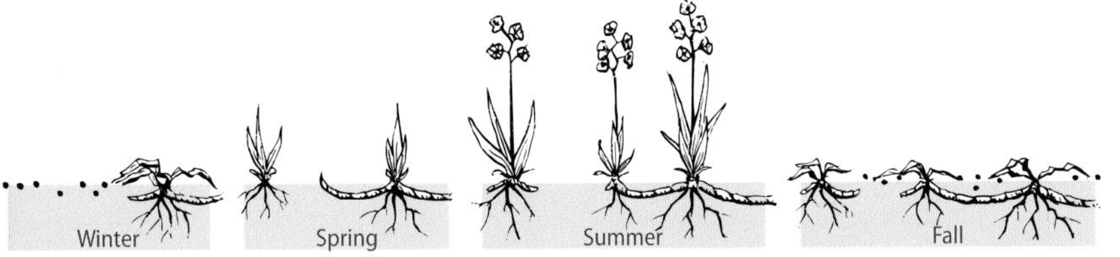

Perennial
Herbaceous perennials grow new plants from seeds or vegetative parts, such as rhizomes, bulbs, tubers, or rootstocks, in the spring. They mature in the summer and die in the fall; seeds and underground parts overwinter.

FIGURE 4-30

Life cycles of winter annual, summer annual, and perennial weeds.

Standard botanical texts often rely on features of flowers, fruit, and mature leaves to identify broadleaf species and grasses. For instance, flowers may occur singly or as compound inflorescences (e.g., panicle, raceme, spike, head, and umbel) (Figure 4-28). However, to manage weeds effectively, they must usually be identified early, so features of seedlings are particularly important, especially seed leaf shape and texture. For good illustrations of weed seedlings and other identifying features, go to the UC IPM Weed Photo Gallery online at www.ipm.ucdavis.edu/PMG/weeds_intro.html.

Two other groups of plants can be classified as weeds in specific circumstances: mosses and algae. Mosses (and liverworts) belong to a unique group of plants known as bryophytes. Mosses lack a vascular system. They are occasional pests in aquatic settings and greenhouses. Algae are nonflowering primitive aquatic plants that often clog streams, lakes, irrigation systems, drainage ditches, and rice fields. Algae lack true stems, leaves, and flowers. They reproduce through cell division or production of spores.

Life Cycle

Most weeds begin as a seed and pass through several stages of development (Figure 4-29). The seed is composed of three basic parts: an embryo, a food source, and a seed coat. Many weeds can remain dormant in the seed stage. Germination occurs at or near the soil surface when the right combination of light, temperature, and moisture occurs.

The first structure to appear during germination is the primary root, followed by the seedling shoot. The first leaves to emerge in dicots are called cotyledons. Primary growth continues from the food stored in the cotyledons until the first true leaves can begin photosynthesis. In grasses, the *coleoptile* is the first leaf to emerge, with the true leaves breaking through the coleoptile. Vegetative growth, or growth of stems and leaves, continues until just prior to the production of flower structures. Plants then enter a reproductive period in which most energy is diverted for flowering, fruit, and seed production. Most annual weeds die soon after the formation of seed, while perennials continue to repeat the vegetative growth and reproductive cycles each year.

Weeds can be grouped according to their life cycles as annuals, *biennials*, or perennials (Figure 4-30). Occasionally, weeds with one type of life cycle may behave as another due to favorable weather conditions or abnormal or unusual changes in the environment. Annual weeds grow, reproduce, and die within one growing season. They sprout from seed, mature, and produce seed for the next generation during this period. Annuals reproduce by seed only.

Depending on the time of year that they begin growth, annuals are divided into two groups, summer annuals and *winter annuals*. Summer annuals germinate in the spring; they flower and produce seed in mid to late summer and die in the fall. Some common summer annuals include pigweed, puncturevine, barnyardgrass, and yellow foxtail. Winter annuals germinate from late summer to early winter, but they may not grow very much until the temperature warms toward the end of winter. They flower and produce seed in mid to late spring and die in the

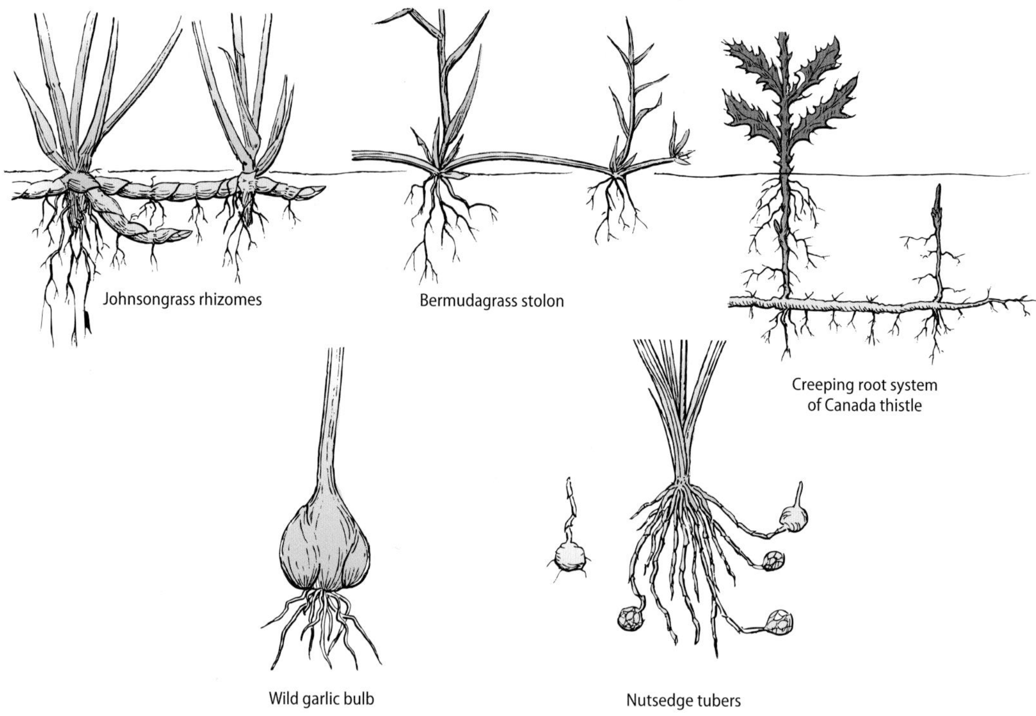

FIGURE 4-31

Vegetative reproductive structures of perennial plants.

early summer. Mustard, wild oats, annual bluegrass, and filaree are examples of winter annuals. The designation of species as summer or winter annuals is not always clear-cut. In cooler areas, some winter annuals may also germinate and grow in summer.

Biennials are plants that usually require two growing seasons to complete their life cycle. Seeds germinate in the spring, summer, or fall of the first year. Biennials overwinter as vegetative growth, generally in a rosette form, often with thick storage roots. Flower stalks and seed are produced after the plant shoot tips have been exposed to cold. After producing seed, biennials die in the fall of the second year. Poison hemlock, wild carrot, common mullein, and scotch thistle are examples of biennial weeds.

Perennials produce vegetative structures that allow them to live for 3 years or longer; some species live indefinitely. Many perennials lose their leaves or die back entirely during the winter but regrow each spring from roots or underground storage organs such as tubers, bulbs, *rhizomes*, or creeping roots or aboveground *stolons* (Figure 4-31). These plants are herbaceous perennials. Most can also reproduce by seed. Many weedy perennials may be introduced as seeds but invade new areas or spread via their vegetative organs. Examples of perennial weeds include yellow nutsedge, johnsongrass, field bindweed, and bermudagrass. Woody plants such as trees and shrubs are perennial and under certain circumstances are considered weeds. Examples of woody plants that are often treated as weeds include poison oak, saltcedar, sagebrush, blackberries, alders, and numerous others.

Successful Characteristics

Plants that have been successful as weeds often possess certain definable characteristics that separate them from other plants. Features including abundant seed production, rapid population establishment, seed dormancy, long-term survival of buried seed (the seed bank), adaptations for seed dispersal, the presence of vegetative reproductive structures, and the ability to invade sites disturbed by people have ensured their survival and dominance. Weeds are competitive, persistent, and pernicious.

Weeds compete with other plants for light, water, and nutrients. Most weed species are successful at germinating early and capturing more of the available nutrients. Rapid root and shoot development gives the weed a distinct advantage by shading out desirable plants.

The ability to produce abundant, long-lived, and highly dispersible seed is essential for the survival and success of weeds. Weeds produce a large number of seed; some produce millions per plant. Some weeds successfully reproduce under conditions of low fertility, minimal water, cool temperatures, and short growing seasons. Most weed seeds exhibit some degree of dormancy. Dormancy ensures survival by preventing the entire population from germinating at one time; it allows the species to wait for external environmental conditions that are suitable for growth. Weed seeds are often viable under some of the most adverse conditions; if buried, some may remain viable for years.

The total number of weed seeds surviving in the soil or on the soil surface of an area is called the weed seed bank. Thousands of seeds can be found in even a cubic foot of agricultural soil, and many of these seeds may have been produced one or more years previously. It is a good idea to assess the seed bank by taking soil samples to determine the type of weed seeds and their abundance in a field. Soil samples can be watered to induce germination to identify weeds. The seed bank can help growers and pest managers make predictions about future weed problems and future management needs.

Perennial weeds that produce seed and vegetative reproductive structures, such as rhizomes, stolons, or tubers, have an even greater chance for survival. Perennial weeds store food in their reproductive structures for overwintering, and on some species, buds become hardened and dormant, ensuring survival in adverse conditions. Reproductive structures produce new shoots and when broken off readily establish as new plants; cutting reproductive structures at certain times of the life cycle may stimulate the production of more plants. In addition, weeds growing from vegetative reproductive structures have more stored energy and can establish more quickly than plants developing from seed.

Dispersal and Movement

Weeds disperse in a variety of ways. Weed seeds may be blown around in wind, washed away in water, or transported by animals that have fed on them. Human activities are an important means of dispersal in managed systems. Weed seeds and other reproductive structures are carried in and out of fields on contaminated equipment, in soil, and by attaching to workers' clothing. Because agricultural systems are disturbed sites, the ability of weeds to establish in new areas is increased.

Some weed seeds have adapted mechanisms that ensure their movement into new areas (Figure 4-32). Barbs, hooks, spines, awns, sticky secretions, or cottony lint can attach to birds, mammals, people, or equipment. Animals also disperse weed seeds after digesting them.

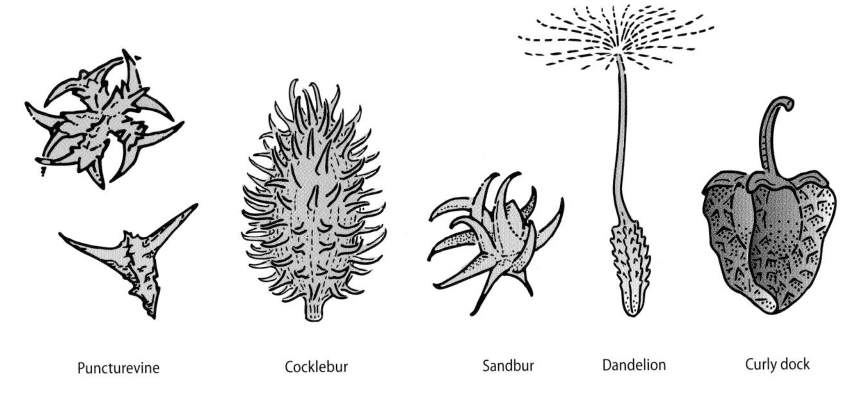

FIGURE 4-32

Some weed seeds have morphological characters that help in dispersal. The prickly outer coating of puncturevine seed becomes entangled in clothing, hair, or fur and cuts tires, shoes, or feet. The seeds of cocklebur and sandbur cling to clothing, hair, or wool. Dandelion seeds have feathery structures that help disperse them in the wind. Curly dock seeds float.

Wind and water movement aid in the dispersal of some weed seeds. Some weed seeds have feathery hairs, parachutes, or wings that carry them through the air. Seeds of other species, such as Russian thistle, are knocked off the plant as it tumbles in the wind. Water transports almost any type of seed. Flood waters and open-canal irrigation systems contribute significantly to the spread of weed seeds. Weeds have adapted other mechanisms that assist in their dissemination. Some species forcefully dehisce seed; others eject them from the mature plant. The ability of perennial weeds to produce seed and propagate vegetatively ensures their dissemination.

The introduction of new weed species from foreign locations is a serious concern in the movement of weed species. It is estimated that two-thirds of today's weeds were introduced into the United States from other countries.

PATHOGENS

Microorganisms that cause disease are commonly called pathogens. Fungi, bacteria, and viruses are the most common pathogen groups causing disease in plants. Other groups that contribute to plant disease include *viroids* and phytoplasmas. Diseases may also be caused by abiotic, or nonliving, factors. To properly resolve disease problems, it is important to distinguish between those that are caused by pathogens and similar disorders and those that are caused by noninfectious abiotic factors.

The Disease Triangle

For a pathogen to attack a plant, the plant and pathogen must come in contact with one another and interact. Plant and pathogen are often present and in contact but disease does not develop, either because the host is resistant to attack or the pathogen is unable to attack due to unfavorable environmental conditions for disease development. These three components—host plant, causal agent, and a conducive environment—are known as the disease triangle (Figure 4-33). One side of the triangle represents each component. If any of the sides is absent or unfavorable, the disease will not develop. It is essential to understand that a pathogen and the disease it causes are not synonymous. The pathogen may be present, but

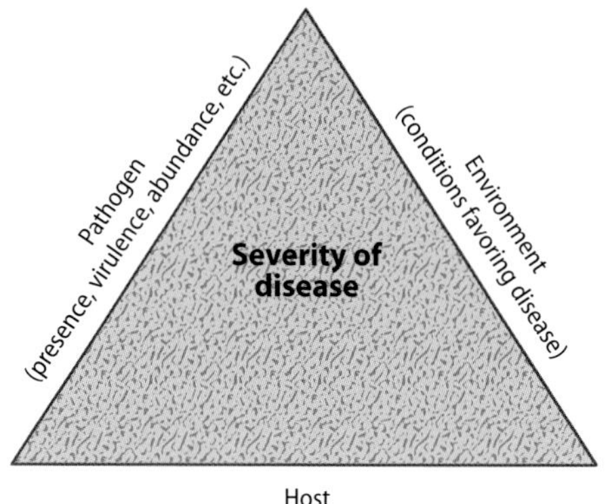

FIGURE 4-33

The disease triangle. All sides of the triangle must be present for disease to occur. The disease becomes more severe as conditions on each side become more favorable.

there may be no disease if either of the other two sides of the triangle is not present.

The Disease Cycle

Specific events unfold that lead to the development and perpetuation of a disease. Together these events are called the disease cycle (Figure 4-34), which refers to the appearance, development, and perpetuation of the disease on a host plant, not just to the life cycle of the pathogen itself. The disease cycle includes changes that occur in the host plant, the plant's symptoms during the growing season, and the survival of the pathogen into the next season.

Inoculation occurs when inoculum, the form of the pathogen that initiates infection, comes into contact with a susceptible plant. In fungi, inoculum can be spores, *sclerotia*, or mycelia. Bacteria, phytoplasmas, viruses, and viroids do not produce specialized structures for survival or dissemination and must gain entry to a host as individual entities. *Primary inoculum* survives the winter and causes the primary infection, or first infection, of the season. In multicycle pathogens, primary infections produce secondary inoculum, which in turn causes secondary infections that can rapidly spread disease within a crop. Sources of inoculum may be nearby plants, seeds, transplants, or other propagative structures. Inoculum may already be present on the host or in the field or may be spread by wind, water, or insects. Some types of inoculum survive in the soil for less than a season; other types survive for many years.

Pathogens enter the host by direct penetration through natural openings or through wounds. Most bacteria enter plants through natural openings, such as stomata, lenticels, nectaries, or through wounds. Viruses, viroids, and phytoplasmas enter through wounds, through feeding by vectors or, in perennial plants, through grafts. Fungi commonly penetrate directly through intact plant surfaces, although some enter through wounds to the plant.

A pathogen has successfully infected a host once the pathogen begins to withdraw nutrients, grow, and multiply in the plant tissues. Successful infections usually result in symptoms such as *necrosis* that are associated with specific pathogens. When symptoms are not evident immediately, the infection is said to be latent. In most diseases, however, symptoms appear within a few days after infection.

Understanding the cycle of a disease can help identify possibilities for control. Opportunities to break the disease cycle exist through the selection of disease-resistant stock, seed treatment, well-timed fungicide application, and many cultural control practices, such as rotation, removal of overwintering inoculum or an alternate host, solarization, and crop-free periods to reduce inoculum or avoid the presence of a *vector*.

Fungi

Fungi are a diverse group of organisms that obtain their nutrients from living or dead plant or animal material. Although once classified in the plant kingdom, fungi are not plants; they do not have chlorophyll or the ability to carry out photosynthesis. Most are in the kingdom Fungi; a few, sometimes referred to as the lower fungi, are now actually classified in the kingdoms Protozoa and Chromista. Most fungi are saprophytes and feed on decaying material, helping to break down dead organic matter and build up soil fertility. Certain fungi live on both dead and living plants; some of these cause plant disease. Still other fungi require living host plants to grow and reproduce; these fungi are called *obligate parasites* and include many of the major plant pathogen species, such as rusts, downy mildews, and powdery mildews.

Identification. Fungi are mostly microscopic organisms, but some have forms, such as mushrooms, that are large and identifiable without magnification. The body of the fungus is made up of tiny,

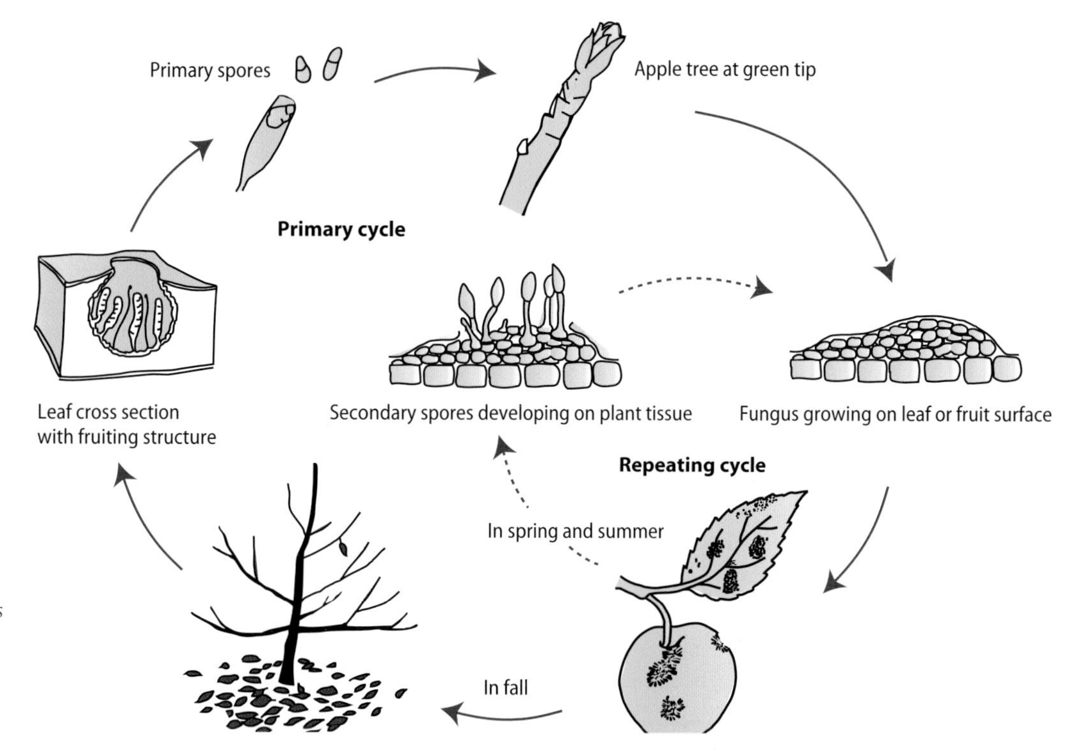

FIGURE 4-34

The disease cycle of apple scab. The disease cycle includes the life cycle of the pathogen and how it interacts with the host to cause disease and perpetuate within the ecosystem.

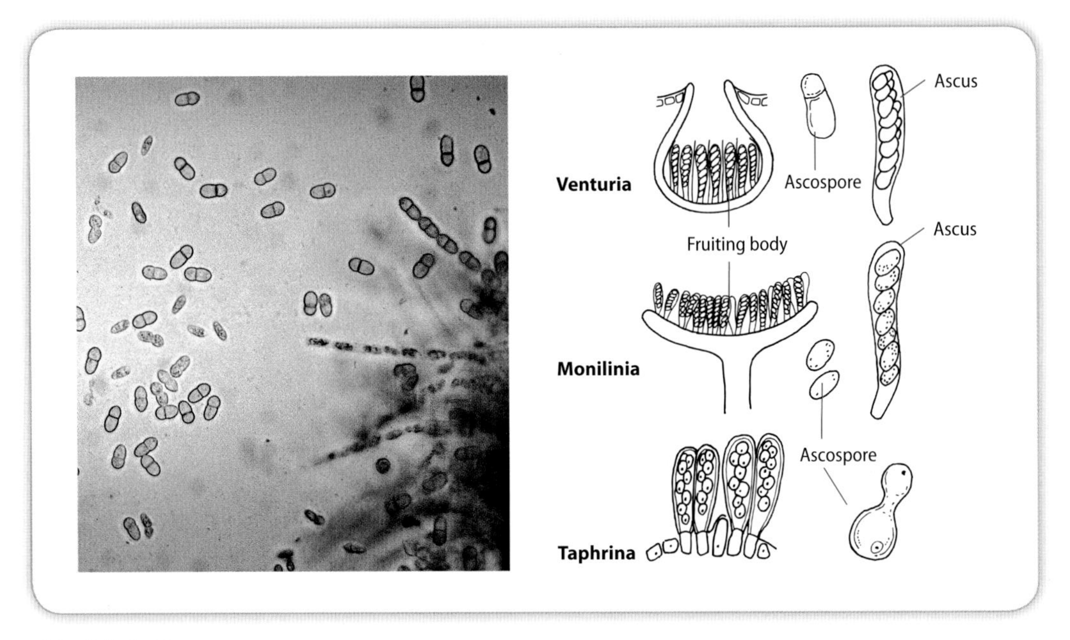

FIGURE 4-35

These footprint-shaped ascospores were produced by Venturia inaequalis, *the causal agent of apple scab. Ascospores are sexually produced spores formed within a sac called an ascus and are often borne on a characteristic fruiting body. Fungi-producing ascospores are classified as* Ascomycetes. *The drawings at right show fruiting bodies, ascospores, and asci for the apple scab pathogen in the genus* Venturia *along with two other types: the genus* Monilinia, *which includes the agent causing brown rot of stone fruits, and the genus* Taphrina, *which includes the agent causing peach leaf curl.*

tubular filaments called *hyphae*. Growth of the fungus occurs at the tips of the hyphae. Hyphae may be colorless or variously pigmented. The mass of hyphae growing together is referred to as the *mycelium*. Reproductive structures called spores may be produced on specialized hyphae or fruiting bodies.

Spores and spore-bearing structures are the most significant characters for identification. Size, shape, color, arrangement of spores, and the shape and color of the fruiting body can be used to help determine the genus and species of the fungus; these characteristics are most often used in taxonomic keys by professional plant pathologists (Figure 4-35).

Field identification of disease symptoms caused by fungi can often help determine the species responsible for the damage. Host indexes and other publications often list the pathogens that are known to affect specific hosts. Although lists may be a good starting point in linking host symptoms to a causal agent, the species responsible must be identified accurately before a control is recommended; when in doubt, enlist the assistance of an expert.

Types of damage that may be associated with fungus infection include leaf spots; blight on leaves, branches, twigs, and flowers; cankers; dieback of twigs; root rot; seedling damping off; stem rots; soft and dry rots; scab on fruit, leaves, or tubers; and the overall decline of the host. All of these symptoms also contribute to the stunting of infected plants.

Fungal infections can also result in distortion of plant parts. Examples of noticeable symptoms include enlarged roots (clubroot); enlarged growths filled with mycelia (galls); warts on tubers and stems; profuse upward branching of twigs (witches'-broom); and distorted, curled leaves (leaf curl). Other symptoms exhibited by fungus infections include wilting or powdery accumulations such as rust and mildew.

In diagnosing disease, plant pathologists distinguish between symptoms and signs (Figure 4-36). The symptoms of a disease refer to changes in the appearance of the infected plant, such as the necrotic, sunken, ulcerlike lesions of an anthracnose infection. The signs of a disease, on the other hand, are structures that the patho-

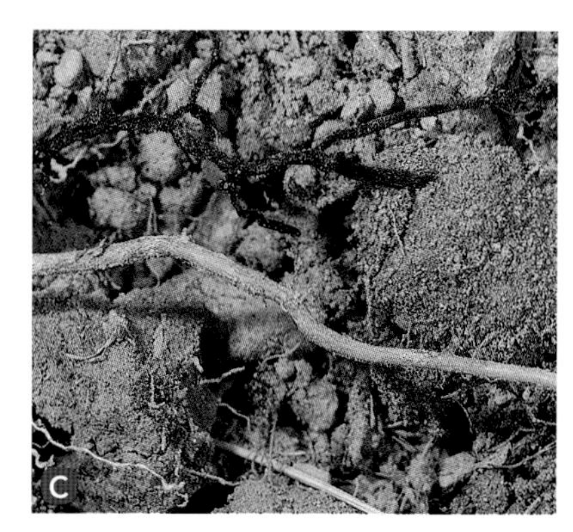

FIGURE 4-36

Distinguishing between signs and symptoms of a disease. One of the symptoms of Armillaria *root rot in tree crops is the yellowing and wilting of foliage, often on one side of the tree (A). Signs of the disease include* Armillaria mellea *mushrooms that may appear around the trunk of infected trees following rain in fall or winter (B) the dark strings of fungal mycelia called rhizomorphs, which can extend from diseased wood to infect the roots of healthy trees. In the photo (C), a rhizomorph is shown above a light brown healthy root.*

gen may produce on the surface of the host, such as mycelia, sclerotia, sporophores, fruiting bodies, and spores (Figure 4-37). Thus, the white cottony mycelia and hard black clusters of dormant sclerotia found on the lower stem of lettuce plants are signs of the *Sclerotinia* fungus; symptoms of Sclerotinia diseases include wilting of the lower leaves and limpness of the entire plant.

Life Cycle. The life cycles of fungi vary. Generally, fungi reproduce through spore production. Spores are specialized reproductive bodies that may be formed sexually or asexually. Almost all fungi have an asexual cycle, and asexual reproduction can occur several times within a growing season. In most of the fungi that have a sexual reproductive cycle, reproduction occurs only once a year. Fungi that are not known to have a sexual cycle are called imperfect fungi.

Although pathogens are usually present on or near their host, they must wait for certain environmental conditions, such as high humidity or warm temperatures, to make disease development possible. Fungi have evolved a number of mechanisms to survive when conditions are unfavorable (Figure 4-38). Fungi overwinter as mycelia or spores in or on infected tissue. Resting spores of some species are resistant to extremes in temperature and moisture. Infected plant debris, bud scales, or bark cankers in trees, shrubs, or soil can provide overwintering habitat for the mycelia, sclerotia, or spores of many fungi. Seeds and other vegetative organs or alternative hosts such as weeds provide other survival sites for overwintering fungi. Some soil-inhabiting fungi survive as saprophytes or by producing resistant oospores.

Dispersal and Movement. Most fungi cannot move long distances without the assistance of people, insects, animals, or environmental factors such as wind or rain. A few fungi are able to move from one host to another through the growth of *rhizomorphs* in the soil. Most fungal spores disseminate in air currents and can be car-

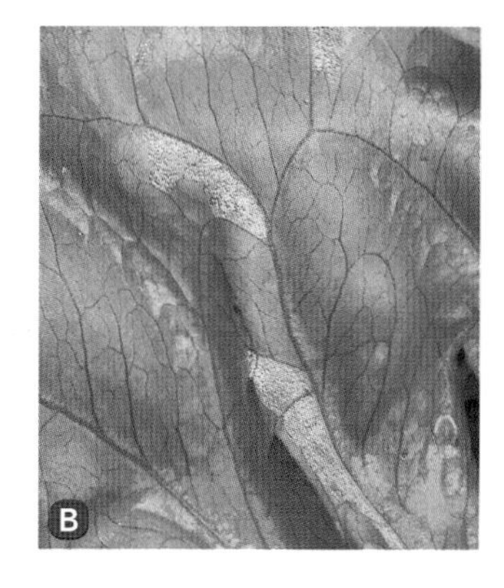

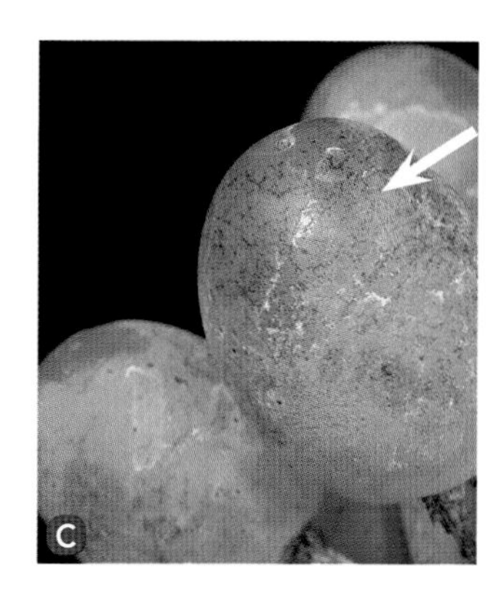

FIGURE 4-37

Some major fungal diseases and their causal organisms that can be identified by the signs (or fungal growth) that appear on the host include (A) leaf rust (Puccinia recondita) *spores on wheat; (B) downy mildew* (Bremia lactucae) *spores on lettuce; and (C) powdery mildew* (Uncinula necator) *spores on grapes.*

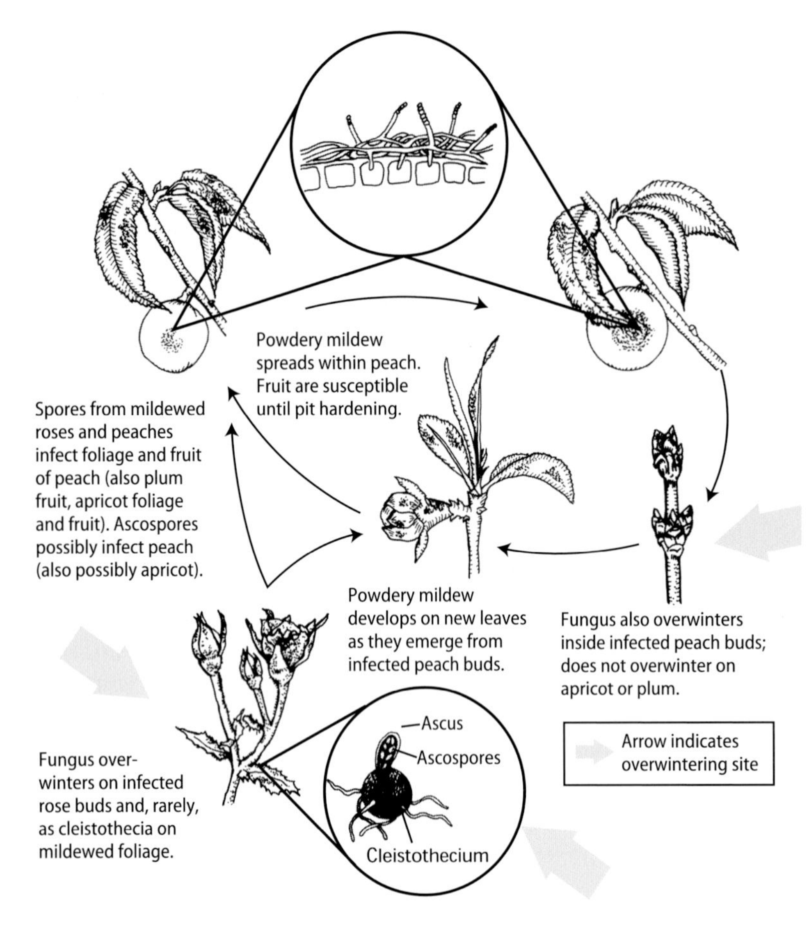

FIGURE 4-38

Seasonal cycle of powdery mildew caused by the Sphaerotheca pannosa *fungus in apricots, peaches, and plums. The pathogen can overwinter as mycelia on rose buds or peach buds or as cleistothecia on rose foliage, but it cannot survive on plum or apricot. Infected shoots and buds can be pruned out during the dormant season on peaches and roses. Removing rose bushes next to plum or apricot may reduce infection of these fruit trees. If fungicides are applied, they will be most effective when applied early in the season as the buds are opening.*

ried over long distances. Spores, sclerotia, and mycelial fragments can be picked up and dispersed by water; some fungi depend on rain for their dispersal. Insects can spread fungal pathogens. For example, the elm bark beetle, *Scolytus multistriatus*, is the primary agent in the transmission of Dutch elm disease (DED) to new areas. Once the tree is infected, DED can continue its spread from one tree to another through root grafts between adjoining trees. Animals, people, and contaminated equipment can carry fungus spores. People also disseminate pathogens through the transport of infected seeds, transplants, nursery stock, and contaminated containers.

Bacteria

Bacteria are microscopic one-celled organisms that are classified as prokaryotes and belong to the kingdom Procaryotae. Bacterial diseases affect almost every kind of plant and can occur wherever moist, warm conditions are present. Most bacteria that cause plant disease are *facultative parasites* and can be readily grown in culture.

Identification. Bacteria are very small, typically under 0.002 mm, or about one twelve-thousandths of an inch, long. They can be rod shaped, spherical, ellipsoidal, spiral, comma shaped, or filamentous. Most bacteria have flagella that allow for some limited movement. The cell wall of bacteria is hard and enveloped by a slime layer. In some species, the slime layer forms a large mass and is referred to as a capsule.

Field diagnosis of bacterial disease relies on recognition of disease symptoms (Figure 4-39). Plants generally have specific reactions to bacterial infections; however, laboratory analysis is required for positive identification. Common symptoms include cankers, galls, wilts, slow growth, distorted fruits, rots, discoloration of plant parts, slow ripening, distorted leaves, brooming, and leaf spots. Some bacterial diseases produce a slimy ooze on the plant surface.

Certain bacteria cause the formation of galls on specific plant parts; examples include olive knot and crown gall. Galls are typified by large, swollen, rapidly dividing cells and disorganized vascular tissues. Galls may interfere with the movement of food and water in the plant. Bacterial wilts generally affect the entire plant; slime is produced that plugs the water-conducting tissue of the infected plant. Destructive wilts include those found in tomato, cotton, cucumber, and corn. Cankers are the result of extensive tissue destruction caused by certain bacteria such as those that cause tomato cankers and fire blight. Localized spotting occurs most commonly on leaves and fruit; potato scab, for instance, appears as local spots on tubers. Rots invade fleshy tissue and often result in a slimy, foul-smelling ooze. Soft rots often enter secondarily, after another pathogen has invaded.

Life Cycle and Growth Requirements. Each bacterial cell is an individual organism. Reproduction occurs by an asexual process known as binary fission: The cytoplasmic membrane grows toward the center of the cell, dividing it, and the bacterium's DNA separates into two equal pieces. When conditions are favorable, a bacterium can reproduce every 20 minutes. Over time, the offspring of a single bacterial cell produces a visible mass called a colony.

Bacteria enter plants through natural openings and wounds. Once inside the plant, some bacteria move along the sap stream; others move about with the flow of fluids between the cells. In early stages of disease, bacteria often develop between the cells of a plant. As cell walls become injured, the bacteria penetrate the cell and continue growing. The disease cycle of fire blight, caused by the bacterium *Erwinia amylovora*, is shown in Figure 4-40.

Disease symptoms become evident at varying times after bacteria enter and develop in plant tissue. Soft rots and leaf spots are usually visible within a day; crown gall infections can take up to 2 years to become evident. The time between the establishment of the bacteria and the appearance of symptoms is referred to as the incubation period.

Survival Characteristics. Bacteria survive mostly within plant hosts as parasites, on seeds, or in plant debris in soil. They do not produce overwintering spores as fungi do; however, many are able to overwinter saprophytically in plant debris. This characteristic contributes to their ability to sustain their populations when the plant host is not present.

Dispersal and Movement. Bacteria can be carried from one place to another in splashing or flowing water and rainwater, by moisture, and by various management practices. Bacteria that survive in organic debris in the soil can be dispersed by any practice that involves moving soil from one place to another, such as cultivation. Bacterial pathogens are frequently carried on infected seeds, cuttings, or transplants. Insects and animals often aid in spreading bacteria. Cucumber beetles, for example, transmit the cucumber wilt organism, and sharpshooters transmit *Xylella fastidiosa*, the bacterium that causes Pierce's disease of grape and oleander leaf scorch.

Viruses

A virus is an obligate parasite of submicroscopic size that is composed of genetic material and is surrounded by a layer of protein. Viruses multiply only in living cells; they are among the smallest and simplest of plant

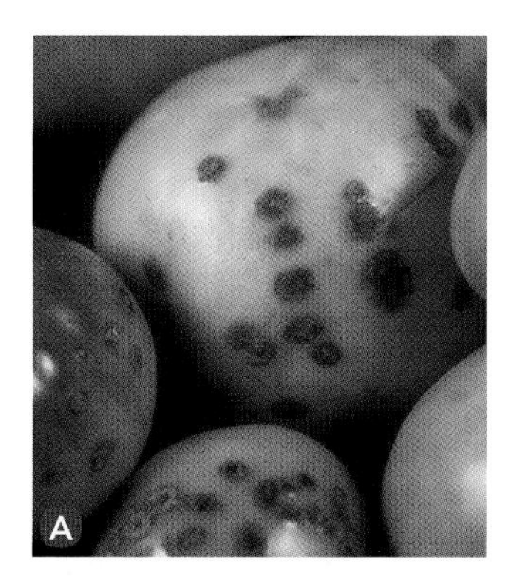

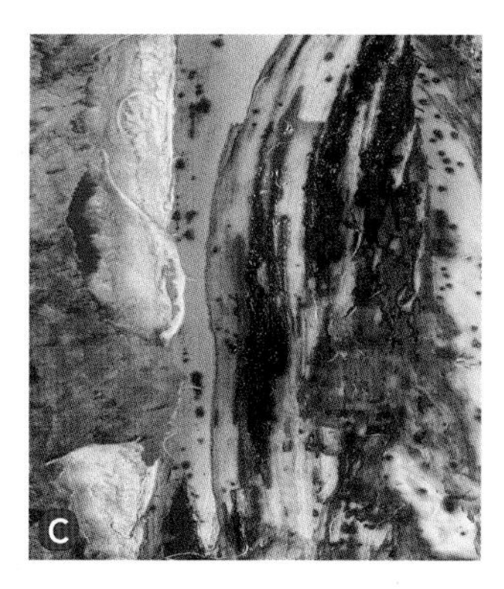

FIGURE 4-39

Disease symptoms of some bacterial diseases. Bacterial spot lesions caused by Xanthomonas vesicatoria *on tomato fruit have a rough, scabby texture (A). When they first develop, the lesions are surrounded by a white halo, similar to bacterial speck. As with all bacterial diseases, laboratory analysis is required for identification. Pear fruit, flowers, and leaves infected with fire blight turn black (B). Tan droplets of ooze are also typically present. A walnut trunk damaged by deep bark canker bacteria,* Erwinia rubrifaciens, *typically shows dark spots underneath the bark (C).*

pathogens. Viruses cause disease by disrupting normal metabolic and physiological processes in plant cells.

Identification. Because of their small size and transparency in the host cells they inhabit, viruses cannot be viewed or detected by the types of methods used for other pathogens. Virus particles in infected plant cells can sometimes be detected with light microscopes in a plant pathology lab, but this detection method is not practical for the field. The main detection clues for the pest manager are plant symptoms.

Viral diseases produce a variety of injury symptoms on plants. Common field symptoms associated with plant viruses include growth reductions, color changes, malformations, and necrosis of tissue; severe stunting and a reduction in yield may be evident. Mosaic patterns (Figure 4-41), a mottling of healthy and discolored tissue on leaves, are a common virus symptom. Some viruses roll or crinkle leaves; sometimes the affected leaves are a deeper green. Vein clearing or leaf yellowing is typical of some viral infections. Some viral diseases cause the plant to become dwarfed, while others stimulate short, sporadic shoots or stunting and rosetting (Figure 4-42). Positive identification of a virus in a plant can only be determined through the use of an electron microscope, serology, or a nucleic acid probe.

Life Cycle and Growth Requirements. All viruses are parasitic on cells and require a host cell for survival and reproduction. They are not cellular organisms themselves. Nucleic acid is released from the virus particle when the virus infects the host; the host cell is then directed to form more viruses. Viruses do not divide, nor do they produce any reproductive structures. Plant viruses have different sizes and shapes, but they are usually elongate, bacilluslike, or spherical.

Viruses enter plants through wounds made by insect or nematode vectors, mechanical injury, or from an infected

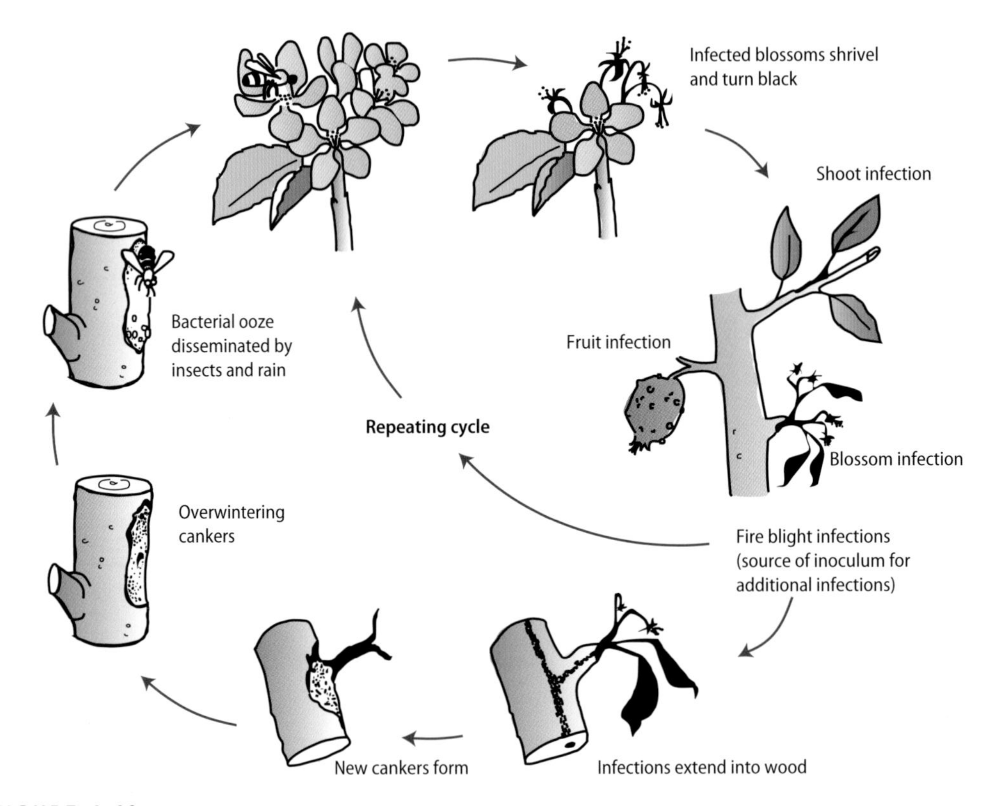

FIGURE 4-40

The disease cycle of fire blight of pear and apple caused by the bacterium Erwinia amylovora. *The bacteria overwinter at the margins of cankers on branches and twigs and are disseminated in spring when trees start growing and sap flows out of the cankers. Splashing rain and insects attracted to the ooze move the bacteria to blossoms. Bacteria multiply rapidly in the blossoms and enter other tree tissues through stomata, lenticels, and wounds.*

FIGURE 4-41

Discolored bands, lines, and ring patterns on leaves, such as these rose leaves infected with rose mosaic virus, are typical of plants affected by certain viruses. Other viruses produce more general yellowing, stippling, or curling that is easy to confuse with damage caused by other agents, such as fungi or bacteria.

FIGURE 4-42

Some viruses can cause severe stunting and tight, bunchy growth (rosetting). This almond tree has been infected with yellow bud mosaic for several years and shows severe stunting and concentration of leaves on terminals.

pollen grain. The first virus particles may appear about 10 hours after inoculation. For infection to occur, the virus must move from cell to cell and multiply in the cells to which it moves. The movement of viruses in the plant varies with the virus and the host.

Survival Characteristics. Viruses cannot survive in dead plant debris or outside living plant tissue. Alternate hosts such as perennial plants, weeds, or volunteer crop plants provide overwintering sites for many viruses and are often suitable hosts for virus vectors as well. While viruses cannot overwinter outside of a plant, living contaminated volunteer plants can be important overwintering sites. Insect vectors also provide an important overwintering mechanism for viruses.

Pest Dispersal and Movement. Viruses can enter plants only through wounds. They are dispersed mechanically from plant to plant through vegetative propagation, sap, seeds, and pollen or by vectors. Vectors of one or more plant viruses include aphids, leafhoppers, whiteflies, beetles, thrips, mites, nematodes, fungi, and dodders. Vectors are specific to certain viruses and hosts.

Viroids

Viroids are low-molecular-weight nucleic acids that can infect plant cells, replicate themselves, and cause disease. Viroids lack a protein coat and contain only RNA. Viroids consist of 250 to 400 nucleotides and exist as free RNA. Symptoms of disease from viroids are similar to viruses. Viroids cannot survive in dead plant matter or outside the host for long periods of time. They overwinter and oversummer in perennial hosts. Viroids are spread by implements, propagation tools, pruning tools, vegetative propagation, and sometimes with seeds. Only a few viroid-caused diseases have been identified. Potato spindle tuber, citrus exocortis, chrysanthemum stunt, chrysanthemum chlorotic mottle, and cucumber pale fruit are known to be caused by viroids.

Phytoplasmas

Phytoplasmas are the smallest cells known to multiply independently of other living cells, and they can survive for extended periods outside plant cells. Phytoplasmas have a cellularlike structure without a true cell wall and may assume a wide variety of shapes and sizes. They reproduce by budding and binary fission and are transmitted primarily by leafhoppers; mites may also transmit phytoplasmas. Phytoplasmas are also readily transmitted among woody crop plants by grafting. Phytoplasmas are responsible for several insect-transmitted diseases, including pear decline, aster yellows, western-X disease of peach, and citrus stubborn disease.

TABLE 4-5.

Common abiotic disorder symptoms and their causes.

Symptoms	Possible cause
Foliage wilts, droops, discolors, and drops prematurely. Twigs and limbs may die back. Bark cracks and develops cankers. Plant may become attacked by wood-boring insects.	water deficiency
Foliage yellows and drops. Twigs and branches die back. Root crown diseases develop. Mineral toxicity symptoms develop.	water excess or poor drainage
Foliage is discolored, undersized, sparse, or distorted and may drop prematurely. Plant growth is slow. Limbs may die back.	mineral deficiency
Foliage turns brown, dry, and crispy. Limbs may die back. Odor of natural gas may be detectable.	natural gas line leak underground
Foliage or shoots turn yellowish, undersized, or distorted. Leaves may appear burned with dead margins and drop.	pesticide toxicity
Yellow, brown, then white areas develop on upper side of leaves, beginning between veins. Foliage may die.	sunburn
Leaves or needles turn yellowish, brownish, or have discolored flecks. Foliage may be sparse, stunted, and drop prematurely.	air pollution
Leaves turn yellowish or brownish, especially along margins. Foliage may drop prematurely. Bark becomes corky.	mineral deficiency
Foliage is abnormally dark.	excess light
Excess growth of succulent foliage. Foliage may appear burned and die. Plant is infested with many mites, aphids, psyllids, or other insects that suck plant juices.	nitrogen excess
Shoots, buds, or flowers curl, darken, and die. Limbs and entire plant may die.	frost
Foliage, twigs, or limbs injured. Cankers may develop.	hail or ice
Bark or wood dead, often in a streak or band.	lightning
Bark is cracked or sunken, often on south and west sides. Wood may be attacked by boring insects or decay fungus.	sunscald

Many of these symptoms can have other causes, including pathogens and insects.

ABIOTIC DISORDERS

Abiotic disorders are noninfectious diseases induced by adverse environmental conditions, often as a result of human activity. These include nutrient deficiencies or excesses; low or high temperatures; toxic levels of salt, herbicides, or other pesticides; air pollution; and too little or too much water (Table 4-5). Activities that compact soil, change soil grade, or injure trunks or roots can also predispose plants to abiotic disorders. In addition to direct damage, abiotic disorders can predispose plants to attack by insects and pathogens.

Some abiotic disorders can be recognized by characteristic symptoms of distortion or discoloring of foliage, roots, stems, fruits, or flowers. Yet abiotic disorders are difficult to identify with certainty. A field history, records of pesticide and fertilizer use, and soil or leaf tissue analysis may help in diagnosis. Patterns of injured plants in the field can also assist in identifying abiotic disorders. In general, symptoms of *abiotic disorders* arise suddenly and do not spread through a plant or to other plants over time as pest damage can.

Inappropriate irrigation practices can cause diseaselike symptoms in plants. Inadequate water causes plants to wilt, discolor, and drop leaves, fruit, or flowers prematurely. Prolonged moisture stress results in stunted growth, dieback, and susceptibility to insects and other pests. Too much water drowns and kills roots and exposes plants to attack from pathogens such as *Phytophthora* spp. As roots die, discolored and dying foliage appear on the aboveground portions of the plant.

Nutrient deficiencies cause foliage to discolor, fade, distort, or become spotted, causing symptoms that are sometimes confused with those caused by pathogens. Fewer leaves, flowers, and fruit may be produced; they can also develop late and be undersized. More severely deficient plants become stunted and exhibit dieback. Excess amounts of nutrients and salts can also be toxic to plants. Toxicity symptoms include marginal leaf chlorosis, necrosis, branch dieback, and increased pests.

Weather conditions can contribute to disorders in plants that can be mistaken for pathogen-caused diseases. Too much or too little sunlight can damage plants. Sunburn, sunscald, and etiolation (bleaching) are associated with improper light levels. Frost damage can cause shoots, buds, and flowers to curl, turn brown or black, and die. Frost damage to foliage often resembles leaf anthracnose diseases. Plant damage from hail can provide entry sites for pathogens such as fire blight on susceptible species. Wind injury may cause scarring similar to insect feeding. If several species of plants in an area show similar damage, the cause is likely to be abiotic because most plant pathogens and many pest arthropods attack only one or a few species of related plants.

5 Management Methods for IPM Programs

IPM programs use a variety of methods to obtain effective, long-term suppression of pests in the least disruptive manner possible. Sound pest management often combines several compatible strategies, such as host resistance or tolerance, biological approaches, cultural controls, mechanical and physical controls, and chemical controls. PCAs must consider the effectiveness of these options and how they can be combined to work together, factoring into each decision the impact on the environment, worker health and safety, and economic considerations. In addition, pest management decision makers should always consider the impact that the control can have on nontarget species, including other pest species. For example, leaving weedy refugees as nectar sources for beneficial insects may increase the incidence of certain insect, plant pathogen, and nematode pests, and also provide a source of weed seeds. The potential for increased natural enemy activity must be weighed against the losses that may be incurred by other serious pests.

This chapter will focus on management tools and their roles in an IPM program. Effective use of IPM as a strategy for managing pesticide resistance will also be discussed. The chapter ends with a discussion of how IPM practices can be incorporated into *sustainable agriculture* systems or pest management programs for *organic farming*.

HOST RESISTANCE OR TOLERANCE

Pest-resistant plants, when available, are one of the most effective and least expensive pest management tools (Figure 5-1). Resistant plant cultivars have inherited characters that result in less pest damage or infestation than other varieties of that species under comparable conditions in the field. Although the terms *variety* and *cultivar* are frequently used interchangeably, plants produced through

FIGURE 5-1

Cotton cultivars resistant to the root knot nematode (outside rows) show little damage from nematodes compared with the susceptible cultivars in the center rows.

breeding programs are most accurately called cultivars; *variety* can also refer to naturally occurring variants in a subspecies. Plant cultivars bred to resist pests are a major tool in management of plant pathogens and nematodes and have also been used effectively against a number of insect pests including grape phylloxera, pea aphid, and spotted alfalfa aphid. Examples of some plant pathogens managed through the use of resistant cultivars in a few California annual crops are given in Table 5-1.

Although extremely valuable, the use of pest-resistant plant cultivars is not foolproof. Occasionally a resistant plant can become more susceptible to pest pressure as a reaction to physical stress such as variations in moisture, evaporation, plant and soil nutrition, and temperature. For instance, Lahontan cultivars of alfalfa are much more susceptible to stem nematode at 25°C (77°F) than at 15°C (59°F). An even more serious concern is the potential for development of *biotypes* or physiological races of pests that can overcome the factors that provide the basis for host resistance. Such races may develop as a reaction to the severe selection pressure exerted by a resistant crop cultivar. Integrating the use of resistant cultivars with other management tools such as sanitation, crop rotation, or *certified seed* can help prolong the useful life of a resistant cultivar.

TABLE 5-1.

Some common plant diseases managed in annual crops in California through the use of resistant cultivars.

Crop	Disease	Causal agent
beans	bean anthracnose	*Colletotrichum lindemuthianum*
	bean common mosaic	bean common mosaic virus (BCMV)
	bean yellow mosaic	bean yellows mosaic virus (BYMV)
	curly top	beet curly top gemini virus (BCTV)
	Fusarium wilt (blackeye and garbanzo beans)	*Fusarium oxysporum*
	halo blight	*Pseudomonas syringae* pv. *phaseolicola*
cole crops	downy mildew	*Peronospora parasitica*
	Fusarium yellows (cabbage)	*Fusarium oxysporum*
corn	common rust	*Puccinia sorghi*
	Fusarium ear rot	*Fusarium moniliforme*
	Fusarium stalk rot	*Fusarium moniliforme*
	head smut	*Sphacelotheca reiliana*
	maize dwarf mosaic	maize dwarf mosaic virus (MDMV)
cucurbits	angular leaf spot	*Pseudomonas syringae* pv. *lachrymans*
	anthracnose (some varieties of watermelon and cucumber)	*Collectotrichum lagenarium*
	downy mildew	*Pseudoperonospora cubensis*
	Fusarium wilt (cantaloupe and watermelon)	*Fusarium oxysporum*
	powdery mildew	*Sphaerotheca fuliginea, Erysiphe cichoracearum*
	Verticillium wilt	*Verticillium dahliae*
lettuce	downy mildew	*Bremia lactucae*
	lettuce mosaic	lettuce mosaic virus (LMV)
	corky root	*Rhizomonas suberifaciens*
small grains	barley yellow dwarf (barley and wheat)	barley yellow dwarf virus (BYDV)
	leaf rust of wheat and barley	*Puccinia* spp.
	leaf scald of barley	*Rhynchosporium secalis*
	net blotch of barley	*Pyrenophora teres*
	powdery mildew (barley and wheat)	*Blumeria graminis*
	Septoria tritici blotch	*Septoria tritic*
	stem rusts of wheat, barley, oats	*Puccinia graminis*
	stripe rusts	*Puccinia striiformis*
tomatoes	Alternaria stem canker	*Alternaria canker*
	bacterial speck	*Pseudomonas syringae* pv. *tomato*
	Fusarium wilt	*Fusarium oxysporum*
	Verticillium wilt (race 1)	*Verticillium dahliae*

Finally, a cultivar that is resistant to one pest may be very susceptible to another pest. For example, a lettuce cultivar that is resistant to lettuce mosaic virus could be very susceptible to downy mildew. Growers and their PCAs must weigh carefully the advantages and disadvantages of each cultivar before planting.

When monitoring pest resistant crops, PCAs must constantly be on the lookout for the breakdown of resistance. For example, most rootstocks resistant to grape phylloxera are not immune to phylloxera attack but are able to tolerate low infestations. Susceptibility to phylloxera increases when plants are stressed; vigorous vines tolerate phylloxera attack better than weak vines. Be aware of differences in soil type, temperature, and drainage that can contribute to the severity of an infestation and decrease a plant's resistance. Good water management, fertilization, and other cultural practices that improve plant vigor may reduce damage.

Pest organisms can overcome host plant resistance when exposed repeatedly to cultivars with the same genetic background. The genetic shift in the population occurs in the same way as the selection of pesticide resistance. Nematode-resistant tomatoes, for instance, are best used in combination with a variety of management techniques, such as crop rotation and fallowing, that reduce the chance that a root knot nematode population will overcome host plant resistance. Some resistant cultivars give only partial control and may be damaged in serious epidemics. Other tolerant cultivars are not damaged by pests but may support populations of the pests, providing reservoirs that may infest subsequent or adjacent susceptible cultivars.

Types of Resistance

Many scientists distinguish between tolerance and true resistance. Tolerant plants are able to endure the presence of a pest with little or no long-term damage; however, tolerant varieties sometimes support populations of a pest that can infest subsequent susceptible crops. Plants with true resistance support few or no pest individuals and have physiological or morphological characters that affect the behavior (antixenosis) or biology (antibiosis) of the pest. Antixenosis (or nonpreference) includes changes in plant color, odor, hairs, or leaf texture that make the plant unattractive or repellent; for instance, certain cultivars of chrysanthemum give off odors that repel the melon aphid. An example of antibiosis would be chemicals produced by the resistant cultivar that inhibit spore germination or mycelium growth or development of an insect. In addition to tolerance and true resistance, a third concept, that of apparent resistance or disease escape, involves plants that fail to become infected with a pathogen because all the requirements for infection or disease development (e.g., environmental conditions, susceptible host stage and quality, or pathogen inoculum) were not present. These same plants, planted under different conditions, could well be susceptible to the disease.

Host resistance is most effective and long-lasting if it relies on more than one gene, and preferably on more than one character, for its mechanism of resistance. Such polygenic, or horizontal, resistance slows the development of individual infections, slowing epidemics and making it more difficult for the pest to develop strains (biotypes) able to overcome the resistance factors.

Plant pathologists refer to resistance that results from the activity of one or a few closely linked genes as *vertical resistance*. Vertical resistance impacts major functions of specific pathogens. As a result, vertical resistance is often very effective and very specific to a pathogen race or group, but sometimes rapidly selected against by the pathogen. In some cases, vertical resistance may inhibit the initial establishment of pathogens, thereby inhibiting the development of epidemics. On the other hand, resistance that is horizontal (polygenic) involves numerous minor processes and defenses working together that affect most strains of a pathogen. Horizontal resistance may be slightly less effective against some pathogen strains but is usually longer lasting (Figure 5-2). In some cases, multilines, or mixes of different varieties with different resistance

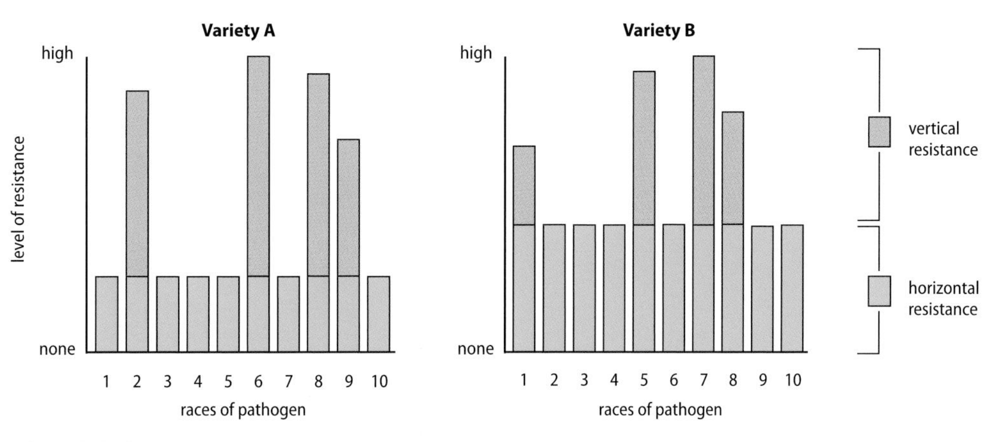

FIGURE 5-2

Levels of horizontal and vertical resistance of two plant varieties toward 10 races of a pathogen. Variety B shows a greater level of horizontal resistance, which reduces damage from a wide range of pathogen races. The two cultivars differ in their level of vertical resistance, which determines the extent to which they are damaged by specific races of the pathogen. When choosing a cultivar, the grower must consider the races of the pathogen most prevalent in the area.
After Vanderplank 1984.

genes, are used to reduce selection in the pest population to overcome host resistance.

Rootstock and Scion Selection

One of the most critical decisions in orchard and vineyard crops affecting management practices, pest damage, and profits throughout the life of the tree or vine is the selection of rootstock and *scion* cultivars. The scion is a shoot or bud that is grafted on to a rootstock to develop into the trunk and branches of the tree or vine (Figure 5-3). Scion choice determines the harvested crop variety, pest resistance above the graft, and many other production and market considerations.

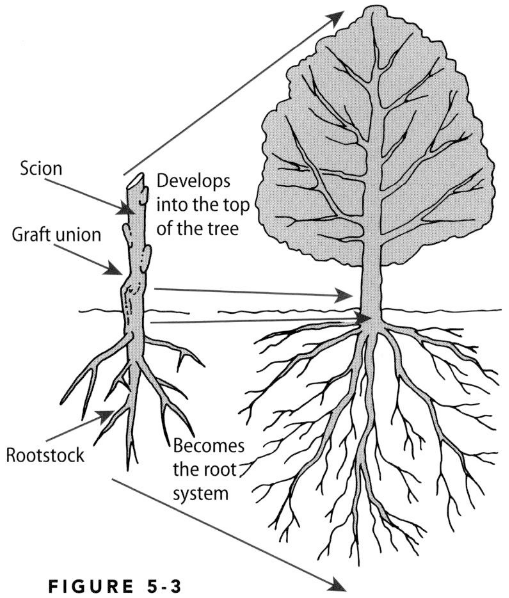

Rootstocks are primarily chosen by growers to control tree vigor and improve fruit or nut size and quality, cold hardiness, and adaptability to climate and soil conditions. Resistance or tolerance of rootstocks to insects, diseases, nematodes, and other pests is another major factor in rootstock selection. If the orchard or vineyard is located in an area where a particular insect, pathogen, or nematode species is known to be damaging, the PCA can recommend that the grower select a rootstock variety, if available, with some resistance or tolerance to that pest. Knowledge of rootstock and scion susceptibility to pests in an existing orchard will help diagnose problems.

Information to help growers and PCAs select pest-resistant and compatible rootstock and scion combinations is available from university publications and Cooperative Extension offices. For example, Table 5-2 lists susceptibility of major almond rootstocks to pathogens, nematodes, and nutritional disturbances.

FIGURE 5-3

In grafted plants, the portion above the graft union develops into the top of the tree from one or more buds on the scion. The root system develops from the rootstock.

TABLE 5-2.

Susceptibility of five major almond rootstocks to pathogens, nematodes, and nutritional disturbances.

	Almond	Lovell peach	Nemaguard	Marianna 2624	Peach-almond hybrid[1]
bacterial canker[2]	SR[3]	SR	SR	VS	SR
Phytophthora crown and root rot	ES	MS	S	SR	VS
Armillaria root rot	S	S	S	SR	S
crown gall	VS	S	S	SR	S
Verticillium wilt	S	S	S	SR	S
root knot nematode	VS	S	R	R	SR
other nematodes[4]	S	S	S	S	S
nutritional disturbances	[5]	[6]	[6]	—	[7]
peachtree borer	S	VS	ES	S	—
pocket gophers and mice	VS	SR	SR	S	—

1. Variable in response depending on hybrid.
2. Susceptibility of scions grown on this roostock.
3. ES = extremely susceptible, VS = very susceptible, S = susceptible, MS = moderately susceptible, SR = somewhat resistant, R = Resistant, — = no data.
4. Other nematode pests include root lesion, ring, and dagger nematode.
5. More resistant to excess boron and lime-induced chlorosis than other rootstocks listed.
6. More sensitive to calcareous soils and high boron or chloride than trees on almond rootstock.
7. More tolerant of lime, sodium, and chloride than Lovell peach and somewhat resistant to excess boron.

Source: Strand 2002.

Techniques for Developing Resistance in Plants

Developing pest resistance in cultivated plants is just one of many objectives of plant breeding and genetic improvement programs. Plant breeders must balance the need for resistance to important pests and diseases with other objectives, such as improved yield and quality. To develop a resistant cultivar, researchers must first locate plants with suitable resistant characters. These are typically found in wild variants of the same species or closely related species. With new bioengineering techniques, resistance characters may now also be derived from different plant genera or even from unrelated organisms.

Traditional Approach. Historically, the development of resistant cultivars has depended largely on finding sources of resistance in host populations and propagating them through sexual reproduction. Various techniques have been used to select genetic material useful in suppressing disease. Occasionally, a disease epidemic in an area will isolate resistant individuals in the field and provide an almost ready-made resistant variety. This was the case with milo (grain sorghum) resistant to the fungus *Periconia circinata*; all surviving plants on a Sacramento Delta island stricken with a severe epidemic of the disease were found to be resistant.

However, far more often a laborious breeding program is required to develop a new resistant cultivar. Such programs require identification of resistant material by infesting plants and selecting the least susceptible individuals, incorporating the resistance into varieties with desirable agronomic attributes through sexual reproduction, and screening offspring for effectiveness of resistance and other desirable qualities. Development of a single cultivar may take many years, although new *tissue culture* and bioengineering techniques can shorten this time substantially. Three primary traditional methods of varietal development are simple selection, pure line selection, and backcrossing (Table 5-3). In actual practice, most breeding programs use a combination of these methods.

Hybrids are the offspring of a cross of two different purebred (homozygous) lines. The heterozygous offspring (F-1) often combine the superior characters of the two lines and also increase the vigor of the crop (hybrid vigor). The introduction of hybrid corn in the 1930s and 1940s greatly increased grain yields. However, offspring of the heterozygous F-1 generation do not retain the vigor or the desirable variety traits of their parents, so the hybridization process must be repeated for each new crop of seed.

Biotechnology

Cell and tissue culture, as well as techniques such as *genetic engineering* and irradiation, allow plant breeders to more rapidly develop new varieties of plants without having to carry out the time-consuming tasks of crossing plants through sexual reproduction and selecting for desirable qualities over many generations. Cell and tissue culture techniques allow breeders to propagate entire plants from a single cell or group of cells (clonal propagation). This approach vastly decreases the time required to propagate and identify new material. Genetic engineering techniques allow breeders to transfer genetic material (DNA) from one organism to another. DNA transfer opens up vast new resources of pest-resistant genetic material.

For woody perennials, budwood may be irradiated to induce mutations that may confer desirable traits. These newly developed tissues will generally be grafted on to rootstocks and grown to maturity to assess their usefulness. However, these modern biotechnological techniques still decrease by years the amount of time required to propagate and identify new plant material.

Transgenic Pest-Resistant Cultivars. Transgenic pest-resistant cultivars result from the transfer of desirable traits from one organism to another. In classical breeding programs, only closely related species can be crossbred. Using transgenic techniques, plant breeders can transfer the genetic material from completely unrelated organisms. Genetic engineering is a tool for moving genetic material, consisting mainly of DNA, while transgenics is a technique for constructing the gene to be inserted into the host.

Tissue culture is initially used to develop the transgenic plant. Once the plant is produced, plant numbers can be increased through normal seed production. Transgenic techniques have been used on rice, cotton, and a variety of vegetable crops, such as lettuce, cabbage, and tomato, as well as on trees and vines to develop cultivars resistant to insect and pathogen pests.

Transgenic technology offers substantial opportunities for developing new cultivars for pest management purposes. For instance, crop cultivars are now available that have transgenic components allowing them to tolerate herbicides to which they were previously susceptible, to produce proteins toxic to certain insects, or to be antagonistic to plant disease. While this technology holds much promise, there are many concerns regarding its use. Introduction of a gene from the bacterium *Bacillus thuringiensis* (Bt) into a plant host, for instance, raises concerns about increasing the chances of insect resistance to the Bt endotoxin. Some seed companies and experts suggest planting a refuge of susceptible cultivars in or around

TABLE 5-3.

Three primary traditional methods of varietal development.

Method	Description
simple selection	Researchers monitor fields where natural infection regularly occurs, selecting seeds from the most highly resistant plants. The selected seeds are planted, and the process is repeated over time. Plants are allowed to cross-pollinate. Generally, resistant varietal development is slow.
pure line selection	Resistant parents (but with less-desirable agronomic qualities) are crossed with another parent that has desirable traits but lacks resistance. Individual offspring from this cross that are highly resistant and have good agronomic characters are propagated under controlled conditions, inoculated repeatedly to test for resistance, and selected over many generations until they are a true breeding cultivar. Offspring of individual parents are kept separate throughout the process to obtain genetic purity. These cultivars are homozygous for the resistant genes and usually have a high degree of specific vertical resistance but a low degree of horizontal resistance.
backcrossing	Cultivars are developed similarly to the method described for pure line selection except several crosses are made back to the parent that has desirable agronomic traits in subsequent generations throughout the varietal developmental process. As a result, there is variation (heterozygosity) within the cultivar for disease resistance, which reduces the pathogen's ability to overcome the cultivar's resistance.

the field to reduce the potential for rapid development of resistance to transgenic cultivars, but the logistics for growers trying to carry out this recommendation can be cumbersome. Some biologists worry that widespread use of plants with altered genes could dilute the gene pool of wild species, threaten diversity, and upset the natural ecological balance of the ecosystem. While much remains to be learned, transgenic technology offers faster, more specific, and more varied options for creating new crop cultivars than ever before.

Nonhost Plants

Host plant resistance is just one of the many tools available for use in an integrated pest management program. In some situations, when heavy pest infestations occur, resistance breaks down, or control options are simply too expensive, the best option is to avoid the pest altogether through the use of alternate nonhost plants. Nonhost plants can be used in crop rotations to reduce pest numbers. In landscapes, nonhost plants providing similar aesthetic qualities are frequently available.

BIOLOGICAL CONTROL

Biological control can be broadly defined as any activity of one species that reduces the adverse effect of another. Biological control agents (natural enemies) include predators, insect parasites (parasitoids), pathogens, antagonists, and competitors. Although all types of pests have natural enemies, biological control is used in pest management systems primarily against mites, insects, specific problem weeds in uncultivated areas, and for only a few specified isolated plant pathogens and nematodes. Refer to the UC IPM Pest Management Guidelines for current recommendations on using biological control. For more in-depth information on natural enemies, refer to *Natural Enemies Handbook: The Illustrated Guide to Biological Pest Control* (see the "Resources" section).

Types of Biological Control Agents

Natural enemies belong to such diverse groups as mammals, birds, amphibians, fish, snails, bacteria, higher plants, and nematodes. Some natural enemies have a minor impact on pest population numbers, killing only a small portion of a pest population even under the most favorable conditions. Other natural enemies keep potentially injurious pests at low levels with no assistance from managers, as long as they are not disrupted by pesticide applications or other factors. PCAs should know the natural enemies of importance in the managed ecosystems and should understand the biology of both the pest and its natural enemies.

Predators. A predator is an animal that attacks, feeds on, and kills more than one prey during its lifetime. Predators are usually larger and stronger than their prey. Some predators are quite specialized and feed only on one or a few closely related prey species. Many predators, however, are generalists, feeding on a variety of similar types of organisms, such as soft-bodied insects or free-living nematodes in the soil. Predators range from large carnivores, such as coyotes and raptors, to microscopic amoebae and other soil-dwelling microorganisms. The most recognized predators in biological control, however, are the predatory arthropods, such as the lady beetle (Figure 5-4) and carabid beetles, lacewings, syrphid

FIGURE 5-4

The convergent lady beetle, Hippodamia convergens, *is a common predator and is often found in gardens, landscapes, trees, and field crops whenever aphids are abundant.*

flies, ants, predatory hemipterans, spiders, and predatory mites. In some insect species, both the adult and immature (larval or nymphal) stages are predaceous. Other species of insects are predaceous only during their immature stages. Adults of many of these predators feed on honeydew and plant nectar or pollen.

Parasites. A parasite feeds in or on a larger host organism. Parasitic organisms have a prolonged and specialized relationship with their host, usually parasitizing only one individual or a few hosts in their lifetime. True parasites often weaken the host but do not kill it outright; in some cases, they may have little negative impact. Only those that significantly weaken or kill their host are important in biological control.

Insects that parasitize and kill other insects are often called parasitoids. Parasitoids are parasitic during their immature stages and kill their host as they reach maturity. Most parasitoids are wasps or flies with adults that feed on insect honeydew and plant nectar and pollen. In certain species, the adult female parasitoid also feeds on hosts.

Because they kill their hosts, insect parasitoids are not considered true parasites. However, in this book and most of the other pest management literature, parasitoids are simply referred to as insect parasites.

Pathogens are microorganisms that cause disease or malfunctioning of host cells and tissues and are a kind of parasite. Pathogens may or may not kill their host. Beneficial pathogenic organisms include certain bacteria, fungi, protozoa, viruses, and nematodes (Figure 5-5) that cause disease in pest insects, weeds, nematodes, and mites.

Herbivores. Herbivores are animals that feed on plants. Herbivores are important natural enemies of weeds, especially when they specialize and feed on only one or several closely related weed species. The most effective natural enemies of weeds limit weed reproduction by feeding on flowers or seeds. Birds feeding on weed seeds after harvest, for example, can help keep the weed seed bank in check. Geese have also been used to control weeds in organically grown cotton, tomato, pepper, and lettuce crops. Livestock such as cattle, sheep, and goats can also play a role in weed suppression, especially on rangelands (Figure 5-6).

Antibiosis. Some organisms release toxins or otherwise change conditions so that the activity or growth of the pest is reduced. Many bacteria and molds secrete antimicrobial substances that inhibit the growth of other microorganisms. The inoculation

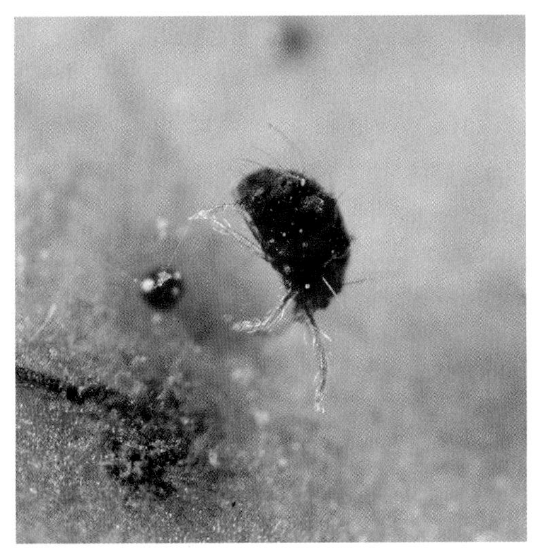

FIGURE 5-5

Its swollen, darkened body suggests that this citrus red mite, Panonychus citri, *is infected by a virus. As the disease progresses, shriveled bodies on plants provide further evidence of infection.*

FIGURE 5-6

The grazing of herbivores such as cattle, when well managed, can contribute to the control of yellow starthistle.

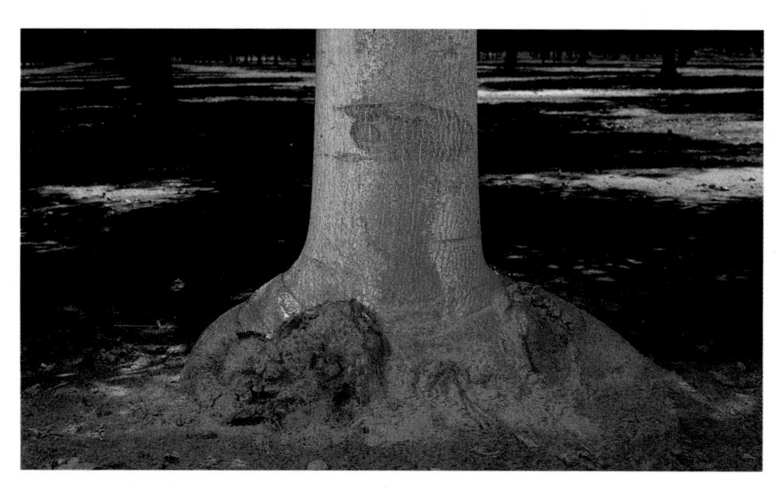

FIGURE 5-7

The crown gall, Agrobacterium tumefaciens, *infection shown here might have been prevented by treating the roots before planting with the antagonist* Agrobacterium radiobacter *strain K84.*

of fruit tree rootstocks with a strain of the bacterium *Agrobacterium radiobacter* K84 can prevent infection by damaging strains of *Agrobacterium tumefaciens*, the pathogen that causes crown gall (Figure 5-7). The K84 strain produces an antibiotic against other *Agrobacterium* bacteria.

Another form of antibiosis is allelopathy, in which one plant releases substances that are toxic to another plant species. Allelochemicals are released into the environment as volatile gases, secreted from living roots, leached from shoots, or released through the decomposition of dead plant material.

Competitors. Competition occurs when two organisms compete for limited supplies of essential resources, such as food in the case of animals and water, nutrients, and light in the case of plants. Due to their superior ability to secure resources, competitors limit populations of pest organisms, sometimes with little negative impact on the crop or other managed resource. Competition can be used in weed control: for instance, the use of highly competitive ground covers in landscape plantings can rapidly outgrow and shade out weedy species (Figure 5-8). Competitors may also be important in the biological control of pathogens, as in the use of strains of bacteria or fungi as protective seed coatings. A soil with a diverse array of harmless nematodes or pathogens may help limit numbers of damaging species.

FIGURE 5-8

Planting ground covers and other plants close together in the landscape uses plant competition to suppress weeds.

Approaches to Use of Biological Control Agents

Three basic strategies are used to increase the effectiveness of biological control agents (natural enemies) in the field: importation, conservation and enhancement, and augmentation.

Importation (Classical). Importation, or classical, biological control involves the deliberate introduction and establishment of natural enemies into areas where they did not previously exist. Importation programs have largely been employed against pests of foreign origin. Usually accidentally introduced, these pests are able to develop high populations in the absence of natural enemies, which may have kept them under control in their native lands. The primary organisms that have been used in classical biological control programs in the United States have been insects that parasitize other insects or insects that prey on pest insects, mites, or weeds. Ideal species belong to groups of organisms that have very restricted host preferences so that nontarget or beneficial species will not be affected.

To implement a successful biological control importation program several steps must be taken. First, the pest organism and its native area must be correctly identified. Next, searches for natural enemies are conducted in the native area and appropriate species shipped back for testing. Shipments are then held in strict quarantine to keep out potential contaminants and to confirm that the natural enemy will have minimal negative impact in the new country of release.

After appropriate quarantine processing and biological testing, the natural enemies can be reared, increased in number, and released. The primary goal is to permanently establish the natural enemy species so that it can exert a regulatory influence on the pest population. Repeated releases or reimportations are often necessary. Imported natural enemies frequently do not thrive in the new area, and biotypes that are better suited to the environment must be introduced. Only specially trained university and government personnel can carry out importation programs. Much of their initial work must be contained in quarantine facilities certified to handle exotic organisms.

Conservation and Enhancement. Conservation and enhancement of established natural enemies include any activities that improve survival, dispersal, and reproduction of resident natural enemies. Many approaches can be taken. Elimination or reduction of pesticides toxic to natural enemies is an important way of improving biological control. When pesticide applications are necessary, selective materials are applied so that natural enemies are least likely to be affected. This selectivity derives from the active ingredient, formulation, method of application, and time of application.

Production practices can also be manipulated to benefit natural enemies. A lack of plant diversity, for instance, can increase a crop's susceptibility to pest attack and reduce its attractiveness to natural enemies. For good survival and high reproduction, natural enemies often require shelter, alternative food sources, water, overwintering sites, and other conditions. Natural enemies of honeydew-producing homopteran insects may also require protection from ants. Certain ant species, such as the Argentine ant, *Iridomyrmex humilis*, feed on honeydew produced by aphids, soft scales, whiteflies, and mealybugs and disrupt biological control by attacking their predators and parasites (Figure 5-9). It may be necessary to take measures to control these ants by pruning branches to deny them access to plant canopies or by applying a sticky material to tree trunks.

Environmental factors such as weather can adversely affect natural enemy populations. Although not much can be done to diminish the ill effects of weather, the microclimates of natural enemy populations can be made more favorable through the manipulation of irrigation, use of cover crops, modification of pruning techniques, or by changes in harvesting practices. For example, border harvesting, where a strip of alfalfa is left standing after each harvest, benefits hay production by maintaining populations of predators and parasites of alfalfa pests in the alfalfa field. In addition, border harvesting also markedly reduces lygus bug migration into cotton and other crops. Similar benefits can be obtained by border harvesting or staggering the cut-

FIGURE 5-9

One way to enhance biological control is to control the ants that protect honeydew-producing soft scales from their natural enemies.

ting of alfalfa hay fields that are nearby or adjacent to each other (Figure 5-10). Figure 5-11 compares the abundance of natural enemy populations in border-harvested alfalfa and conventional alfalfa.

The majority of biological control agents at work in agricultural and urban environments are naturally occurring, providing excellent regulation of many potential pests with little or no assistance from people. The importance of such agents has often become apparent when pesticides applied to control a key pest destroy the natural enemies of another species, elevating that species to pest status (a secondary pest outbreak). Because of their greater stability, perennial cropping systems often provide a better habitat than annual crops, and thus IPM programs in many perennial crops place a greater reliance on natural enemies.

Augmentation. *Augmentation* involves supplementing the numbers of naturally occurring biological control agents with releases of laboratory-reared or field-collected natural enemies (Figure 5-12). Because it is so much more expensive, augmentative approaches are usually attempted only when importation and conservation techniques are not promising.

Two approaches used in augmentation are *inoculative releases* and *inundative releases.* In inoculative programs, released natural enemies reproduce in the field and build up their population so that their progeny provide control for several generations. Inoculative releases may be used to build up populations of natural enemies earlier in the season than usual or to establish a natural enemy in an area where it was not previously present. Inoculative releases are appropriate when the population of an otherwise effective natural enemy is severely devastated by pesticide applications, unfavorable weather conditions, cultivation practices, or a lack of seasonal hosts.

Inundative releases are aimed at achieving immediate biological control through the activities of the released individuals. Offspring of released individuals are not expected to survive to assist in control. Additional releases may be required throughout the season should pest populations approach damaging levels. Inundative releases are most effective against pests that cause economic damage only during a limited period of the

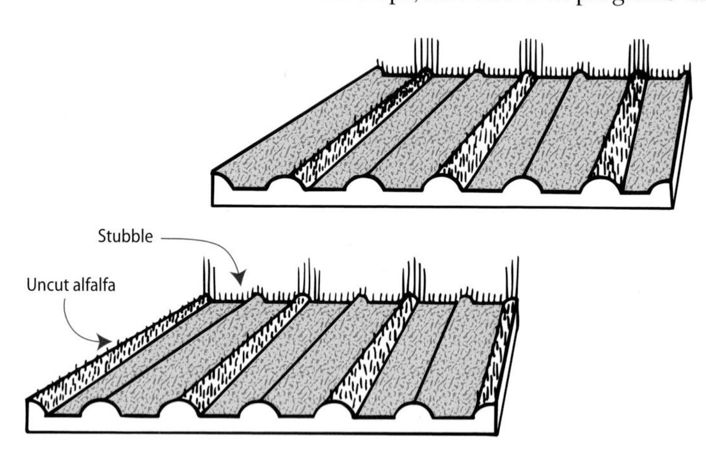

FIGURE 5-10

Border harvesting can be used to reduce lygus populations and maintain predator and parasite populations. In border harvesting, a row of alfalfa is left uncut along every other border. The next time that field is harvested, the uncut borders are mowed and the alternate borders are left standing.

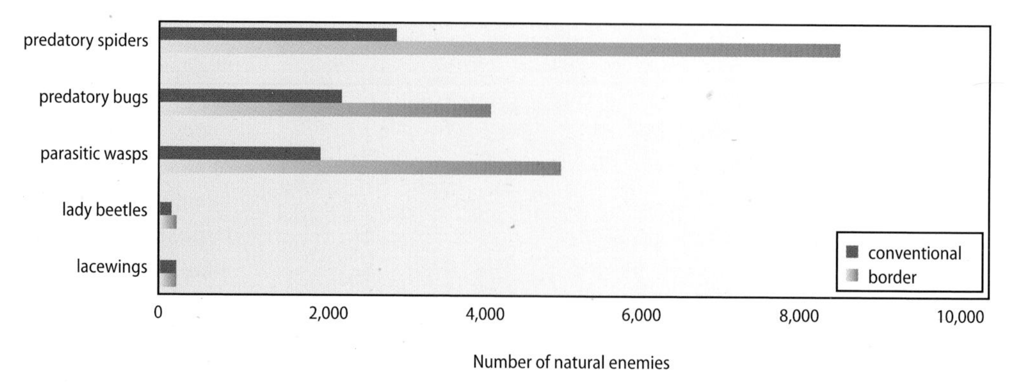

FIGURE 5-11

Comparison of the natural abundance of natural enemies in a border-harvested versus a conventionally harvested alfalfa field over a 4-month period from May through September. Data from Summers 1976.

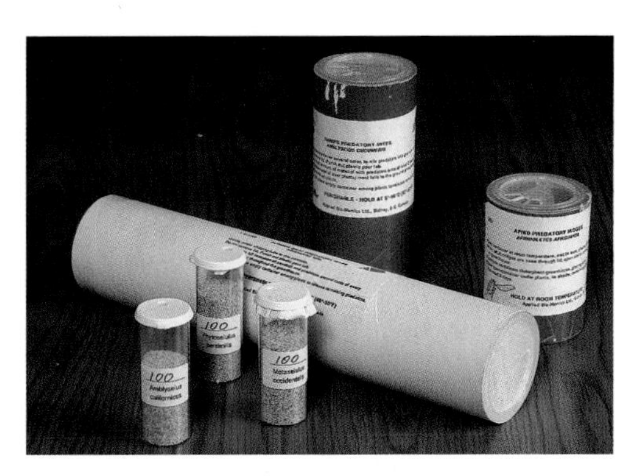

FIGURE 5-12

Many natural enemies are available for purchase and are shipped through the mail in containers such as these. If resident natural enemies are insufficient, releases can augment biological control in certain situations.

year and against pests in a controlled environment. A wide variety of predators and insect parasites have been used in inundative programs and are the primary method of using microorganisms and nematodes for biological control.

Although their use is still quite limited, natural enemy augmentation and, in particular, inundative releases have become increasingly important. However, many release programs have only a minimal impact on the target pest. These programs require a thorough understanding of the managed system; significant differences in efficacy may exist among uses in different crops. There are many reasons for the deficiency of this technology, including factors related to quality of the natural enemy, choice of released species, and application method (Table 5-4).

Research on inundative releases has not yielded enough information to make many of these types of programs successful. There are only a limited number of situations where inundative releases of natural enemies have been shown to be effective. The release of the egg parasite *Trichogramma* against codling moth, *Aphytis* parasites against California red scale, and predatory mites to control spider mites are three examples that show that properly designed inundative releases can be effective tools in IPM programs.

The quality of commercially available natural enemies is critical to the success of a release. PCAs can take steps to aid in a successful release. Natural enemies should be obtained from a quality supplier. Check to see that the organisms were shipped properly and evaluate the quantity and quality of each shipment upon arrival. Learn how to store them until they are ready to be released, and release them when conditions are right.

Effectively releasing natural enemies or applying microbial or biological control agents requires knowledge, experience, and imagination. Look for information in the UC IPM Pest Management Guidelines and other pest management publications on the number of natural enemies to release for effective control and the appropriate development stage of the host and natural enemy to be released. The control action threshold in some pest management guidelines takes into account the abundance of natural enemy populations (Chapter 6). Monitor for the presence of natural enemies as well as the population of the pest and follow recommended thresholds. Any release of natural enemies should be followed up with a

TABLE 5-4.

Common reasons for the failure of inundative releases of natural enemies in augmentative biological control programs.

- ☐ **Inappropriate biological control agent released:** General versus specific predators, or terrestrial versus arboreal natural enemies. For many pests there are no effective natural enemies.
- ☐ **Poor-quality predators or parasites released:** Low vigor, poor reproduction, diseased, or poor dispersal ability.
- ☐ **Released natural enemies rapidly consumed by other natural enemies**: For example, ants and other general predators can consume lacewing eggs.
- ☐ **High cost** relative to other control options such as pesticides.
- ☐ **Poor release technology:** Release rate too low or too infrequent, resulting in poor coverage; release too late for effectiveness; agent injured during release.
- ☐ **Life stage of pest not susceptible to agent released:** Most beneficial organisms feed on or parasitize only certain life stages.
- ☐ **Released natural enemies killed** by pesticide applications or residues.

monitoring program to assess the effectiveness of the release. Field trials with untreated controls can also be conducted to assess the success of natural enemy releases, but be cautious. Refer to Chapter 7 for information on conducting successful field trials.

Biological Control of Insects and Mites

There are far more applications of biological control agents for the management of insects and mites than for any other pest group. Natural enemies can play a role in the management of virtually every arthropod pest. Predators and parasites are among the best-known enemies of insects and mites. Other biological control agents include *entomopathogenic nematodes* and microbial agents such as bacteria, fungi, and viruses. Vertebrates, including birds, bats, and fish, may have a significant impact on some insect populations. During the past century, more than 40 insect species have been partially, substantially, or completely controlled through classical biological control efforts in California alone.

Predatory Arthropods. Insects and mites serve as prey to an enormous variety of predatory arthropods, with representatives from every insect order as well as spiders, several mite families, and centipedes. Important insect predators include green and brown lacewings, syrphid flies, aphid flies, robber flies, damsel bugs, big-eyed bugs, assassin bugs, minute pirate bugs, lady beetles, soldier beetles, predaceous ground beetles, wasps and ants, and predaceous mites and spiders (Figure 5-13). Many predatory arthropods eat each other in addition to pests. Thus their contribution to pest management in a given situation may be beneficial or detrimental, depending on the complex web of trophic relationships in a particular location.

Parasitic Insects. Most insects that parasitize other insects belong to the orders Hymenoptera (wasps) and Diptera (flies). Insect parasites (parasitoids) are free living as adults but parasitic in their larval stage. Adult females lay their eggs inside or attached to the outside of the host's body.

FIGURE 5-13

Predators of insects and mites are found in almost every insect order. A few examples are shown here. (A) Hemiptera: a minute pirate bug. (B) Neuroptera: a lacewing larva eating a lace bug. (C) Thysanoptera: sixspotted thrips and a spider mite egg. (D) Hymenoptera: gray ant attacking a peach twig borer larva. (E) Diptera: a syrphid fly larva attacking an aphid. (F) Coleoptera: a twice stabbed lady beetle feeding on walnut scale.

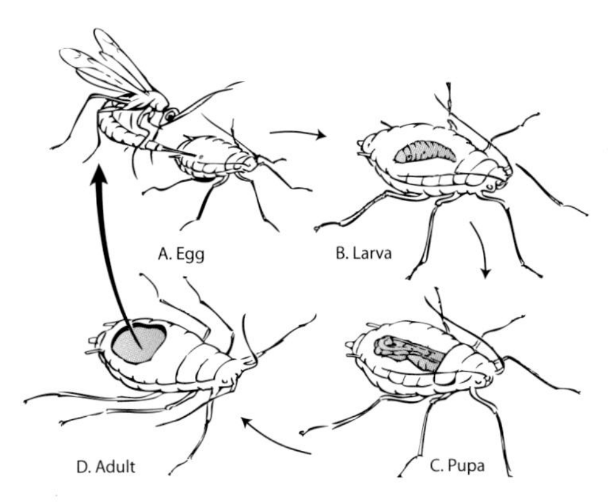

FIGURE 5-14

The life cycle of a typical insect parasite, or parasitoid, illustrated using a species that attacks aphids: (A) an adult parasite lays an egg inside a live aphid; (B) the egg hatches into a parasite larva that grows as it feeds on the aphid's insides; (C) after killing the aphid, the parasite pupates into an adult wasp; (D) the wasp chews a hole and emerges from the dead aphid; (A) the wasp then flies off to find and parasitize other aphids.

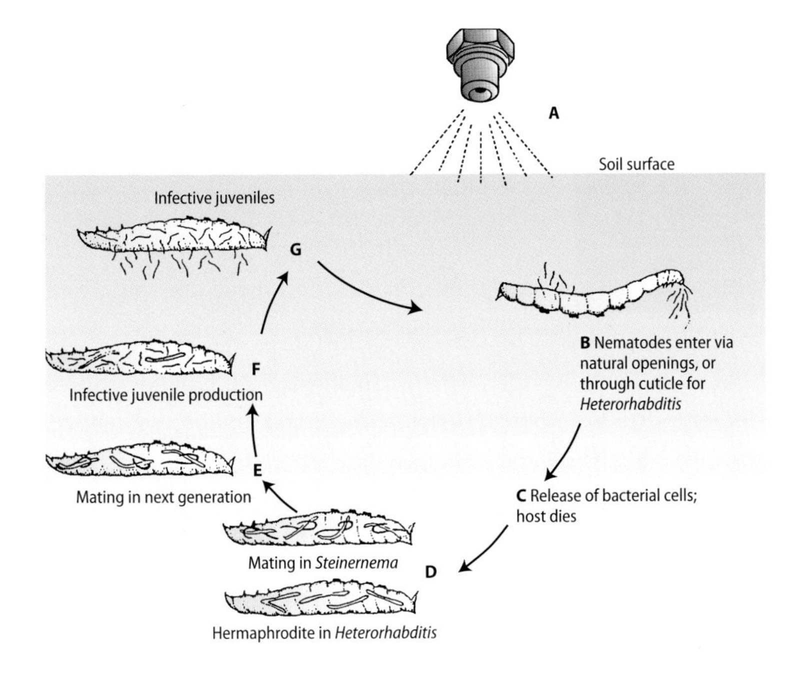

FIGURE 5-15

Life cycle of beneficial nematodes: (A) infective stage nematodes are applied to soil; (B) the nematodes seek a host and enter it; (C) once inside, the host is killed by the nematodes and the mutualistic bacteria carried by the nematodes; (D) nematodes feed, grow, mature, and reproduce (the initial development of nematodes in the host differs because there are no separate male and female nematodes in the first generation of Heterorhabditis*—they are hermaphrodites); (E) all generations of* Steinernema *and subsequent generations of* Heterohabditis *in the host produce both males and females that mate; (F) females then produce infective-stage juvenile nematodes inside the dead host; (G) infective nematodes exit and seek hosts. The entire life cycle from infection of the host to release of the new infective generation takes 7 to 14 days. Adapted from Kaya 1993.*

After hatching, the immature stage of the parasite completes its development on or in the host, killing the host just before the parasite pupates or emerges as an adult (Figure 5-14). Only one host individual is killed during the parasite's development. Some parasites attack other parasites (hyperparasites) or predatory arthropods. These parasites may be detrimental to biological control of pests.

Nematodes. The role of naturally occurring nematodes that attack insects is not well understood, but their potential in biological control is receiving increased attention. Nematodes used in insect pest management are called entomopathogenic nematodes and are packaged, sold, and applied for release inundatively, similar to insecticides. Most currently available entomopathogenic nematodes are in the families Steinernematidae and Heterorhabditidae. Depending on the species and the formulation in which it is packaged (gel, clay matrix, or liquid flowable), the shelf life for heterorhabditid-based formulations extends to 3 months, and up to 5 months for steinernematids at room temperature.

Entomopathogenic nematodes serve as vectors of pathogenic bacteria (Figure 5-15), and it is these bacteria that actually kill the host. The nematodes can be applied to the foliage, soil, or in insect galleries and are most effective against insects that feed in enclosed areas where moisture levels can remain high. Applications of nematodes for soil-inhabiting insects should be watered in to increase efficacy. Table 5-5 lists some commercially available nematodes and the pests they control.

Although entomopathogenic nematodes appear to act as broad-spectrum pathogens and can kill a wide range of insect species, their spectrum is limited by their environmental requirements. The most important factors are adequate soil moisture and avoidance of temperature extremes. Nematodes survive best when applications are made late in the day; soil surface temperatures above 86°F (30°C) decrease nematode effectiveness. Modification of the environment can increase nematode efficacy; changes in soil acidity,

TABLE 5-5.

Selected commercially available nematodes and the pests they can control. Follow the recommended application procedure.

Nematode scientific name	Pests controlled	Habitat	Application procedure
Heterorhabditis bacteriophora	Japanese beetle larvae, turf grubs, white grubs	turf	Apply as soil drench or conventional spray to the surface at a rate of 23 million per 1,000 square feet in 5 gallons of water. Soil must be warm (60°F) and moist. Irrigate every 2–3 days for 2 weeks after application.
	weevils, flea beetles, root maggots, and certain other soil-dwelling insects	soil	
Steinernema carpocapsae	carpenterworm, clearwing moth larvae	woody plants, typically beneath the bark	Apply with a squeeze-bottle applicator or 20-ounce oil can at a concentration of 1 million or more per 1 ounce of water. Clear tunnel entrance, insert applicator, and inject suspension until gallery is filled.
Steinernema feltiae	fungus gnat	soil	Apply as soil drench or surface application.

FIGURE 5-16

The two silverspotted tiger moth caterpillars, Lophocampa argentata, *hanging beneath this Monterey pine twig have been killed by a naturally occurring disease. A healthy larva is on top.*

tilling and irrigation practices, and livestock grazing can increase the rates of parasitism. In addition, nematodes have various natural enemies, such as nematophagous mites, Collembola, and fungi, that can have a detrimental effect on nematode survival.

Pathogens. Pathogens play an important role in the control of insect pests. Insect pathogens include viruses, bacteria, fungi, and other microorganisms that produce disease in their insect hosts. Many pathogens occur naturally in nature and frequently keep insect pests below damaging levels. For example, many caterpillar species are kept under control by natural outbreaks of disease. Of those that occur naturally, viruses are most common. Caterpillars killed by viruses and bacteria often turn dark, and their bodies become soft and limp, eventually degenerating into a sack of liquefied contents (Figure 5-16). When the bodies are broken, new viral particles or bacterial spores are released, which infect other caterpillars.

Naturally occurring fungal disease is an important contributor to the control of aphids. In humid conditions, aphids are very susceptible to fungal disease. Fungus-killed aphids turn orangish or brown and have a fuzzy, shriveled texture; sometimes fine white mycelia can be seen growing over their surfaces.

The use of microorganisms for the management of insects and mites is a special category of augmentation. There are several microbial pesticides, including bacteria, fungi, viruses, and protozoa, currently registered for the management of insects and mites. Because these formulations are developed to be applied as pesticides, such microbial agents will be discussed more thoroughly in the section on pesticides.

Biological Control of Weeds

Worldwide, biological control organisms have been responsible for the control of a number of introduced weed species. In California, major successes include the importation of a cochineal insect, *Dactylopius opuntiae*, to control two species of prickly pear cactus and the control of the European native klamathweed with the imported European leaf-feeding beetle *Chrysolina quadrigemina* (Figure 5-17). Many other *noxious weeds* have been targets of biological control programs but with limited success to date.

Historically, classical biological control of weeds has been attempted primarily in pastures, forests, rangelands, and aquatic systems. These ecosystems usually have a low level of disturbance that enhances natural enemy survival. Also, unlike a cultivated crop, these systems can be compatible with the slow rate of control often produced by biological control and the need to maintain low populations of target weeds to retain natural enemies. Most weeds targeted for biological control are specific problem weeds that are toxic to livestock or have some other undesirable quality. Once target weeds are removed, the area must be managed so that desirable wild plants will take their place.

In crop production systems, weed management is directed at complexes of weed species that compete with crop plants. If one species is removed by biological control, another troublesome species is likely to take its place. As a result, classical biological control has less potential for weed management in agricultural or landscape systems.

Insects. Most natural enemies used in classical weed biological control programs have been insects. This is due in large part to their size, high rate of reproduction, dispersal, host-searching abilities, and considerable degree of host specificity. Insects control weeds by feeding on plants or by transmitting disease organisms that injure plants. Insects in such diverse groups as moths, thrips, mealybugs, scale insects, wasps, chrysomelid and other beetles, leafminers, gall midges, and others have been successfully used in biological control programs for weeds.

In California, the CDFA and the USDA coordinate most biological control programs for weeds in cooperation with local county agricultural commissioners' offices. Table 5-6 lists some biological control agents that have been released in California against weeds. Contact your county agricultural commissioner's office if you are interested in participating in a biological control program for weeds.

Pathogens. Plant pathogens offer some potential applications in biological control of weeds. Pathogens have extremely rapid

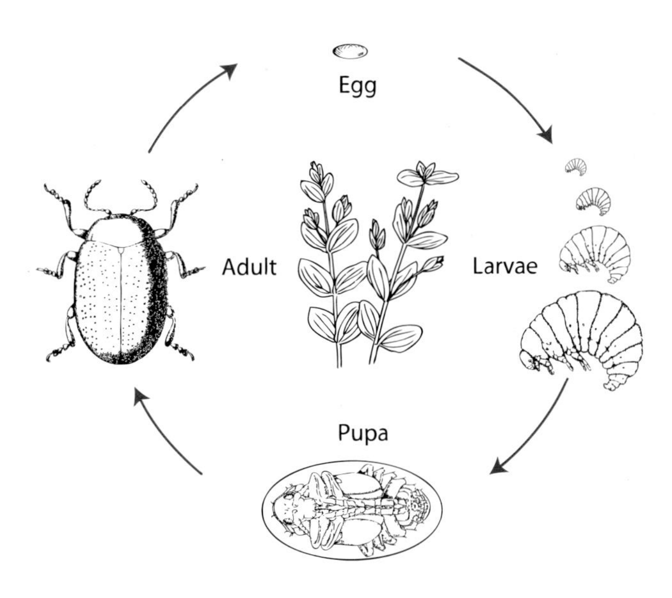

FIGURE 5-17

Klamathweed, Hypericum perforatum, *is an excellent example of classical biological control of weeds in the western United States. The female klamathweed beetle lays eggs singly or in small batches on young* Hypericum *leaves. The grayish larvae feed on foliage and develop through four instars before dropping or crawling to pupate in a cell they make just below the soil surface. Adults feed on foliage year-round, except during the hot, dry summer. Most damage occurs from larvae feeding during the spring and early summer. Klamathweed beetle commonly has one generation per year, although multiple life stages may occur at one time and development varies with location and weather.* Source: *Flint and Dreistadt 1998.*

generation times, can cause severe damage rapidly under proper environmental conditions, have methods for efficient dispersal, and exhibit a high degree of specificity toward the plant species they attack. However, because many weeds are closely related to desirable species, any release of pathogens must be done with care. One of the few successful cases of the introduction and establishment of a pathogen for biological control of a weed involved the introduction of the rust pathogen *Puccinia chondrillina* against rush skeletonweed in Australia in the 1970s. The rust was also introduced into California and the Pacific Northwest but achieved less success. Unlike release of insects for biological control, any pathogenic microorganisms applied augmentatively for biological control of weeds must be registered as pesticides. Although presently no pathogens are being used as bioherbicides in California, two fungi have been registered for use in other states against stranglevine and northern jointvetch.

Vertebrates. Plant-eating vertebrates are used to manage weeds in some situations. Vertebrates are largely nonspecific in their feeding habits and rely heavily on seeds, fruit, roots, seedlings, or other plant parts as major components of their diets.

TABLE 5-6.

Selected weeds for which biocontrol agents have been imported and established in California through 2003

Common name	Scientific name	Natural enemy and date introduced
broom, Scotch	*Cytisus scoparius*	*Apion fuscirostre* (beetle), 1964 *Leucoptera spartifoliella* (moth), 1960
gorse	*Ulex europaeus*	*Exapion ulicis* (beetle), 1953 *Tetranychus lintearius* (mite), 1994
hydrilla	*Hydrilla verticillata*	*Bagous affinis* (beetle), 1991 *Hydrellia pakistanae* (fly), 1994
klamathweed	*Hypericum perforatum*	*Chrysolina quadrigemina* (beetle), 1946 *Zeuxidiplosis giardi* (fly), 1950
knapweed, various species	*Centaurea* spp.	*Bangasternus fausti* (beetle), 1994 *Larinus minutus* (beetle), 1995 *Spheoptera jugoslavica* (beetle), 1980 *Urophora affinis* (fly), 1976
loosestrife, purple	*Lythrum salicaria*	*Galerucella calmariensis* and *G. pusilla* (beetles), 1998 *Hylobius transversovittatus* (beetle), 1996 *Nanophyes marmoratus* (beetle), 1997
prickly pear	*Opuntia* spp.	*Dactylopus opuntiae* (cochineal scale), 1951
puncturevine	*Tribulus terrestris*	*Microlarihu lareynii, M. lypriformis* (beetles), 1961
ragwort, tansy	*Senecio jacobaeae*	*Longitarsus jacobaeae* (beetle), 1969 *Pegohylemyla seneciella* (fly), 1966 *Tyria jacobaeae* (moth), 1959
sage, Mediterranean	*Salvia aethiopis*	*Phrydiuchus tau*, 1976
skeletonweed	*Chondrilla juncea*	*Cystiphora schmidti* (fly), 1975 *Eriophyes chondrillae* (mite), 1977 *Puccinia chondrillina* (fungal pathogen), 1976
starthistle, yellow	*Centaurea solstitialis*	*Bangastermus orientalis* (beetle), 1985 *Chaetorellia australis*, 1988 *Eustenopus villosus* (beetle), 1990 *Urophora sirunaseva* (fly), 1984
thistles, Italian and others	*Carduus* spp.	*Rhonocyllus conicus* (beetle), 1973
thistle, Russian	*Salsola tragus*	*Coleophora klimeschiella* (moth), 1975 *Coleophora parthenica* (moth), 1973
water hyacinth	*Eichhornia crassipes*	*Neochetina bruchi* (beetle), 1982

Source: M. Pitcairn, CDFA.

FIGURE 5-18

Weeder geese have been effectively used to manage grasses and sedges in cotton and other crops.

When used as biological control agents, they require active management to confine their feeding to target weeds. Among vertebrates, fish, birds, and mammals have the greatest potential for biological weed control.

Fish have been successfully used for the control of aquatic weeds, particularly submersed weeds. The sterile triploid grass carp, *Ctenopharyngodon idella*, for example, has been effective at reducing hydrilla in the canals of the Imperial and the Coachella irrigation districts. Additional fish species are being studied to control other aquatic weeds and may become available in the future. However, there are numerous concerns with the introduction of exotic fish species due to the potential for displacement of native fish and consumption of nontarget species.

Geese have been used as selective grazers to control weedy grasses in cotton, orchards, strawberries, and organically grown vegetables (Figure 5-18). Geese prefer grasses, but they will turn to broadleaves when grass species have diminished. Feeding preferences have been observed in geese, showing a noticeable distaste for weed species with hairy leaves. If not properly managed, geese will attack the crop when there are no more weeds to eat.

Grazing animals such as cattle, sheep, and goats can be used for weed control, especially on rangelands. Each type of animal has dietary preferences that can be applied to manipulate predominating vegetation in a grazing area. Cattle primarily consume grasses but will also ingest some forbs. Sheep are primarily grazers of herbaceous vegetation and will partially control such noxious weeds as leafy spurge, dyers woad, Russian thistle, and tansy ragwort. Goats are primarily browsers of woody vegetation and are particularly useful in areas of rough terrain.

Controlled grazing has proven to be an effective vegetation management tool in the control of yellow starthistle. Timed to specific stages of yellow starthistle development, several years of intensive grazing management can effectively diminish large stands. Grazing does not eliminate yellow starthistle; rather, it is one component of a long-term management program integrating the use of several weed control practices.

Allelopathy. Allelopathy occurs when a plant releases chemicals (allelochemicals) that impair growth of other plants nearby. An example of a plant with allelopathic qualities is the black walnut tree, which produces a toxin that inhibits growth of most plant species around the base of the tree. Other crop plants with well-demonstrated allelopathic effects on weeds include barley, rye, and sudangrass. Some weed species also have the potential to express allelopathic effects, including some *Cirsium*, *Cynodon*, and *Cyperus* spp. Some organic mulches have also been used for their allelopathic effects. Natural toxins leaching from uncomposted fresh bark or foliage of certain species, such as pine and eucalyptus, can temporarily retard young weeds but have little effect on more mature weeds. Although the concept of allelopathy has been known for years, there are currently few specific recommendations for its use in weed control. Lack of effective weed control with allelochemicals may be partly attributed to the poor stability, high volatility, and strong adsorption of these chemicals to soil particles, which makes them unavailable to weed seedlings.

Competition. Weeds decrease crop yields and negatively influence neighboring plants in landscapes by competing for light, water, and nutrients. Tilting any of these resources in favor of the host plant can help it outcompete weeds. For instance, using transplants can give the host plant an advantage by shading certain weed species and giving the host a head start. Managing water and nutrients can also effectively exploit crop competitiveness. Banding and side-dressing of fertilizer applications, for instance, is more favorable for crop growth than for weed growth.

Biological Control of Plant Pathogens

Organisms that kill or inhibit growth of plant pathogens and other microorganisms are called antagonists. They may act as predators, parasites, pathogens, and competitors, or they may have repellent or antibiotic effects. Although many naturally occurring organisms may antagonize plant pathogens, little is known about most of the biological control agents involved, and pest managers generally cannot identify them in the field. In some cases, soil solarization, crop rotation, or incorporating compost, green manure, or other amendments can increase activities of beneficial soil organisms and reduce disease occurrence. Solarization has been used to increase fluorescent pseudomonad and *Bacillus* spp. bacteria known to be natural enemies of soilborne pathogens, for example, in the root zone of lettuce (Figure 5-19). However, relatively few specific recommendations for using these techniques in commercial management situations can be made at this time.

Mycopesticides are commercially available beneficial microorganisms or their by-products that control plant pathogens. Other than disease-suppressive composts and amendments, mycopesticides are currently the main strategy growers can use to implement biological control against plant diseases. For instance, the competitive bacterium *Pseudomonas fluorescens* A506 has been used effectively in combination with antibiotics to control fire blight disease of pears in commercial pear orchards in California. Table 5-7 lists some commercially available mycopesticides.

Disease-Suppressive Soils. Disease-suppressive soils are soils in which disease incidence remains low even though a pathogen, a susceptible host, and environmental conditions that favor disease development are present. Most soils possess some ability to limit disease, but *disease-suppressive soils* are known to have substantially lower incidence of a disease.

Suppression may involve a small number of microbial organisms antagonistic to specific pathogens. Physical and chemical characteristics of the soil may also be involved in suppressing the pathogen or its antagonists. In general, the low incidence of disease in suppressive soils is due to the antagonistic effects of soil microorganisms and their ability to limit the activity of the pathogen. However, plant pathologists have not been able to transfer disease-suppressive characteristics from one site to another, and the exact mechanisms of suppression are not known. Among the best-known examples of disease-suppressive soils are soils suppressive to Fusarium wilt and take-all decline of wheat.

FIGURE 5-19

Properly applying clear plastic to bare soil for several weeks during warm, sunny weather superheats the soil and can control pathogens, nematodes, and certain weeds near the surface. Soil solarization also favors the survival and reproduction of heat-tolerant beneficial microorganisms that are antagonistic to soilborne pathogens. Pest control can sometimes be increased by solarization in combination with incorporation of crop residues.

Biological Control of Plant-Parasitic Nematodes

Effective use of biological control agents for plant-parasitic nematodes is complicated by the dynamics of the soil environment that most pest nematodes inhabit. Soil varies physically, chemically, and biologically. Nematodes and their antagonists are influenced by interactions with the host plant, the physical environment, and the soil microflora and microfauna. Nematodes have many types of natural enemies that regulate their numbers in nature. These include predatory insects and nematodes, pathogenic microorganisms, competitors, and organisms that secrete harmful or repellent substances. Some of the best-studied natural enemies are fungi that capture nematodes in various forms of traps (Figure 5-20). However, very little is known about how most natural enemies control nematodes or how they could be manipulated in the soil environment. At present, there are no practical applications of biological control for nematodes.

TABLE 5-7.

Selected commercially available mycopesticides.

Mycopesticide organism	Pathogens controlled	Example application methods
Agrobacterium radiobacter strains	crown gall (*Agrobacterium tumefaciens*)	preplant preventive as cutting, root, or seed dip
Ampelomyces quisqualis	powdery mildew	preventive foliar spray
Bacillus subtilis	damping-off pathogens, such as *Pythium* spp., and foliar diseases such as powdery mildew	seed inoculant or foliar spray
Burkholderia cepacia	Fusarium and Rhizoctonia root rots and certain nematodes	preventive application prior to planting as cutting, seed, or seedling dip
Candida oleophila	postharvest fruit decay	preventive application to harvested fruit
Gliocladium virens	*Pythium* and *Rhizoctonia* spp.	incorporation in soil or media and incubation before planting
Pseudomonas cepacia, *P. fluorescens*	fire blight and frost of pears	foliar spray
Pseudomonas syringae	postharvest fruit decay	postharvest preventive application to certain fruits before storage
Streptomyces griseoviridis	root decay fungi, e.g., *Alternaria*, *Fusarium*, and *Phomopsis* spp.	preventive as dip, drench, or spray for seeds and container-grown plants
Streptomycin from Streptomyces griseus	bacterial blights, cankers, leaf spots, and wilts; crown gall	curative as cutting dip or foliar spray
Trichoderma harzianum, *T. polysporum*	*Pythium* spp. and other soilborne pathogens	seed and bulb dip, soil drench, tree wound dressing

Mycopesticides must be labeled and registered in accordance with pesticide regulations of the U.S. EPA (and in California with DPR). Check labels for permitted uses and methods. Many of these microorganisms are also available in commercial amendments or inoculants, which are largely unregulated, and information on their use may be less reliable than the information for registered mycopesticides.

CULTURAL PEST CONTROL

Cultural practices can sometimes be used to modify the environment, making it less favorable to pest invasion, reproduction, survival, or dispersal. For instance, in some cases simple changes in the timing of irrigation may reduce disease incidence or weed germination; or, a change in planting or harvest time can allow a crop to escape pest invasion.

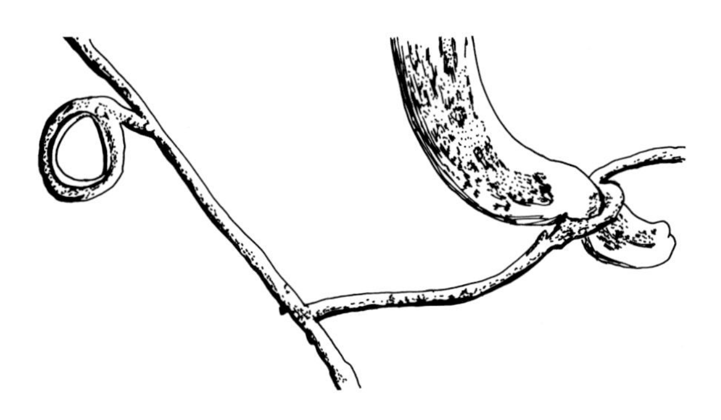

FIGURE 5-20

A ring trap fungus snares a nematode at right. Another trap waits for prey at the upper left. Certain fungi (such as Arthrobotrys brochopaga *and* A. dactyloides*) produce these looplike structures in spaces among soil particles where nematodes commonly travel. When a nematode enters, the loop contracts like a noose. Fungal hyphae then grow into the captured nematode and consume its body. Other trapping fungi use sticky nets (*A. oligospora*), sticky knobs (such as* Dactylaria haptoyla *and* Nematoctonus *spp.) or sticky spores (*Drechmeria coniospora *and* Hirsutella rhossiliensis*).*

Cultural pest controls complement other control options and are an essential component of IPM programs. Cultural controls are generally familiar, simple, and inexpensive, and they can often be incorporated into management systems with only minor modifications. Some of the oldest pest control practices used by growers, foresters, and other resource managers include cultural methods. In fact, some practices have been so routinely used and are so closely associated with management practices that they are not clearly recognized as pest control strategies.

Cultural pest control is most often used as a preventive pest management tool. To properly implement cultural control measures, it is essential to have a good understanding of crop and pest biology, ecology, and phenology. Special attention should be paid to the weak links in the life cycle of the pest. There are hundreds of different cultural practices, and their use varies among the different pest types and management systems. The most common and potentially useful are described below.

FIGURE 5-21

In landscape situations, many pest problems can be avoided by choosing plants that are adapted to local conditions and by grouping plants that have similar requirements. The ceanothus and flannel bush above are well adapted to landscapes in dry areas of central and southern California.

Site Selection

Sometimes pest problems can be prevented by selecting a site that is pest-free, or by choosing a crop, plant species, or variety that is particularly well suited to the site. Before planting, evaluate whether the site and the resource being planted are a profitable match or whether the combination will create or aggravate pest problems. For example, trees or shrubs are often planted in landscapes without regard to the environmental conditions favorable to that plant (Figure 5-21). Plants poorly adapted to conditions at their site are more prone to insect and disease damage and generally do not perform well. Likewise, it is good management to avoid fields with weeds that are potentially troublesome and hard to control in a crop. Instead, plant an alternative crop where available herbicides, cultural practices, or crop competitiveness will help reduce troublesome weeds. Sampling for nematodes, propagules of potentially damaging pathogens, and weed seeds, and looking at the field history prior to planting can help determine whether the site is suitable for a given type of crop.

Sanitation

Sanitation is an important management tool that is applicable in nearly every habitat and for most kinds of pests. Good sanitation practices minimize pests by removing or destroying their breeding, refuge, and overwintering sites or the pest species themselves. For instance, some plant pathogens cannot survive without a plant host; removal of all residues from previous crops can limit future infestations. Some sanitation practices are effective enough that they eliminate the need for pesticide applications. For example, removal and destruction of *mummy nuts* from almond trees eliminates the overwintering sites of the navel orangeworm, *Amyelois transitella*. Table 5-8 lists some important sanitation techniques of general applicability.

Destruction of Alternate Hosts

Destroying alternate hosts can effectively suppress certain pest populations. For example, some weeds or other plants act as alternate hosts for insect and pathogen pests; destruction of these reservoirs can aid in suppressing pest populations. The control of curly top virus in sugar beets involves the destruction of weeds that are alternate hosts of the beet leafhopper, *Circulifer tenellus*. Lettuce root aphid, *Pemphigus bursarius*, overwinters in galls on poplar trees. Lettuce growers have successfully reduced populations of this aphid below damaging levels in their fields by eliminating poplar windbreaks.

Habitat Modification

Pest problems occur when conditions essential for survival, that is, food, shelter, alternate hosts, and proper environmental conditions, are favorable. Habitat modification intentionally changes the environment to limit availability of one or more of these requirements, thus reducing the suitability of the host to pest populations. Habitat modification is very important in vertebrate control. For example, weeds, ground cover, and litter provide food and cover for meadow mice. Eliminating these areas in and around crops, turf and landscape, and cultivated areas reduces the potential of these areas to support these and other vertebrate pests. Habitat modification is effective in limiting numerous insect pests as well. For example, draining areas containing standing water reduces breeding sites for mosquitoes and is an important management technique.

Habitat modification is also a critical element in the management of pest flies in poultry houses and dairies. Flies have three basic developmental requirements: food, warmth, and moisture. Eliminating any of these in the poultry house, barn, or barnyard can break the life cycle of the fly and reduce the problem. Although eggs may hatch, larvae cannot develop into flies without food, warmth, and moisture. Warm weather combined with moist organic matter will produce large numbers of flies. Manure, waste feed, silage stored on a cement base, and contaminated animal bedding provide ideal sources for feeding, egg laying, and development of pest flies. Barns and poultry houses must be designed for easy removal of manure, and barn floors should be surfaced and sloped to facilitate drainage (Figure 5-22). Moist bedding, old feed, and other decomposing organic matter should be regularly removed. Manure-breeding flies have many natural enemies, and certain practices can enhance their activities. For instance, in poultry houses, when fresh manure is being removed, a single pad of manure 6 to 8 inches

TABLE 5-8.

Important sanitation techniques for reducing pests.

- ☐ **Use certified** seed, tubers, or rootstock to prevent the spread of nematodes, weed seeds, and viral, bacterial, and fungal pathogens that may be carried in contaminated plant material.
- ☐ **Thoroughly clean** with steam or pressurized water all equipment as it moves between different sites to prevent the spread of pathogens, nematodes, and weeds.
- ☐ **Use clean irrigation water.** Do not irrigate with tailwater from fields infested with root knot nematodes or other soilborne pathogens or pests; do not use water that may carry runoff containing harmful herbicides.
- ☐ **Install screens** in pipes bringing irrigation water from canals or ditches to filter out seeds, rhizomes, and other weed parts.
- ☐ **Remove or clean up** unharvested crops or areas that might provide overwintering habitat for pests.
- ☐ **Eliminate weedy borders.**

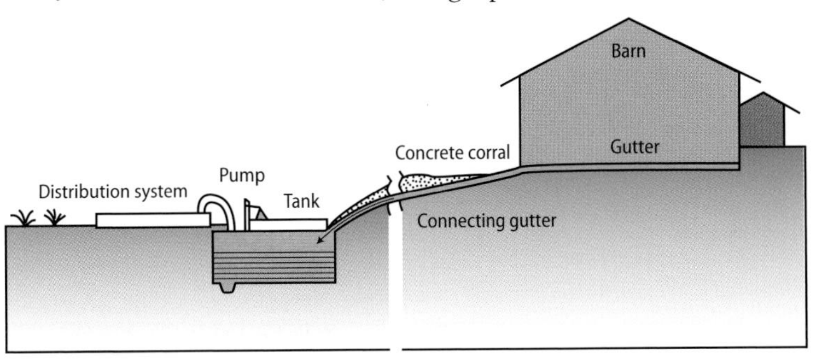

FIGURE 5-22

Flushing manure regularly out of dairy barns and properly disposing of it eliminates habitat for manure-breeding flies.

(15 to 20 cm) thick can be left on the house floor to provide habitat for natural enemies such as ants, mites, beetles, and parasitic wasps so they can quickly invade adjacent fresh droppings.

Smother Crops and Cover Crops

Smother crops are grown for their ability to suppress weeds and for their cash value. They are effectively employed in crop rotations following the main crop and are planted at high densities to rapidly occupy a site. Cereals, sorghum, safflower, field corn, and domestic sunflowers have been effectively used as smother crops.

Cover crops are noncrop plant species grown either concurrently with the host crop (usually perennial plants) or in rotation with annual crops; they are generally not harvested. Cover crops can suppress weeds; provide nutrients to the soil; and provide food and shelter to beneficial insects, mites, and spiders (Figure 5-23). They are also valued for their ability to improve soil texture, increase organic matter, increase water infiltration rates, reduce pesticide runoff into surface water, and reduce soil erosion. Successful cover crop varieties often planted in orchards and vineyards in California include strawberry clover, annual clover, cereal grasses, annual grasses, and vetches, or a combination of species selected for their ability to improve soil fertility or attract beneficial insects.

FIGURE 5-23

Native ground covers are used as cover crops in many vineyards to minimize weed growth, improve soil water penetration, and reduce erosion.

An example of a cover crop improving biological control is the use of a vetch cover crop in vineyards to improve biological control of mealybugs. Vetch supplies ants with adequate amounts of nectar and keeps them from moving into vines, thereby reducing ants' interference with the natural enemies of grape pests—-especially parasitic wasps that can control the mealybugs. However, in some vineyards cover crops can increase early populations of the caterpillar pest omnivorous leafroller by providing it with an alternate food source.

In some cases, cover crops may compete with crops or may become weeds if not properly managed. When using cover crops, it is important to select a species or mix of species that provides the specific benefits desired. Cover crop selection should take into consideration soil type, water availability, method of irrigation, cropping sequence, and cultural practices. While properly selected cover crops can provide numerous benefits in addition to weed suppression, there can also be disadvantages associated with their use (Table 5-9). Take care to choose species that will not increase nematodes, rodents, or pathogens, or deplete nutrient supplies. More information on cover crops can be found in the "References" and "Resources" sections.

Intercropping. Intercropping involves growing more than one crop in a field at the same time: multiple crops are planted in alternating strips or intersown into a main crop. Intercropping is sometimes used to reduce pests. For instance, older stands of alfalfa can be overseeded with oats to reduce weed pressure, and the harvested alfalfa-oat mix is sold as a forage crop. In cotton, a trap crop of alfalfa can be planted in strips to keep damaging lygus bugs out of the crop (Figure 5-24), since lygus does little damage to alfalfa hay grown for forage. In new orchard plantings, intercropping is sometimes used to produce income prior to the maturing of the trees.

TABLE 5-9.

Advantages and disadvantages of cover cropping.

ADVANTAGES
☐ decreases soil erosion
☐ increases organic matter
☐ improves soil structure and water infiltration
☐ decreases water and pesticide runoff
☐ may add or conserve nitrogen
☐ may suppress weed growth
☐ may attract or provide nectar source for beneficial insects, spiders, and mites
DISADVANTAGES
☐ depletes soil moisture
☐ may decrease availability of plant nutrients, especially nitrogen
☐ may require additional irrigation and nitrogen applications
☐ may increase weeds
☐ may attract arthropod and rodent pests
☐ may increase nematode populations
☐ increases danger of frost damage to trees or vines
☐ increases associated costs

Adapted from Ingels et al. 1998.

FIGURE 5-24

Alfalfa intercropped in cotton fields attracts lygus bugs, which are harmless to alfalfa and would otherwise feed on and damage cotton.

Problems can arise in intercropping systems if pesticide applications are needed. A pesticide registered for one crop in the system may not be registered for the other crop. Care must be taken to select pesticides that are compatible with both crops.

Crop Rotation

Crop rotation is the intentional planting of specific crop sequences to improve crop health. It is one of the oldest cultural practices in use and is widely practiced to increase organic matter, improve soil properties, conserve water, and manage pests. Crop rotations have provided effective control for certain host-specific plant pathogens, nematodes, and insects by disrupting the pests' life cycle and by changing environmental conditions to deter certain species.

Pests that can be successfully controlled with crop rotation are those that originate in the field and are not likely to move in from adjacent areas. These pests tend to be soilborne and immobile. The host range of the pest cannot be so extensive that a suitable alternate crop cannot be found. Pests that attack only one or a few closely related crops are the best candidates. And, the pest population should not be able to survive after a 1- or 2-year absence of the living host.

Good candidates for management by crop rotation include soil- and root-dwelling nematodes and soilborne pathogens that do not produce airborne spores and have limited host ranges. When rotation is used for disease and nematode management, alternate weed hosts must be controlled as well. Weeds are sometimes controlled with a combination of rotation and herbicide management by rotating to a crop that allows the use of a more effective herbicide. Rotating to a flooded crop, like rice, or a highly competitive crop, like alfalfa, can also help reduce a variety of troublesome pest species. Small grains are routinely used as part of a crop rotation sequence to reduce nematode and soil pathogens as well as provide additional agronomic benefits in broadleaf crops (Table 5-10).

Planting and Harvest Dates

Planting and harvest dates should be chosen to favor crop development and discourage pests. For weed management, choose a planting time that favors germination of the crop over key weed species. For instance, in many parts of California, fall alfalfa planting can be adjusted to avoid both late-germinating summer weeds and late-fall-germinating winter weeds. However, crops planted too late in fall will germinate and grow slowly, allowing winter weeds to become well established. If fields are infested with field bindweed, perennial grasses, or nutsedge, planting in early fall will ensure that alfalfa is established and vigorously growing when these perennials start growing in spring.

Adjustments in the timing of planting can also aid in insect and disease management. Winter-grown carrots and potatoes in California can be planted when the potato cyst nematode is not active, avoiding exposure in early crop development stages when the pest is most damaging. In sugar beets, beet yellow virus and beet mosaic virus have been kept under control by following strict planting programs that allowed no early-spring plantings in or adjacent to growing districts where sugar beets were overwintered in the field. Certain seedling pathogens, such as *Thielaviopsis basicola* and *Pythium ultimum*, can be avoided in cotton by planting later to reduce the chance of infection. Timing spring row crop planting to allow sufficient time for winter row crop residues to decompose is an important cultural practice used in coastal row crop production to prevent seedcorn maggot damage to seeds and seedlings. Using a chain drag behind the seed shanks on the planter also helps to camouflage the seed row from adult egg-laying flies.

Harvest can also be timed to limit or avoid pest damage. Early harvest can reduce the number of generations of nematodes that can damage a root crop. In northern California potato fields, early harvest effectively reduces the incidence of nematode blemishes on tubers by shortening the time nematodes have available to reproduce. Early harvest is also beneficial in the management of various insect pests. Early harvest of coastal avocados

TABLE 5-10.

Selected crops in which rotation with small grains is combined with other management practices to reduce nematode and disease populations.

Crop	Pest	Rotation (years)	Other management practices
barley	barley root knot nematode	1	Rotate with resistant cultivars of oats; onion and potato rotations also reduce populations.
beans	Anthracnose	1	Use disease-free seed.
celery	late blight	2	Use disease-free seed or transplants; use fungicides.
cole crops	cabbage cyst nematode	3–10	Control mustard family weeds; can rotate with crops other than small grains.
cotton	Verticillium wilt	1	Plant resistant cultivars; soil solarization.
potato	scab	1	Provide adequate soil moisture during tuber development; completely decompose crop residue.
sugar beets	sugarbeet cyst nematode	3–10	Control broadleaf weeds.
tomato	Phytophthora root, fruit and seedling rot	1	Irrigate carefully.

Source: Strand 1990.

can help control greenhouse thrips on tough-skinned varieties by minimizing crop-to-crop overlap and removing much of the insect population before it has time to move to a new crop. In some situations, early harvest of alfalfa hay can eliminate need for treating alfalfa weevil or alfalfa caterpillar.

Irrigation and Water Management

Poor water management is a major contributor to many pest problems. Excess soil water excludes oxygen that plant roots need to survive and is a primary factor in the development of many root and crown diseases, such as Phytophthora root rot. The fungi causing these diseases are present in many soils but become damaging when excessive moisture favors them. Excess water and poor drainage in low areas of fields or landscapes also favor numerous hard-to-control weeds such as nutsedge and barnyardgrass.

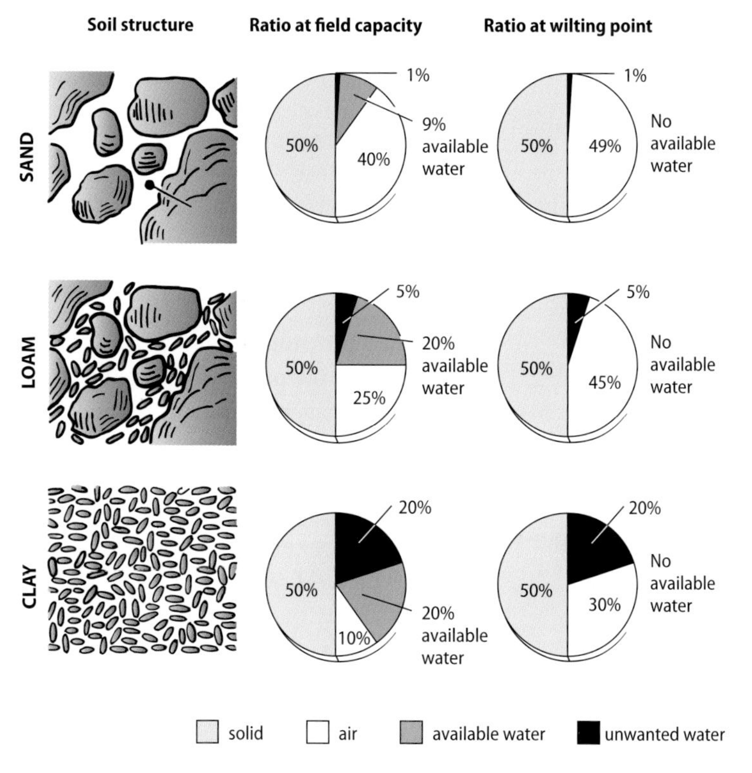

FIGURE 5-25

Pore spaces are the openings between soil particles that contain water and air. Sandy soils have big pore spaces that contain large amounts of air and allow water to drain quickly. Clay soils have smaller pore spaces and retain more water, resulting in poor drainage and often insufficient oxygen for roots. Loamy soils are a mixture of sand, silt, and clay that are not compacted; loam provides the best balance between water-holding ability and adequate air.

Underwatering, or drought stress, on the other hand, may cause wilting, sunburn, sunscald, and branch cracking that can allow invasion of pathogens and attract boring insects, or it can stress plants to the extent that they cannot overcome attack by pests. This is especially true of conifers in the landscape. Crop competitiveness is reduced and weed invasion may occur.

Certain types of irrigation systems are both positively and negatively associated with pest populations. Droplets from overhead irrigation can dislodge, drown, or drive off some insect and mite pests, reduce successful mating of moth pests such as the diamondback moth, and reduce powdery mildew of grapes. However, the increased moisture on the plant surface can also encourage other types of fruit or foliar diseases. Flood irrigation, while inexpensive, can often contribute to root diseases and weeds. Drip irrigation systems deliver water only to sites where it is needed, reducing root diseases and weeds.

To avoid problems associated with over- or underwatering, learn the water requirements of the plants and find out how much available water the soil can hold. Available water is that portion of the soil water that can be withdrawn by plants (Figure 5-25). During the growing season, irrigation is needed when a certain proportion of the available water, the allowable depletion, has been used. Factors affecting allowable depletion are soil type, stage of plant growth, total amount of available water, weather conditions, and irrigation cost.

Schedule the frequency and amount of irrigation by using weather-based methods or by measuring soil moisture, or by using a combination of reliable methods. Reliable devices for monitoring soil moisture include gypsum blocks, neutron probes, and tensiometers.

An indirect method of monitoring soil moisture around the plant root zone is to develop a water budget (Figure 5-26) to estimate how much water plants use from day to day under prevailing weather conditions. After soil has drained to field capacity, further loss of soil water occurs mainly through transpiration by the plant and by evaporation from the soil surface. This process is called evapotranspiration (ET). Knowing how much available water is in the plant rooting depth at field capacity and how much water is lost through ET each day helps to estimate the amount of available water remaining at any time by adding up the daily ET values, which are based on long-term average conditions. They are available for many crops and are often provided through local California Irrigation Management Information System (CIMIS) weather stations, on the UC IPM website, at www.ipm.ucdavis.edu, in local newspapers, on the radio, or by irrigation districts. For more information on irrigation scheduling and ET, see the California Irrigation Management Information System (CIMIS) website, www.cimis.water.ca.gov.

Water budgets provide estimates of crop water use. However, disease, weeds, insects, physical characteristics of individual fields, and management factors affect actual water use. It is a good idea to follow up with a direct measurement of soil moisture to make the final decision about when to irrigate.

FIELD NO. 14

ALLOWABLE DEPLETION 40% of 2.0 in. = 0.80 inch

DATE	ET IN INCHES PER DAY	CUMULATIVE ET	
7/10			irrigation
7/11	0.23	0.23	
7/12	0.25	0.48	
7/13	0.27	0.75	check field
7/14	0.26		irrigation
7/15	0.27	0.27	
7/16	0.25	0.52	
7/17	0.24	0.76	check field
7/18	0.26		irrigation
7/19	0.28	0.28	
7/20	0.27	0.55	
7/21	0.25		irrigation

FIGURE 5-26

With a source of evapotranspiration (ET) data, you can use a water budget to estimate the interval between irrigations. Keep a running ET total starting the day after irrigation. When the total approaches the allowable depletion, check soil moisture in the field to make a final decision on when to irrigate.

Fertilizers and Soil Amendments

Fertilizers and soil amendments are applied to promote healthy plants and increase yields. Although not a pest management practice, the use of fertilizers and soil amendments can influence the activity of many pest species to the advantage or detriment of the host plant. It has been argued that healthy plants have a greater tolerance to pest attack and damage. While this may be true, overfertilization may attract or enhance development of many pest species. Excess nitrogen on nectarines, for example, results in increased levels of brown rot (*Monilinia fructicola*), oriental fruit moth (*Grapholita molesta*), and peach twig borer (*Anarsia lineatella*). Excessive nitrogen levels are also a contributing factor in the incidence of brown patch, *Rhizoctonia solani*, a common disease in fescue lawns. Aphids, leafhoppers, lace bugs, and other sucking insects are associated with the lush growth that results from excess nitrogen application. Adding the proper amounts of soil nutrients when the plant needs them greatly assists pest management practices and limits environmental contamination from excess nitrogen, especially nitrogen lost from leaching to groundwater.

MECHANICAL AND PHYSICAL CONTROL

Mechanical and physical control methods include practices that destroy pests or present a barrier to pest infestation by creating conditions unsuitable for their entry, dispersal, survival, or reproduction. Such methods include regular management practices such as cultivation and mowing. Other methods used specifically for pest control include solarization, flaming, barriers, and traps. Some mechanical and physical methods of pest control have been in use since the dawn of agriculture. Others have become more important with the renewed emphasis on nonchemical control measures.

Land Preparation

Adequate land preparation prior to planting can avoid many potential pests. Well-prepared fields and landscapes are easier to irrigate and manage, resulting in better weed control and fewer diseases.

Proper site preparation reduces soil compaction and contributes to good drainage. When planning how to prepare the site, consider soil depth, texture, and topography. Survey the site for the presence of troublesome weeds and potential vertebrate pests. Evaluate the surrounding habitat as well; adjacent sites may harbor vertebrates and other pest organisms that could contribute to future pest problems. Many pests, if detected before planting, can be overcome by preplant treatments including cultivation or pesticide application.

Soil Tillage

Tillage, or cultivation, contributes to pest management by killing weeds, disrupting the life cycle of some insect pests, and burying disease inoculum. Tillage has many purposes and is often combined with other management practices to turn under crop debris, incorporate fertilizer, improve water penetration, or enhance growing conditions for the crop. Cultivation is the most important and widely used weed management tool in many crops and, with proper timing, kills annual weeds, biennial weeds without a taproot, and seedlings of annual and perennial weeds. Mature perennial weeds can be sufficiently controlled by repeated cultivations under dry soil conditions. Improper timing, however, can increase perennial weeds. Tillage can also bring additional weed seeds to the surface, resulting in germination flushes after each cultivation, while some perennial weeds, such as johnsongrass, bermudagrass, and field bindweed, can increase due to regrowth from chopped-up underground stems.

Different types of cultivation are often used to manage different pests. For instance, moldboard plowing is used in specific situations because it buries weed seeds deeply, reducing germination and establishment. The French plow is used in vineyards for weed management and to bury the overwintering larvae of the omnivorous leafroller, *Platynota stultana*, and their overwintering diet of dried berries and vineyard debris (Figure 5-27). Sclerotinia drop of lettuce, caused by *Sclerotinia minor*, can be managed, in some instances, by burying the propagules 10 to 12 inches (25 to 30 cm) beneath the soil by deep plowing, reducing the number of sclerotia that can germinate. However, in fields with high inoculum density, deep plowing may actually increase disease incidence by spreading sclerotia and increasing the number of plants likely to be infected.

All cultivation techniques have their advantages and disadvantages; their impacts on the ecosystem should be carefully weighed. Tillage can destroy soil structure and contribute to soil erosion, loss of fertilizer, increased compaction, disruption in the life cycle of beneficial organisms, and air pollution. For this reason, some growers have adopted *conservation tillage*. Conservation tillage is any conservation practice that retains a plant residue cover of at least 30% from the previous crop on the soil surface and includes no-till, ridge-till, strip-till, mulch-till, and other tillage systems that

FIGURE 5-27

Row plowing in vineyards in early spring before new growth begins buries the mummies of the omnivorous leafroller, a major pest of grapes.

meet this requirement. However, conservation tillage has its disadvantages as well: soils do not warm as rapidly due to the insulating effect of the residues, and weeds and plant diseases can increase.

Mowing

Mowing is an effective weed management tool, particularly when used in combination with other management methods. It is the most common method of nonchemical weed control used along rights-of-way and orchard floors (Figure 5-28). Proper timing and site conditions are very important. Mowing should be completed before weeds set seed or before seeds mature, and it should be done when soil moisture is low. Also, mowing a cover crop can result in mass migration of arthropod pests such as thrips or mites to trees or vines, so mowing should be avoided when these pests can be most harmful, such as during bloom. In most situations, follow-up mowing is necessary, depending on the growth and flowering pattern of the weed species present.

As a component of a long-term management plan, mowing has been particularly effective in managing yellow starthistle populations. After 2 years of mowing treatments, flower head and seedling production are significantly reduced when mowing is timed for the early flowering stage (when about 5% of the flower heads are in bloom).

Correct mowing height and frequency of mowing are critical for preventing weed invasion in turf. Different turf species have different mowing height requirements. Mowing Kentucky bluegrass too short (below 1.5 inches, or 4 cm) weakens the turf and encourages weed growth. Conversely, mowing bermudagrass too long (above 1 to 2 inches, or 2.5 to 5 cm) results in a buildup of thatch, which reduces the competitive ability of the grass.

Flaming

Flaming is a weed management technique that has been used in row crops, orchards, roadsides, and industrial sites. Flaming commonly uses special propane burners, but equipment employing hot water, steam, or infrared light is also available. Flaming requires only brief plant contact and extremely high temperatures to disrupt cell membranes and cause cell walls to burst; treated weeds wilt and die within a few days. Proper use of flaming should not heat weeds so long that they smolder, char, or burn.

Flaming is most effective on young annual dicot weeds. Young perennial weeds are also susceptible but require more than one treatment. Repeated flaming eventually starves the roots, killing the weeds. Flaming has also been successfully used in the landscape (Figure 5-29) as a spot treatment for dodder, and to control weeds and weevils in alfalfa. In late fall or winter, when alfalfa plants are dormant, flaming destroys the adult and egg stages of weevils.

Burning

Prescribed, or controlled, burns have been a long-standing practice in crop production, forest, and range ecosystems, particularly when exotic weed species invade. Burns conducted in late spring over consecutive years can dramatically reduce yellow starthistle populations. Controlled burns have also provided effective management of stem rot and sheath spot in rice. Recent legislation, howev-

FIGURE 5-28

Mowing or flailing several times during the year is one method of managing weeds in orchards.

er, calls for burning to be substantially curtailed because of air pollution generated by this practice. The benefits of burning in any crop must be weighed against cost and potential environmental hazards. Many rice growers are successfully using alternatives such as rolling rice residue and keeping fields flooded throughout the winter. These flooded areas also serve as winter wildlife habitat for migratory birds along the Pacific flyway.

Mulches

A mulch is a layer of material covering the soil surface (Figure 5-30). Mulches have been used extensively in landscape plantings for weed control for years and are also used in some vegetable, orchard, and strawberry plantings. The use of plastic mulches is a standard practice in strawberry production. Mulches discourage weed growth, conserve soil moisture, enhance the water-holding capacity of light sandy soils, and help maintain a uniform soil temperature.

Composts used as mulches in nursery crops have been shown to reduce Phytophthora root rots and other diseases. Equally important in the landscape is the aesthetic value that the appropriate mulching material can provide. A number of mulching materials are available, including bark and wood chips, composted green waste, and plastics such as polyesters and polyethylenes (Table 5-11). Different materials are sometimes used together. For instance, a woven weed mat or plastic sheet may be laid down and wood chips spread on top.

Polyethylene mulches have been used extensively in strawberry production. Besides providing weed control, they reduce decay by limiting fruit contact with soil and irrigation water. The type of mulch used and the timing of application depend on cultivar, planting and harvest seasons, and other management practices. Various types of mulches are beginning to be used in vegetable, vineyard, and orchard crops as well (Figure 5-31). Silver polyethylene mulches used in the production of cucurbits can repel aphids and whiteflies and reduce the incidence of the nonpersistent virus diseases they carry.

FIGURE 5-29

Flaming kills weeds in bare soil, along fence rows, in pavement cracks, and in certain mulched areas like this, where weeds are being killed before reapplying mulch to an area where mulch is too thin.

FIGURE 5-30

A layer of wood chip mulch in a landscape planting discourages weed growth and conserves soil moisture.

Where weeds are severe, organic mulches are most successful when applied in the spring after the soil is weed-free to a depth of 2 to 6 inches (5 to 15 cm). It is also important that mulching materials be weed-free. Fertilizer can be added to organic mulches that are not completely decomposed to prevent the material from robbing the soil of nitrogen.

Mulches should be regularly inspected; they can provide hiding places for other pest species. Snails, slugs, earwigs, ants, sowbugs, and other invertebrate pests can be found hiding in mulched areas. In addition, mice, gophers, and other vertebrate pests seek out mulches for the protection and food they provide. Plastic mulches don't allow air or water penetration, so special care must be taken to use drip irrigation beneath them and avoid waterlogging of roots that can lead to root disease.

Soil Solarization

Soil solarization involves covering moist soil with clear plastic and allowing the soil to heat up. This practice reduces or eliminates many soil-inhabiting pests by raising the temperature in the top 2 to 3 inches (5 to 7.5 cm) of soil to levels lethal to many soil pest organisms. Solarization favors beneficial organisms in the soil by creating changes in the soil microflora that the beneficials are able to exploit. In some situations, solarization contributes to increased yields and improved crop quality following treatment.

Soil solarization has been effective in controlling certain soilborne pathogens and many weed species, and in partially controlling many other pests (Table 5-12). To be effective, a clear plastic tarp is placed over bare, moistened soil for 3 to 6 weeks during the hottest part of the year. Weed control is enhanced if fields are irrigated prior to being covered because moisture helps conduct

TABLE 5-11.

Common materials used as mulches in the landscape.

Plant-Based (Organic) Mulches	
compost	May harbor weed seeds if not properly composted. Holds water, so if placed too close to tree trunks may promote crown disease. Weeds grow easily in it. Apply at 2-inch depth. Best in annual beds or vegetable gardens for short-term mulching.
grass clippings	May contain weed seeds. Mats and reduces water penetration, especially if applied too thick and not dried out before applied. Not generally recommended.
ground bark	Attractive but decomposes rapidly unless used with landscape fabric. Weed seeds grow easily in it and must be pulled out. Can tie up nitrogen as it decomposes when mixed in soil. Don't apply more than 2 inches deep. Often free and is a good source of organic matter. Best for short-term mulching in annual beds or for use on top of landscape fabric.
medium-sized bark chips	Longer lasting than smaller particles, like ground bark. Needs to be replenished as it decomposes. Excellent as a topping for landscape fabric. Best overall choice for a plant-based mulch to be used without fabric. Apply 3 to 4 inches deep and keep replenishing.
peat moss	Not a good mulch. Resists wetting. May blow away. Expensive. Better as a soil amendment for alkaline soils when worked into the soil.
wood chips	A good topping for landscape fabric. Where there is a lot of runoff, it may float away. May decompose faster than bark chips.
Nonplant Mulches	
landscape fabric	Excellent mulch for permanent plantings of woody landscape plants. Allows for air and water penetration. Many products are available. May last up to 5 years when properly maintained with plant-based mulch on top.
newspaper/cardboard	Two or three sheets of newspaper can be placed under organic mulches in landscape beds on top of drip irrigation; remains effective for the whole season. Cardboard can also be used.
plastic	Inexpensive but breaks down rapidly. Not permeable to air and water. Requires drip irrigation. Weeds can grow through tears. Use black plastic, since clear plastic encourages weed growth. Unattractive unless covered with plant-based mulch. Landscape fabric or plant-based mulches are generally a better choice.
rock	Attractive as a top mulch for landscape fabrics. Tends to become weed infested if used alone. May get too hot and injure roots. Hard to clean. Time consuming to remove.

FIGURE 5-31

A mow and blow operation in plums. A Sidewinder mower blows the cover crop mulch into the tree row for strip weed control.

heat under the tarp. In cooler areas, solarization may not be as effective or the required treatment period may increase.

Temperature Manipulation

Temperature manipulation is widely used in greenhouse and nursery operations for the control of many insect, weed, nematode, and pathogen pests. Heating or steaming soils to temperatures around 70°C (158°F) kills most plant pathogens; nematodes are killed at temperatures of 50°C (122°F) for 30 minutes (Table 5-13). Long exposure to high temperatures, however, can produce undesirable side effects, such as the formation of toxic breakdown products and lethal effects on beneficial organisms, that increase the chances of success for later invasion by opportunist pest organisms. To avoid these problems, soil pasteurization is sometimes used. In compar-

TABLE 5-12.

Selected pathogens and pests controlled by soil solarization.

Type of pest	Scientific name	Common name or disease
	Pathogens or Pests Largely Controlled	
fungi	*Fusarium oxysporum*	Fusarium wilt
	Phytophthora spp.	Phytophthora root rot
	Rhizoctonia solani	seed or seedling rot
	Verticillium dahliae	Verticillium wilt
bacteria	*Agrobacterium tumefaciens*	crown gall
	Clavibacter michiganensis	canker
	Streptomyces scabies	potato scap
nematodes	*Criconemella xenoplax*	ring nematode
	Heterodera schachtii	sugarbeet cyst nematode
	Paratylenchus spp.	pin nematode
	Pratylenchus spp.	lesion nematode
	Xiphinema spp.	dagger nematode
weeds	*Abutilon theophrasti*	velvetleaf
	Amaranthus spp.	pigweed
	Amsinckia douglasiana	fiddleneck
	Brassica nigra	black mustard
	Convolvulus arvensis	field bindweed (seed)
	Cynodon dactylon	bermudagrass (seed)
	Malva parviflora	cheeseweed
	Solanum spp.	nightshade
	Sorghum halepense	johnsongrass (seed)
	Pathogens or Pests Partially Controlled	
fungi	*Macrophomina phaseolina*	charcoal rot
bacteria	*Pseudomonas solanacearum*	bacterial wilt
nematodes	*Meloidogyne incognita*	southern root knot nematode
weeds	*Convolvulus arvensis*	field bindweed (plant)
	Cynodon dactylon	bermudagrass (plant)
	Cyperus spp.	yellow and purple nutsedge
	Sorghum halepense	johnsongrass (plant)
	Malva niceanis	bull mallow

Adapted from Elmore et al. 1997.

TABLE 5-13.

Recommended soil temperature and exposure time necessary for the control of selected plant pathogens and pests by solarization.

Pest	Soil temperature (°C)	Soil temperature (°F)	Exposure time (minutes)
most bacteria	60–70	140–158	10
Botrytis cinerea	55	131	15
Fusarium oxysporum	60	140	30
Pythium spp.	53	127	30
Rhizoctonia spp.	52	125	30
Sclerotium rolfsii	50	122	30
Verticillium dahliae	58	136	30
most pathogenic fungi	60	140	30
most actinomycetes	90	194	30
foliar nematodes	49	120	15
Meloidogyne incognita	48	118	10
Pratylenchus penetrans	49	120	10
most viruses	100	212	15
worms, slugs, centipedes	60	140	30
most weed seeds	70–80	158–176	15

ison with sterilization, pasteurization uses less-intensive or shorter-duration treatment, such as heating to a lower temperature. Pasteurization kills most pests but does not greatly change the chemical composition of the soil and allows some heat-tolerant beneficial microorganisms to survive.

Temperature manipulation is also used in the control of many stored-product pests; for example, cold storage destroys apple maggot and plum curculio in apples, and some postharvest diseases in kiwifruit can be effectively managed by prompt cooling of firm fruit after harvest.

Chaining and Dredging

Chaining and dredging are mechanical methods of weed control used primarily in aquatic systems and forests. Chaining physically tears the plant off the root, providing immediate but temporary suppression of weeds. Since large masses of plants are released into the waterway, chaining should be conducted only at sites that provide access for removal of the plant material. Chaining is typically performed in canal systems as a part of annual maintenance.

Dredging, on the other hand, removes the entire plant, including the root system, decreasing the possibility of spreading the weed species. The effectiveness of dredging depends on the depth of mud dredged from the water and the water depth after dredging. Dredging is typically performed in systems that have become inundated with undesirable plant species, organic matter, and silt. Dredging offers a higher degree of control than chaining, with longer intervals between controls.

Traps

Mechanical traps are practical devices for controlling many vertebrate and invertebrate pests. Traps are especially important in the management of vertebrate species such as ground squirrels, moles, meadow voles, pocket gophers, rats, mice, and some species of large animals. Insect traps used for control include fly traps, roach traps, and other types of sticky traps. Some of these traps are regularly used for monitoring but can also be used to assist in management efforts.

Traps are of various designs and are specific to a pest species. Some traps are designed to kill, while others catch the offending individuals until they can be transported and released elsewhere. An alternative trap type for mice is the glue board. Glue boards work much in the same way as flypaper; as mice travel across the glue board, they get stuck. Mice trapped on glue boards do not die immediately and must be disposed of. Table 5-14 lists various types of traps used in pest control.

When using traps, placement is very important (Figure 5-32). For vertebrates, the most effective locations are usually in natural travel ways and near or in runways, depending on the species. Be aware of legal requirements when trapping vertebrate pests. Some species are classified as nongame mammals by the California Fish and Game Code and can be controlled whenever they injure crops, while other species are classified as game animals and fall under special provisions of the Fish and Game Code. Check with the county agricultural commissioner's office for specific information on trapping.

USING PESTICIDES IN AN IPM PROGRAM

Pesticides, as defined by the Federal Insecticide, Fungicide, and Rodenticide Act (FIFRA), include "any substance or mixture of substances intended for preventing, destroying, repelling, or mitigating any insects, rodents, nematodes, fungi, or weeds, or any other forms of life declared to be pests; and any substance or mixture of substances intended for use as a plant regulator, defoliant, or desiccant." Many pesticides kill pest organisms outright; others control pests by repelling them or by inhibiting reproduction or normal development.

TABLE 5-14.

Various traps used for pest control and some of the pests they control.

Trap	Pest controlled
automatic multicatch house sparrow trap; double funnel trap	house sparrows
box-type squirrel trap; multicatch box-type squirrel trap	ground squirrels, tree squirrels
conibear trap	ground squirrels
snap trap	mice, rats, meadow voles
Macabee trap	pocket gophers, moles
box trap	pocket gophers
electronic trap	rats, house mouse
glue boards	mice
live traps	rats, mice, birds, skunks, opossums, other small animals
bait traps	Indian meal moth, fruit flies
sticky traps	flies, some cockroaches, white flies, thrips
walk-through traps	horn fly
cone-shaped bait traps	yellowjackets, flies

FIGURE 5-32

Conibear traps are placed over the entrance to ground squirrel burrows and are secured with a stake. Follow restrictions to protect endangered species, such as covering the trap with a box that has a small entrance to keep out the San Joaquin kit fox.

Pesticides are an important part of many IPM programs. For some pest organisms, pesticides offer the most effective control; sometimes, they may be the only control available. In most situations, pesticide use is one component of an integrated program that may include biological, cultural, and mechanical pest management techniques.

The selection and use of a pesticide can be quite complex. To control a single pest in a crop, the pest manager is expected to choose the most effective material from a list that includes as many as a dozen pesticides registered for the pest and site and that are packaged in a variety of formulations. The manager must understand how the material will react with the plant or crop, the target pest, nontarget organisms, and the environment. In making the selection, hazards and risks of the material to humans and application safety must also be considered.

This section discusses the factors to consider when recommending the use of pesticides in an IPM program. Issues concerning pesticide safety, such as environmental concerns and impacts on living organisms, are covered in Chapter 8. Regulatory issues are covered in Chapter 1, and liability issues are covered in Chapter 9. For more discussion on specific application techniques, see the "Resources" section and refer to *The Safe and Effective Use of Pesticides*, UC ANR Publication 3324.

Factors to Consider

Pesticides differ from most other pest management tools because of their potential to injure people or nontarget organisms on and off the application site. When using pesticides, PCAs must first look at the pest, its life stage, and the beneficial species present. The efficacy of the pesticide in controlling the pest must be considered, as well as whether it is selective enough to allow important beneficial species to survive. The location of the application, the potential hazards, the commodity being treated, harvest timing, and cultural practices, as well as subsequent crops, must be part of the decision-making process. Finally, the philosophy of the client regarding pesticide use must be considered. The ultimate goal is to select the least toxic material that will satisfactorily and economically control the pest in the least disruptive but most effective manner possible.

The Philosophy of the Client. When making pest control recommendations, the client's philosophy regarding management of the commodity and pests is an important consideration. For instance, most growers have cost limitations that drive their pest management decisions. Some may have production contracts that put limits on the pesticides they can use, and others may wish to market their product as organic or free of pesticide residues.

In urban situations, the need for pesticides varies according to landscape uses. Turf in a golf course, for instance, may require a number of herbicide applications to control invading weed species. The same weed species may be tolerated in turf in a park setting because economics, aesthetic requirements, and public concerns about herbicide use might not favor turf herbicide treatments.

The Location and the Commodity. The suitability of the pesticide for the application site is an important consideration in the selection of the material. Prior to writing a pesticide recommendation, identify all important structures and environmental conditions on and around the application site. The location of dwellings, schools, hospitals, parks, playgrounds, commercial areas, and other places where people may work, live, or play imposes limitations that prohibit the use of certain pesticides. The location and type of adjacent crops can also be determining factors in the selection of the material. The possibility of *drift* and the sensitivity of an adjacent crop could eliminate certain pesticides as feasible alternatives. Noncrop areas should also be inspected; the proximity of endangered species, wildlife areas, lakes, streams, ponds, wells, and irrigation canals to the application site can influence pesticide selection if the pesticide has the potential to harm organisms in these areas.

The appropriateness of the pesticide for the commodity is a primary consideration. The pesticide must be registered for use on the commodity and crop growth stage for which the application is being considered. Some herbicides, for example, can only be applied during nonbearing years, after harvest, or during dormant stages in deciduous orchards. There are many different crops with many of the same pests. Although a pesticide may be registered in one crop for use against a pest, it may not be legal to use against the same pest in another crop. Be sure to consider subsequent crops and plantback restrictions. Some pesticides leave residues that harm subsequent crops or may be illegal. Always check the label to confirm legal uses.

The Pesticide. Pesticides are available in many different formulations, including liquids, dusts, granules, and baits (Table 5-15). A *pesticide formulation* consists of the active ingredient (a.i.) that controls the target pest, and the inactive, or inert, ingredients added to the product. The *inert ingredients* dilute the pesticide, enhance its activity, or make it safer and easier to mix and apply.

When more than one formulation is available for control of a pest, consider how the formulation may affect the host plant, people, nontarget organisms, and the environment. Formulations that have the longest residual activity would more likely control a larger portion of a pest population. However, these formulations tend to be more destructive to natural enemies and other nontarget organisms or potentially contaminate the environment.

Different formulations of the same pesticide and different types of packaging, such as water-soluble bags (Figure 5-33), also vary in cost, safety, and the type of control desired. They present different challenges for worker safety and exposure to people, livestock, and pets in the area of the application. The potential for *phytotoxicity* (damage to the crop plant) can be a concern with some formulations. For example, pesticides formulated with oil tend to cause more phytotoxicity, especially to water-stressed plants. On the

TABLE 5-15.

Common formulations of pesticides.

Type (abbreviation, if used)	Description	Comments
wettable powders (W, WP)	Active ingredient combined with finely ground dry carrier, usually clay; contains between 15 and 75% active ingredient.	Less phytotoxic; mixes well with other pesticides and fertilizers; dust inhalation hazard.
dry flowables (DF); water-dispersible granules (WDG)	Active ingredient incorporated with emulsifiers and other enhancers; must be mixed with water before use; higher percentage of active ingredient per unit of weight.	Fewer mixing and loading hazards; less phytotoxic; abrasive to application equipment.
soluble powders (SP)	Active ingredient, carrier, and all other ingredients completely dissolve in water to form true solution; only a few pesticides available in this formulation.	Agitation not required; dust inhalation hazard.
emulsifiable concentrates (EC)	Petroleum-soluble pesticides formulated with emulsifying agents and other enhancers.	Better at penetrating porous materials such as soil, fabrics, paper, and wood.
flowables (F, FL)	Finely ground pesticide particles mixed with liquid, along with emulsifiers, to form a concentrated emulsion; used when active ingredient is an insoluble solid and does not dissolve in water or other solvents.	Less inhalation hazard than powders; hazard potential from spills and splashes; abrasive to application equipment.
water-soluble concentrates or solutions (S)	Liquid pesticide formulations that dissolve in water; only a limited number of pesticides dissolve in water and are formulated as solution.	Does not require agitation.
low-concentrate solutions (S)	Small amount of active ingredient dissolved in organic solvent.	Ready to use; useful for structural and institutional pests.
ultra-low-volume concentrates (ULV)	Highly concentrated—between 80 and 100% active ingredient.	Applied with little or no dilution.
slurry (SL)	Thin, watery mixture of finely ground dusts.	Bordeaux mixture is commonly used slurry; applied to plants as fungicide.
fumigants	Solid, liquid, or gas in form; solids and liquids volatilize into gas after application; gases at room temperature are packaged under pressure in steel cylinders and metered into treatment area through valves and hoses.	Used to treat pests of stored products; soil pests such as weeds, insects, nematodes, and microorganisms; and to control vertebrate pests such as ground squirrels and gophers. Also used in structural pest control.
invert emulsions	Liquid formulations having small water droplets suspended in oil.	Used to reduce drift; also serves as spreader-sticker.
dusts (D)	Finely ground pesticide combined with inert dry carrier; contains from 1 to 10% active ingredient.	Effective in hard-to-reach indoor areas; used in seed treatment, around homes and gardens, and on pets, livestock and poultry; presents drift hazards in agricultural applications.
granules (G)	Pesticide and carrier combined with binding agent; range in size from 4 to 80 mesh (the size of screen that particles can pass through).	Used for control of aquatic pests such as algae, weeds, or fish; applied to soil for control of weeds, nematodes, and soil insects; used for systemic insecticides.
pellets (P, SP)	Identical to granules except manufactured into specific uniform weights and shapes.	Same applications as granules; applied with precision planters.
microencapsulated materials	Pesticide particles (liquid or dry) enclosed in tiny plastic capsules.	Residential, industrial, and institutional applications; hazard to bees in agricultural applications.
baits	Pesticides combined with food, attractant, or feeding stimulant.	Used indoors for control of rodents, ants, roaches, and flies; outdoor use includes control of slugs, snails, insects, and vertebrates such as birds, rodents, and larger mammals.
impregnates	Pesticides incorporated into household and commercial products.	Used in pet collars, livestock ear tags, adhesive tapes, and plastic pest strips.

(Table continued on next page)

Type (abbreviation, if used)	Description	Comments
aerosol and lotion formulations	Repellents.	Used as insect repellents, applied to skin or clothing; others mixed with water, used as deterrents.
animal systemics	Administered to pests and livestock as food additive; sometimes applied to skin of animal.	Used for controlling animal parasites such as insects, fleas, and mites.
fertilizer combinations	Insecticides, fungicides, and herbicides combined with fertilizer.	Used in landscape situations and by homeowners.
water-soluble packets	Preweighed amount of formulated pesticide powder packaged in water-soluble plastic bags.	Used to reduce mixing and handling hazards.
attractants	Various substances, including pheromones, sugar and protein hydrolysate syrups, yeasts, and rotting meats.	Used to attract pests in sticky traps or combined with pesticides for control.
aerosol containers	Pesticide combined under pressure with a chemical propellant in disposable or refillable self-dispensing can.	Used in residential, industrial, and institutional pest control.

other hand, wettable powders are among the safest formulations to use when phytotoxicity is a concern, but the potential hazard for inhaling dust during mixing can be very high. An alternate choice would be dry flowable formulations, which, like wettable powders, are less likely to be phytotoxic, but which do not have the dust associated with wettable powders. Handling and mixing are easier because dry flowables are packaged in easy-to-pour containers and can be measured by volume as well as weight.

When selecting a pesticide, drift, runoff, wind, and rainfall must be considered, along with soil type and characteristics of the surrounding area. Also, be sure that the formulation is compatible with available application equipment. Most of this information is on the label. Prior to considering the use of a pesticide, READ THE LABEL.

Mode of Action. The mode of action is the mechanism by which the pesticide kills or controls the target organism. Most pesticides act by interfering with a metabolic process. Some pesticides cause physical damage to the target organism; for example, desiccants remove the waxy coating on certain insects, resulting in water loss. Generally, pesticides in the same chemical class have similar modes of action on specific types of pests. For example, all organophosphate insecticides attack the nervous system of insects.

Knowing the mode of action will help you assess which pesticide will work best in a particular situation. It can also provide a clue as to which nontarget organisms may be affected along with other pests. When trying to reduce selection for pesticide resistance, it is helpful to rotate among pesticides that have different modes of actions (see the section "Resistance Management Strategies"). Different life stages of a pest commonly have different susceptibilities to pesticides due to biological and physical characteristics. Applying a pesticide at a time when the most vulnerable life stage of the pest is present increases the effectiveness of control.

Site of Action. The site of action is the primary location of the enzymatic, metabolic, or physical reaction caused by the pesticide. The site of action is important in determining which pests may be controlled and when they may be most vulnerable to an application. However, for many pesticides, the exact site of action is not known.

Persistence. Persistence indicates the amount of time it takes for a pesticide to degrade. Persistence is measured in terms of *half-life*—the amount of time it takes for the material to be reduced to half the amount of active ingredient originally applied. The half-life of a pesticide may be affected by its chemical nature or formulation and environmental factors such as soil microbes, ultraviolet light, heat, and the pH of water used in mixing. In some situations where reinfestation is likely to occur or long-term pest control is desired, persistence may be advantageous, but prolonged persistence can also increase

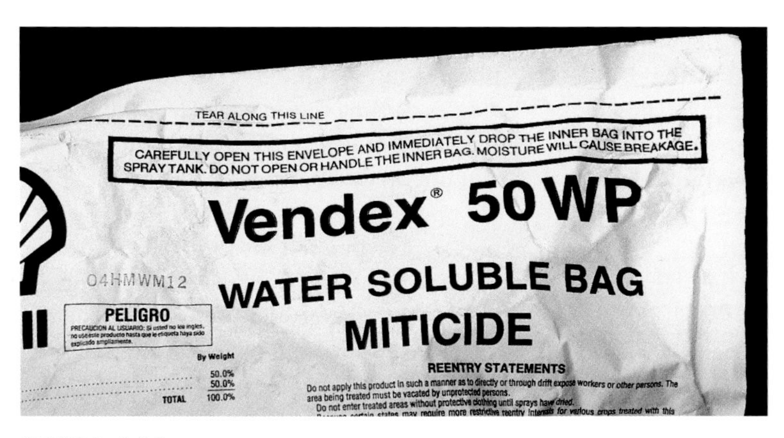

FIGURE 5-33

Water-soluble bags protect handlers during mixing of some types of highly toxic or hazardous pesticides. A preweighed amount of formulated pesticide powder is contained in a plastic bag inside this paper envelope. The bag is removed from the envelope and dropped into a water-filled spray tank. The plastic bag will dissolve in the water and release the powder.

hazards to people, wildlife, beneficial insects, and honey bees. Highly persistent pesticides select for resistance faster.

The breakdown of a pesticide does not always mean that hazards associated with it have been eliminated. Some pesticides break down into other more toxic materials. For instance, the low-toxicity insecticide acephate may break down into the more toxic methamidophos.

FIGURE 5-34

Broad-spectrum persistent pesticides can sometimes be used selectively. Spot spraying a band around the trunk (bark banding) kills elm leaf beetle larvae as they crawl down to pupate near the trunk base.

Selectivity. Another criterion to consider in pesticide use is selectivity—the range of organisms and life stages of organisms affected by the pesticide. A broad-spectrum pesticide kills a wide range of pests and nontargets, whereas a selective pesticide controls a smaller group of closely related organisms. Selective pesticides are generally desirable in IPM programs because they often have less impact on beneficial organisms and lower risks for humans and wildlife. Selectivity that prevents damage to a crop plant is a key feature of any herbicide applied while the crop is in the field.

Selective pesticides target chemical processes unique to one pest or pest group. Selectivity also is influenced by the rate of penetration of the toxicant, the binding of the toxicant to the organism's tissues, and the speed with which the organism breaks down the toxicant.

Selectivity can also be achieved through the use of application techniques that cause a pesticide to come into contact with the target pest and not with nontarget organisms. For example, spraying only the trunks of elm trees to control elm leaf beetle larvae as they crawl down from the tree canopy to pupate in the soil leaves the beneficial species in the foliage unharmed (Figure 5-34). Another example is the use of shields and wipers to direct herbicide applications away from susceptible plants.

Pesticide Toxicity. Toxicity is the capacity of a material to cause injury to organisms. All pesticides are toxic to some organisms; each pesticide has a toxicity rating that suggests the pesticide's relative hazard to people and other organisms in the environment. The toxic effect of a pesticide in any given situation is influenced by various conditions including temperature, humidity, and exposure to sunlight, wind, and rain. The age and general health of the pest or other organism can also influence how a pesticide affects it. For example, water-stressed weeds may tolerate some herbicides better than nonstressed weeds. Environmental factors sometimes change the chemical composition of pesticides, making them more or less toxic than the original pesticide. Some pesticides may be highly toxic to some organisms but have low toxicity to others. The ideal pesticide would be toxic only to the target pest. Chapter 8 provides a detailed discussion of pesticide toxicity to humans and other nontarget organisms.

Selecting the Right Pesticide for Use in an IPM Program

Properly used, pesticides can provide effective and economical protection from pests that otherwise would cause significant damage. In an IPM program, pesticides are used only when field monitoring or other site-specific information indicates they are needed to minimize economic losses. Learn what materials are available and the properties of each.

Pesticides are categorized in several ways. The most common is according to type of pest controlled—such as insecticides, herbicides, fungicides, rodenticides, or nematicides. They may also be classified according to mode of action or according to chemical class (such as organophosphates, pyrethroids) or source of material (such as *botanicals*, microbials, inorganics, fixed coppers). Some pesticides are categorized as systemic or nonsystemic according to how far they move within the treated plant or animal from the site of application. Contact pesticides have no systemic activity and must touch the insect or plant part to injure it. Pesticides with translaminar activity are absorbed short distances into tissues and have activity there.

In choosing a pesticide, consider not only its effect on the target pest, but also the effects it may have on other pests, natural enemies, honey bees, people, the environment, and the protected commodity. Consider special attributes of the site such as soil type, surrounding crops, rivers, wildlife, human habitation, and the potential for worker exposure. This section discusses specific criteria to consider when choosing different types of pesticides—insecticides, herbicides, nematicides, fungicides, *bactericides*, and vertebrate control materials. Chapter 8 provides more in-depth information on *toxicology* and protection of human health and the environment.

Insecticides. Key to selecting an insecticide for an IPM program is identifying the species causing the damage, determining the life stage most effectively controlled, and timing the application for that window of opportunity. Many insecticides are fairly broad-spectrum poisons. It is therefore equally important to consider the potential impact of the insecticide on beneficials, nontarget organisms, and people in the area, as well as its potential for moving off-site and posing hazards elsewhere.

All insecticides should be selected based on the site and pest specifics of the situation. Emphasis should be placed on choosing the least-toxic material that will effectively manage the pest population. In an IPM program, preservation of natural enemies is a high priority, so selective materials should be chosen, whenever feasible, to improve overall pest control. Formulations, application methods, and timing can further improve control and selectivity, and can reduce hazards to people and the environment.

Certain insecticides are often referred to as *contact poisons* or *stomach poisons*. A contact insecticide provides control when target pests come in to physical contact with it. Stomach poisons must be ingested to affect the pests. For instance, if a pest feeds on the underside of leaves, an application to the upper surface will not be effective unless the material is a systemic that can move in the plant. Many insecticides have both contact and stomach activity.

Systemic insecticides are taken up by the crop, plant, or animal and move, after application, to other tissues. Systemic insecticides may also be fed or applied to pets and livestock for control of external and internal parasites. On plants, systemics may be applied to the soil to be taken up by the roots or applied to foliage and transported to the leaves and stems, where they kill feeding insects. Sometimes beneficial insects feeding on nectar from treated plants may be affected as well.

The differences in the chemistry of insecticide compounds should be considered. Chemical classes reflect the chemical makeup of insecticides. Over the last 3 decades of the twentieth century, the most widely used insecticide chemical classes were the organophosphates and the N-methyl carbamates. Organophosphates and N-methyl carbamates disrupt nerve activity and are toxic to a broad spectrum of animal species. However, different materials vary substantially in their persistence in the environment and in their toxicity to people and other nontarget organisms. Partly because of environmental and health concerns associated with many of the organophosphates and carbamates, new types of insecticides have been increasingly used.

The pyrethroid class of insecticides includes many widely used insecticides such as resmethrin, permethrin, fluvalinate, and many other compounds. These materials are synthetic compounds based on the chemical and physiological action of natural pyrethrins, which are plant-derived insecticides. Pyrethroids disrupt nerve activity by stimulating nerve cells and eventually causing paralysis. They are much more toxic to insects and generally more persistent in the environment than the natural pyrethrins. These materials are generally effective at much lower rates than organophosphate materials.

The neonicotinoids are an expanding insecticide chemical class. These products are nerve poisons with a similar mode of action to the botanical chemical nicotine. Many of these materials are systemics that move through the plant. Common active ingredients include imidacloprid, dinotefuran, acetamiprid, and thiamethoxam. These products are generally broadly toxic against many arthropods, including honey bees.

The chlorinated hydrocarbons, or organochlorines, once widely used, is another well-known chemical class. Very few organochlorine insecticides are currently registered for insect control in the United States. Most have been removed from the market because of their long persistence and negative impact on wildlife. Well-known organochlorine insecticides that are no longer available include DDT, chlordane, aldrin, dieldrin, endrin, mirex, and toxaphene. The few remaining organochlorine insecticides include endosulfan and lindane; both pesticides are restricted-use pesticides, and most uses of these materials are likely to be curtailed in coming years.

The chemical class, which provides clues to the material's specificity, toxicity, and mode of action, has been a traditional way of classifying insecticides. However, a number of insecticides, especially many of the newer biorational types of materials, do not fit into chemical class groups because they are not closely related chemically to other insecticides. (For classification of insecticides according to chemical class and mode of action, see the Insecticide Resistance Action Committee website, www.irac-online.org.) On the other hand, some materials can be grouped according to other commonalties in their origin or mode of action. Some commonly designated insecticide groups include soaps and oils (fatty acids), abrasive dusts, botanicals, *insect growth regulators*, and microbials.

Insecticidal oils, soaps, abrasives, and desiccants act as physical toxicants: they smother or desiccate the pest. They are broad-spectrum in that they kill many soft-bodied insects and mites on contact. Oils and soaps are also considered somewhat selective because they do not leave toxic residues.

Botanicals are natural products that plants produce to protect themselves from insect attack. Some of these plants are cultivated to harvest natural botanical insecticides. Well-known botanicals include natural pyrethrins from the pyrethrum daisy, nicotinoids from tobacco, rotenoids from many plants including *Derris* and *Lonchocarpus* spp., sabadilla from *Schoenocaulon officinale*, and azadirachtin from the neem tree. The modes of action of many of these natural products are not completely understood.

Insect growth regulators (IGRs) are synthetically produced chemicals used to control insects by interfering with the insects' normal development. Some inhibit development by mimicking hormones in the insect's body or inhibiting chitin synthesis needed to provide skeletal structure when the insect molts. These insecticides disrupt egg hatch and molting. Examples include hydroprene (cockroaches), pyriproxyfen (fleas and scale insects), and diflubenzuron (caterpillars).

Microbial insecticides arise from microorganisms that cause disease in insects (or other pests) and have been combined with other ingredients to form pest control products. The most commonly available microbial insecticide is *Bacillus thuringiensis* (Bt) (Figure 5-35). Different strains of Bt are available to control moths and butterflies (*B. t. aizawai, B. t. kurstaki*) and mosquitoes and fungus gnats (*B. t. israelensis*) in agricultural crops and ornamentals. Development of new strains of Bt for use against other pests is continuing. Codling moth granulosis virus is another microbial insecticide that has recently become widely available for use in orchards and home gardens. Other bacteria, viruses, and fungi are also registered as insecticides and may become more widely used in the future. Because each pathogen attacks only a group of related species, microbials are highly selective insecticides.

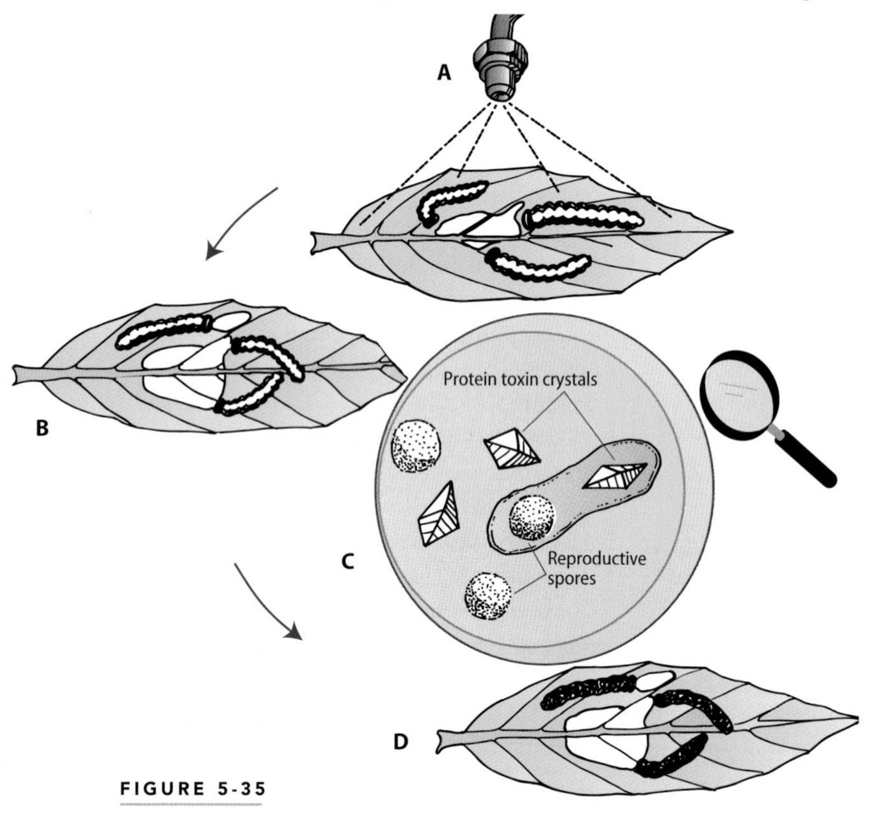

FIGURE 5-35

Several subspecies of Bacillus thuringiensis *(Bt) are widely used to control certain insects. The Bt subspecies* kurstaki *controls moth and butterfly caterpillars: (A) Bt must be sprayed during warm, dry conditions to thoroughly cover foliage where young caterpillars are actively feeding. (B) Within about 1 day of consuming treated foliage, caterpillars become infected, relatively inactive, and stop feeding. (C) An enlarged view of Bt in the gut of a caterpillar. The natural bacteria are rod shaped and contain reproductive spores and protein toxin crystals (endotoxins). The spores and protein crystals are separate components and one whole bacterium (greatly enlarged) is shown here. (D) Within several days of ingesting Bt, caterpillars darken and die, and their carcasses eventually decompose into a dark, liquid, putrid mass.* Source: *Flint and Dreistadt 1998.*

Some new types of microbially derived insecticides have recently become available. These products, including the avermectins (abamectin) and the spinosyns (spinosad), are natural fermentation products of bacteria. Both these products affect the neurological systems of arthropods but are selective for certain pest groups, with minimal impact on most natural enemy species. Abamectin has been used especially against thrips, mites, pear psylla, and leafminers. Spinosad is used primarily against caterpillars and thrips. While the mammalian toxicity of spinosad is quite low, the mammalian toxicity of abamectin is much higher, and abamectins are also highly toxic to honey bees.

Each chemical class or insecticide group is unique in terms of the target pests it controls and its impacts on the environment. However, some generalities may apply. The following factors are important in the selection of any insecticide:

- mode of action
- site of action
- persistence
- selectivity
- mammalian toxicity
- movement and impact in the environment

Herbicides. Herbicides are chemicals that kill plants. When combined with good cultural practices, herbicides control many weed species and are an important component of an integrated weed management program. Application method and timing are critical for the success of an herbicide application. For instance, in field or vegetable crops, herbicides may be mechanically mixed into the soil (preplant incorporated) before planting, or irrigated into it after planting before (preemergence) or after (postemergence) the crop or weed penetrates the soil surface. Figure 5-36 shows the different types of herbicide applications and the stages of weed growth on which they are properly applied.

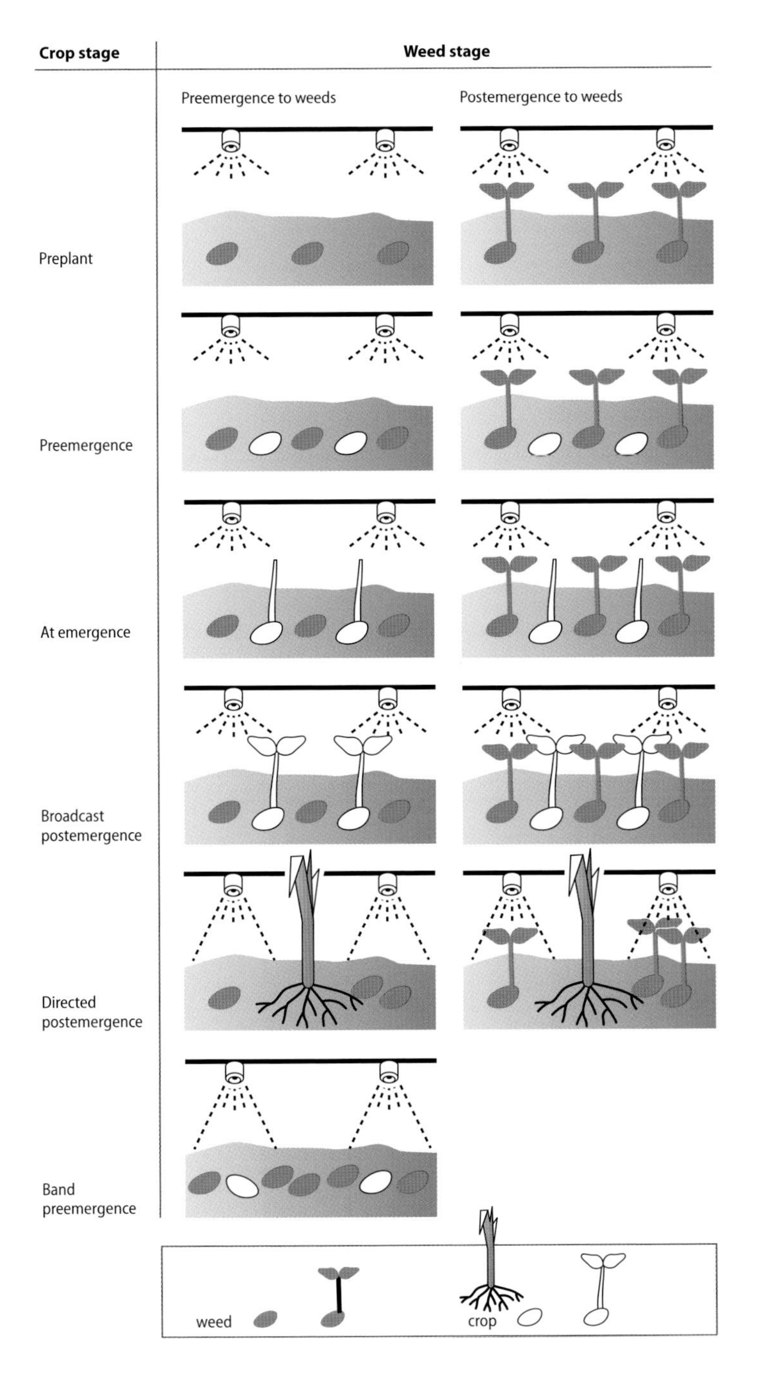

FIGURE 5-36

Terms used in describing application timing of herbicides in relationship to crop growth and weed emergence. Adapted from Fryer and Evans 1968.

Herbicides are classified in several ways, generally relating to how they affect plants or which types of plants they affect. Selective herbicides kill certain types of plants (such as broadleaf or grass weeds) without killing the crop or desired plant; nonselective herbicides kill all types of vegetation. Contact herbicides kill primarily the plant parts on which the herbicide is applied; *translocated herbicides* are absorbed by the roots or aboveground plant tissue and move throughout the plant, affecting the entire plant.

Proper choice of herbicide depends on the weed species or species complex, the susceptibility of the weeds to the material, and the tolerance of the crop and surrounding desirable vegetation to the material. Weed infestations are usually composed of a complex of species. Use a weed susceptibility chart to help select an herbicide or mix of herbicides that offers the most effective control. Figure 5-37 is an example of a weed susceptibility chart for weed species found in stone fruits. Plantback restrictions should also be considered when selecting herbicides. Because herbicide soil residuals can limit the growth of sensitive rotational crops, plantback restrictions are legal label mandates that must be adhered to. Follow the label instructions regarding use and plantback restrictions.

Before finalizing the herbicide selection, always consider the pesticide mode of action, selectivity, persistence, site of action, mammalian and nontarget organism toxicity, resistance, and environmental concerns. In addition, soil type and preparation affect the action of many herbicides. Specifics for individual herbicides can be found in the Weed Science Society of America's *Herbicide Handbook* (see the "References" section).

Mode of Action. Herbicides kill plants by interfering with various vital plant processes. For example, they may interfere with photosynthesis, inhibit root or shoot growth, or induce abnormal plant tissue development. Often, closely related herbicides exhibit the same mode of action, but there are exceptions. Knowing the mode of action helps in recognizing damage symptoms caused by a specific herbicide on target weeds as well as on nontarget plants. Modes or mechanisms of action for specific herbicides can be found on the Herbicide Resistance Action Committee website, at www.hracglobal.com, or from the Weed Science Society of America.

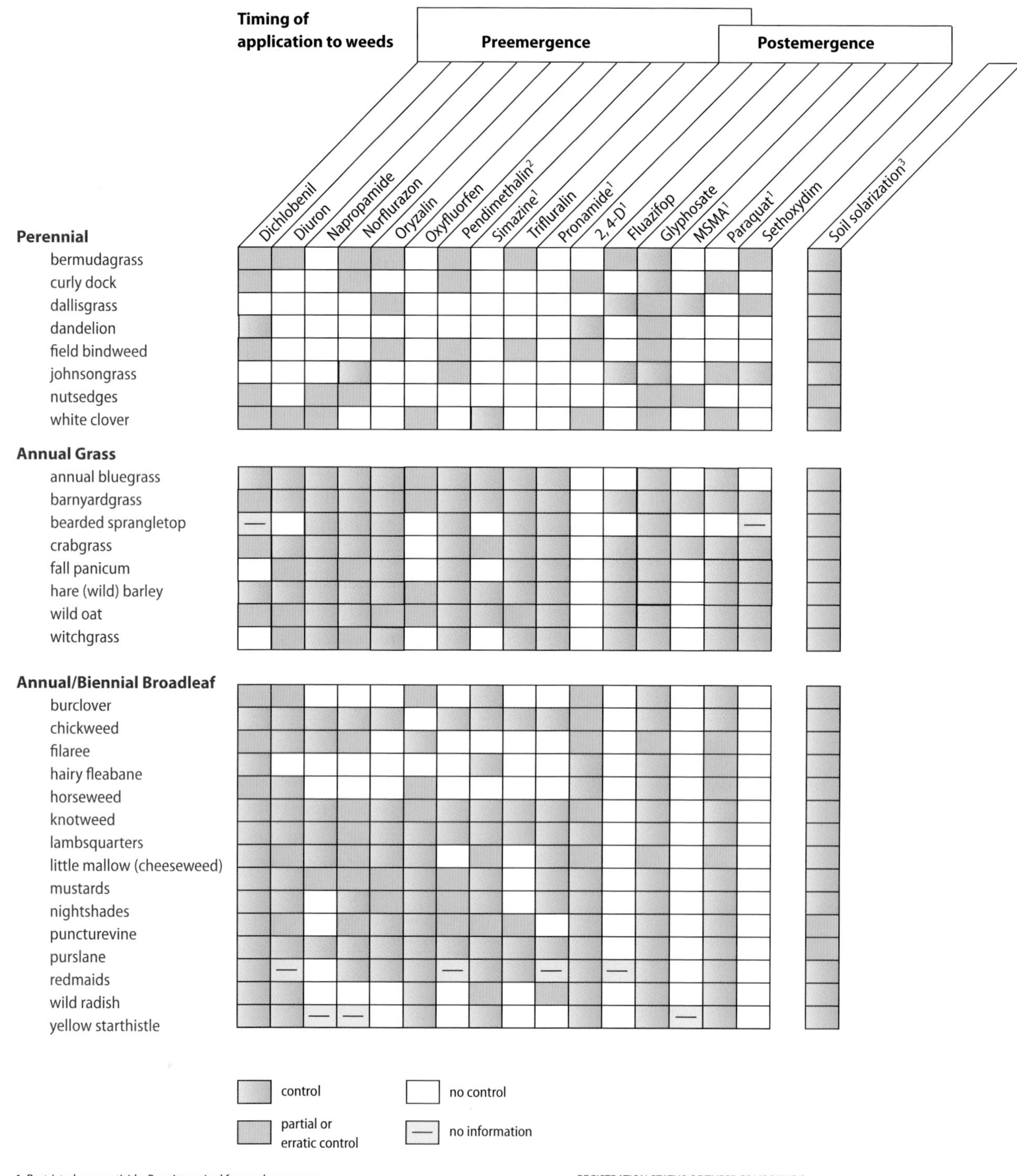

1. Restricted-use pesticide. Permit required for purchase or use.
2. Nonbearing orchards only.
3. Solarization controls perennial structures in upper few inches of soil.

REGISTRATION STATUS OF THESE COMPOUNDS MAY CHANGE. CHECK WITH LOCAL AUTHORITIES BEFORE USING.

FIGURE 5-37

Example herbicide selectivity table for stone fruits. Herbicide efficacy tables are regularly updated online at the UC IPM website, www.ipm.ucdavis.edu.

Selectivity. Herbicides that kill weeds but leave desirable plants unharmed are called selective. For instance, certain herbicides kill most broadleaves but leave grassy plants such as turf or grain crops unharmed. Know the identity of the weed species to take advantage of herbicide selectivity.

Selectivity can also be achieved by placing materials away from crops (Figure 5-38) or applying them before crop plants emerge or after they are vulnerable to damage. Proper timing and placement, therefore, is essential for the successful use of herbicides. Applying no more than the prescribed rate is also important since many herbicides lose selectivity when applied at too high a rate, and applying herbicides above label rate is illegal.

Persistence. Herbicide persistence varies substantially. Many soil-applied herbicides are designed to remain toxic to weeds for several weeks or months after application to provide weed control over an extended time. Persistence of an active residue of such a material may pose a hazard if susceptible crops are to be planted before the material degrades or if the herbicides accumulate in groundwater or water supplies.

Soil type can affect both the persistence and the effectiveness of many herbicides. Some herbicides are held tightly in soils that have high clay content or high levels of organic matter and thus are not as available for weed control. As a result, higher rates may be required on these soils for effective weed management or their use may not be recommended at all (Table 5-16). Herbicide labels specify these differential rates.

Site of Entry and Translocation. Soil-applied herbicides enter plants through roots, shoots, seeds, rhizomes, bulbs, or tubers. The primary site of entry for foliar-applied herbicides is the leaf. Once inside the plant, the herbicide must reach a specific susceptible site. Contact herbicides typically kill only the tissue that is sprayed.

Translocated herbicides move in the plant apoplast, the system of spaces between cells and cell walls (including the xylem) that conducts water and nutrients, or they move in the symplast (phloem) with the sugars (Figure 5-39). Apoplastically translocated herbicides move upward and outward (they cannot move down the plant) and affect older leaves first. Symplastically translocated herbicides become distributed throughout the plant. Immediate effects include contact burn and distortion, and chronic effects

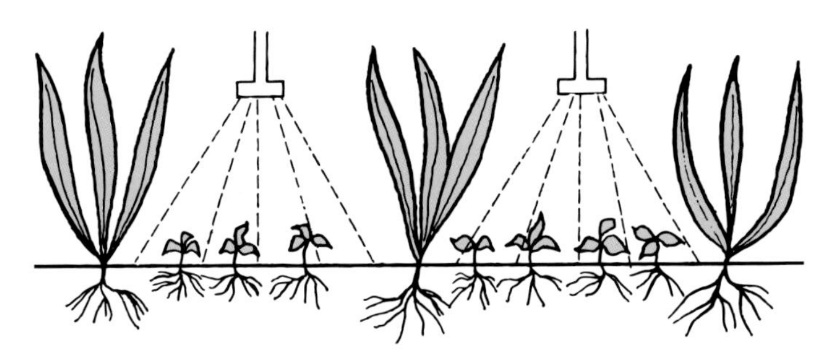

FIGURE 5-38

Selectivity in herbicide applications can be achieved by directing the spray away from the desired plant toward the weed species. Source: Ashton and Harvey 1987.

TABLE 5-16.

Preemergence broadcast application rates per acre for a sample herbicide. Some herbicide labels recommend different application rates based on soil texture and the percentage of organic matter found in the soil.

	Brand X 80W herbicide applied (lb) taking into account % organic matter in soil					
Soil texture	**Less than 1%**	**1%**	**2%**	**3%**	**4%**	**Over 4%**
coarse (sand, loamy sand, sandy loam)	do not use	1.5	2.0	2.5	3.0	3.5
medium (loam, silt loam, silt, sandy clay, loam)	2.0	2.5	3.0	3.5	4.0	4.5
fine (silty clay loam, clay loam, sandy clay, silty clay, clay)	2.5	3.0	3.5	4.0	4.5	5.0
peat or muck	not recommended					

Source: Ross and Lembi 1999.

include inhibition of new shoot tips, young leaves, and buds. Understanding the site of entry and whether (and how) herbicides are translocated is essential when choosing herbicides and also aids in determining appropriate application timing and herbicide placement.

Mammalian Toxicity. Most herbicides have quite low acute mammalian toxicity; they are relatively safe to the user, wildlife, and the environment. However, there are notable exceptions, such as paraquat. The acute mammalian toxicity of paraquat is very high, and, as a result, it is a restricted-use pesticide.

Environmental Concerns. A hazard associated with most herbicides is the potential for drift or runoff onto other areas, where nontarget plants may be damaged. Some herbicides are quite persistent or are easily leached into groundwater, rivers, or other bodies of water where they may pose a hazard to plants or wildlife. Leaching is influenced by the characteristics of the herbicide and the soil type. Coarse-textured soils, such as sandy loam, are prone to leaching. The California Department of Pesticide Regulation's Ground Water Protection Program may restrict the application of certain herbicides in some growing areas where leaching to groundwater has been a problem. Always consider the potential for runoff, *volatilization*, and drift when selecting a material and application method and check with regulatory officials for local regulations.

Fungicides and Bactericides. Pesticides used for disease control fall primarily into three categories: bactericides, fungicides, and *fumigants*. A few bactericides are available to control bacterial plant diseases. They consist primarily of copper compounds and certain antibiotics, such as streptomycin. Natural bactericides are also available, such as *Agrobacterium radiobacter* strain K84 for the control of crown gall in fruit trees, *Pseudomonas fluorescens* strain A506 for the control of fire blight in pears, and *Bacillus subtilis* used against a range of fungal pests including those causing powdery mildew and Alternaria leaf blights.

Many fungicides are available to aid in managing a variety of fungal diseases (Table 5-17). New products are developed every year. An up-to-date list of fungicides and modes of action is maintained by the Fungicide Resistance Action Committee at www.frac.info.

Fungicides can be grouped into two basic classes: *eradicants* and *protectants*. Most fungicides are surface protectants that must be

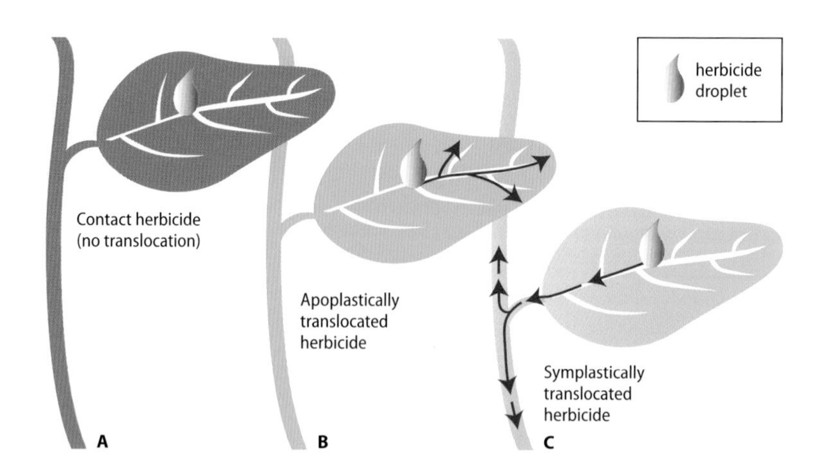

FIGURE 5-39

(A) Contact herbicides, when applied to a leaf, kill the contacted tissue immediately and do not move from the application site. (B) Apoplastically translocated herbicides move outward to the edges of the leaf and up the plant. (C) Symplastically translocated herbicides move back into the stem and are distributed to the points of active growth (meristematic tissue) or food storage (underground structures). Source: *Ashton and Harvey 1987.*

applied before the fungal spores germinate and enter the plant. Their presence usually prevents the spores from germinating or kills the spores once they germinate. If infection occurs, protectants cannot prevent disease.

Eradicants directly affect pathogens after they have invaded the plant tissue by killing the fungus inside the host or by suppressing sporulation of the fungus. Many eradicant fungicides are systemics that are absorbed into plant tissue and translocated to other parts of the plant; they can control established infections to a limited extent. Systemics are not as susceptible to weathering as surface protectants, but because they usually act on specific metabolic processes, they are more susceptible to resistance. Because of their effectiveness as eradicants, systemic fungicides are beginning to replace protectants.

Antibiotics are substances produced by microorganisms that inhibit growth and are toxic to other microorganisms. A number of

TABLE 5-17.

Common types of chemicals used for plant disease control

Chemical Type	Examples	Activity
Antibiotics		
	streptomycin, terramycin	systemic
Biologicals		
	Bacillus subtilis, Bacillus pumilis. Trichoderma spp., Pseudomonas fluorescens A506	contact
Protectants		
inorganic chemicals	copper compounds: Bordeaux mixture, fixed copper	contact
	sulfur compounds: sulfur, lime sufur	contact
	carbonate compounds: potassium bicarbonate	contact
	film-forming compounds: mineral oils	contact
organic chemicals	dithiocarbamates: thiram, maneb, mancozeb	contact
	aromatic hydrocarbons: dicloran, chlorothalonil	contact
	heterocyclic compounds: captan, iprodione, vinclozolin	contact
Eradicants		
benzimidazoles	benomyl, thiabendazole, thiophanate-methyl	systemic (local)
oxanthins	carboxin	systemic (local)
phosphonates	fosetyl-Al	systemic
pyrimidines	fenarimol	systemic (local)
triazoles	difenoconazole, myclobutanil, propiconazole, triadimefon	systemic
miscellaneous systemics	chloroneb, imazalil, propamocarb, triforine	systemic

antibiotics, such as streptomycin and tetracycline, have been formulated as bactericides and are registered only on a few crops. They act on the host or pathogen. Antibiotics are absorbed by the plant and, to a limited extent, may be translocated systemically.

When selecting a fungicide or bactericide, it is important to choose the most effective material for the given situation. Protectants have limited usefulness once an infection has started. Precise timing of pesticide applications is probably more important in the control of plant pathogens than with any other type of pest. Keep track of weather, especially temperature and moisture, because disease epidemics can build rapidly after changes in the weather. Repeated applications are frequently required, depending on the persistence of the material. Systemic fungicides reduce the need for repeated applications.

To date, fungicides have not been identified as sources of environmental problems nearly as much as have insecticides and herbicides. This is partially because of their generally rapid breakdown in the environment, with the exception of copper compounds. Also, fungicide modes of action against microorganisms make them less likely to impact mammals, fish, birds, and higher plants. Few fungicides have high acute mammalian toxicity. However, some have been identified as potential carcinogens and mutagens and have raised consumer concerns because they can persist on harvest food crops for some time. Processors and certain markets do not allow the use of some of these materials.

Nematicides. The number of nematicides available is very limited. Effective nematicides must be able to move throughout the soil profile where nematodes reside. Soil texture, moisture, organic matter, pH, temperature, soil profile variability, adsorptive characteristics of the toxicant, application method, and application rates also influence successful control. As broad-spectrum pesticides, most of the available nematicides have substantial potential for environmental and human hazard if not used with great care. Some nematicides have been associated with groundwater contamination, and others are volatile organic compounds (VOCs) that can contribute to smog.

Nematicides can be generally classified as fumigants or nonfumigants. Most fumigants must be applied preplant, before the crop or tree is planted, and they are generally much more effective at controlling nematodes than nonfumigants. Common fumigants historically used for nematode control include methyl bromide, metam sodium, and 1,3-dichloropropene; however, these materials have been increasingly restricted because they can be sources of volatile organic compounds (VOCs) and also pose health concerns. Nonfumigants, such as fenamiphos and sodium tetrathiocarbonate, may be applied at or after planting. Irrigation or rain to disperse the material into the soil profile must follow application or materials can be applied directly in irrigation water.

Before deciding to use a nematicide, it is critical to know the nematode species present and to have estimates of their populations. Species of nematodes differ in their sensitivity to nematicides, and within a single species, certain stages are more sensitive than others. Juveniles and stages during a molt are more sensitive than adults to fumigation, while eggs are less sensitive than adults.

Rodenticides and Other Chemicals to Control Vertebrate Pests. The most commonly used pesticides available for vertebrate pest control include burrow fumigants, strychnine, anticoagulants, chemosterilants, and repellents. Chemicals used as fumigants for vertebrate control include several different types of toxic gases. Formulated as gas cartridges or volatile tablets, they are designed to control species that burrow or have dens. Strychnine is a single-dose poison widely used to control gophers. It is one of the fastest-acting rodenticides and is sometimes formulated as bait.

Anticoagulants interrupt the target organism's blood-clotting ability, causing death. Chemosterilants are used to control the population of the target species by affecting their ability to reproduce. Repellents are available for rodents, cats and dogs, deer, rabbits, and birds, but are not widely used in commercial agriculture.

Many types of pesticides are formulated as baits to control a wide variety of vertebrate pests. Baits are primarily dry formulations consisting of whole or pelleted grain mixtures treated with a toxicant. Some baits are formulated to kill in one feeding. Some are formulated to be consumed during several consecutive feedings over a period of 5 days or more in order to kill the rodent, which reduces the risk of nontarget animals being killed.

Key to selecting and using the appropriate pesticide for vertebrate control is accurate identification and knowledge of the habits and biology of the species causing the damage. Often, the pest is not present when damage is observed, and identification must be made from damage symptoms such as chewed bark or from other signs of the pest's presence such as droppings, tracks, burrows, nests, or food caches.

Pesticide selection should be appropriate to the location, time of year, and the environmental conditions specific to the site. Consider the potential for poisoning people and nontarget species and secondary poisoning of predators and domestic animals. When using pesticides to control vertebrate pests, consider the effectiveness of the material compared to other methods. Pesticides should be combined with appropriate habitat modification techniques whenever possible. Check with the local county department of agriculture to find out about vertebrate pest control materials that can legally and effectively be used in each site. Vertebrate control materials can pose special hazards for endangered vertebrate species. In California the Department of Pesticide Regulation provides online maps indicating what endangered species may be located in a given area. Consult this database at www.cdpr.ca.gov/docs/endspec/.

Adjuvants. Adjuvants are materials added to a pesticide formulation to enhance its performance, customize the application to site-specific needs, or compensate for local conditions. Formulation adjuvants are mixed with the pesticide active ingredient during packaging and formulation. Spray adjuvants are packaged separately from pesticides and can be added to the spray mix before application.

Spray adjuvants may act in two ways: as activator adjuvants or utility adjuvants. Activator adjuvants directly enhance pesticide performance once the spray hits the plant. They include humectants (moisture retention promoters), penetrants, stickers, and wetter-stickers. Utility adjuvants help the spray application process. These adjuvants include acidifiers, buffers, colorants, defoamers, deposition aids, drift control agents, and water conditioners. A single *adjuvant* product can be both an activator and a utility adjuvant.

When recommending the addition of a spray adjuvant to a pesticide formulation, become familiar with the different types of adjuvants and how they should be used. Understand the function that the adjuvant is to perform, then check the pesticide and adjuvant labels to ensure compatibility and suitability to the application site, target pest, and application equipment. In California, spray adjuvants are considered pesticides and must be registered and recorded in pesticide use reports.

Table 5-18 lists various adjuvants and compares their intended effect. Many adjuvants function in more than one way. For instance, buffers are commonly added to lower the pH of water to enhance the effectiveness of the pesticide solution. They also retard the breakdown of the solution. In some situations, this may be desirable; in others, it may not. As with the selection of the pesticide, what occurs after the application (restricted-entry intervals, nontarget exposure) could be affected by addition of a buffer.

Adjuvants are also often used to reduce drift. Some also reduce volatilization, while others do not. This can be an important consideration in the selection of an adjuvant. Special attention to the placement of the pesticide and the application equipment can also effectively reduce or eliminate many drift problems.

TABLE 5-18.

Comparsions of adjuvants.

Function	Surfactant	Sticker	Spreader-sticker	Extender	Activator	Compatibility agent	Buffer	Acidifer	Deposition aid	Defoamer	Thickener	Attractant
reduce surface tension	•		•		•							
improve ability to get into small cracks	•		•									
increase uptake by target	•		•		•			•				
improve sticking	•	•	•									
protect against wash-off/ abrasion	•	•	•	•								
reduce sunlight degradation		•	•	•								
reduce volatilization	•	•	•									
increase persistence			•	•					•		•	
improve mixing						•	•	•				
lower pH							•	•				
slow breakdown							•	•				
reduce drift									•	•	•	
eliminate foam										•		
increase viscosity									•		•	
increase droplet size									•		•	
attract pests to pesticides												•

Selective Application Techniques

The choice of application equipment, the placement of the pesticide, and the timing of the application are almost as important as the selection of the pesticide in the effectiveness of a pesticide application. To use a pesticide safely and effectively, the material must be applied to the treatment area in the proper amount using the proper equipment. Coverage must be uniform and applied at the optimal time for control.

Specific application techniques can be used to improve coverage, reduce drift, and achieve better control of pests. Spot treatments, band applications, and alternate-row treatments (Figure 5-40) can be used instead of spraying an entire area to enhance selectivity and reduce the amount of pesticide used (Table 5-19). The amount of active ingredient applied can also be reduced as long as the lower rate is effective against the target pest and is consistent with the label directions.

Equipment. Pesticide application equipment varies from simple hand-held sprayers to complex self-propelled ground applicators (Figure 5-41), fixed-wing aircraft, and helicopters. Some equipment is designed for specific uses such as weed control or for specific situations such as orchard, vineyard, row crop, field crop, right-of-way, turf, or ornamental applications. Different pesticide formulations, such as dusts, granules, and liquids, require different application equipment. The type of equipment chosen to perform the application can be as important as the selection of the pesticide. In some situations, the equipment available will be the deciding factor in the selection of the pesticide. Check the pesticide label for equipment recommendations. Always know the type of equipment, its availability, the condition it is in, and its capability to perform the task before making a pesticide recommendation.

Placement. The ability of the application equipment to adequately place the pesticide where it is needed is an important consideration in pesticide selection. Aerial applications, high-pressure hydraulic sprayers, low-pressure sprayers, and air blast sprayers (Figure 5-42) are typically used in broadcast applications. In many situations, however, broadcast applications are not necessary and

TABLE 5-19.

Application methods that reduce pesticide use.

Method	Description
spot treatments	Rope wick applications for weed control. Compact hand-held sprayers for application in infested areas only. Treating clumps of perennial weeds as opposed to total field. Treating field edges or landscape areas.
band treatments	Applying herbicides as a band in tree or vineyard rows.
treating alternate rows or blocks	In orchards, vineyards, or field crops, spraying alternate rows when frequent applications are needed.
low-volume application	Improves efficiency: pesticide application volume can be reduced by as much as one third.
reducing dosage level of pesticide	Reduces amount of active ingredient for control of certain insect pests and mites. Should only be used when recommended on the label or by university specialists.

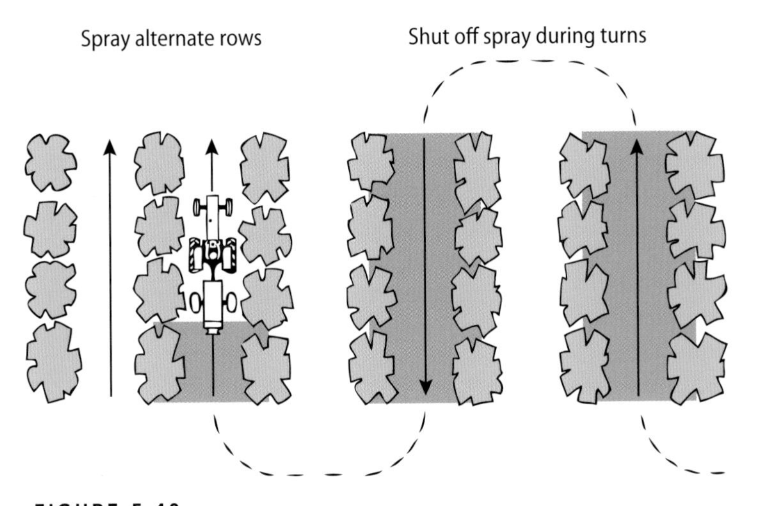

FIGURE 5-40

Applying pesticides to alternate rows or alternate blocks is sometimes used when several sprays are required for control of the same pest. This technique reduces the amount of pesticide used each time by 50%. It also provides untreated locations in the crop for the protection of natural enemies.

FIGURE 5-41

A mechanical bait applicator makes an artificial burrow into which it dispenses poison bait.

often not desirable. Many pest species are located in specific locations in the field or on the host. Spot treatments can be used to selectively place the pesticide where it will be most effective, eliminating unnecessary coverage to uninfested areas, lessening impacts on natural enemies and other nontarget species, and reducing the amount of pesticide used by as much as 90%. For example, in orchards where mite activity is concentrated on the lower leaves, certain nozzles on air blast sprayers can be turned off, directing the application to where the mites are most abundant.

Band treatments are also valuable to direct the pesticide to where it is needed most. Herbicides are often applied as bands within tree or vine rows in orchards and vineyards, reducing the actual amount of herbicide used. Banding of herbicides has also been used effectively in some row crops. Applications of insecticides or acaricides are also made on alternate rows or blocks in orchards, vineyards, and field crops, effectively reducing the amount of pesticide applied and protecting natural enemies in untreated sections.

Application equipment can be adapted or special equipment can be used to selectively apply pesticides. Different nozzle sizes on a sprayer produce different spray patterns. By recommending the use of specific nozzles, the placement of the pesticide can be more precisely directed and drift can be reduced.

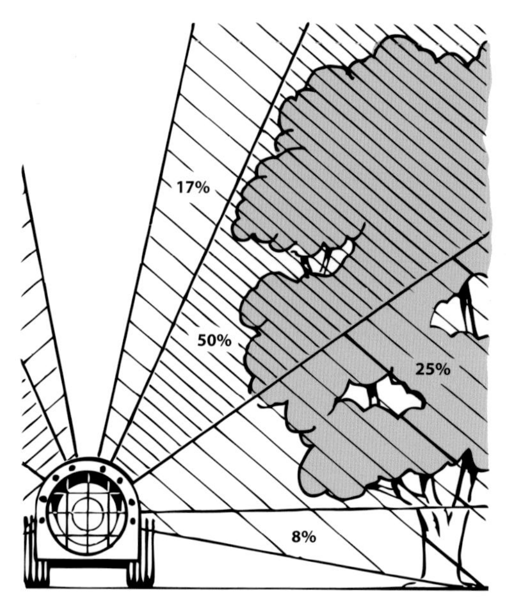

FIGURE 5-42

An orchard air blast sprayer may be adjusted to apply greater amounts of pesticides to some parts of the tree by using different nozzle sizes or increasing the number of nozzles.

Backpack sprayers, wick applicators, and ultra-low-volume (ULV) sprayers can also be used to direct smaller amounts of pesticides to more specific locations. In the case of ULV sprayers, the pesticide solution is highly concentrated; accurate calibration is especially critical and extra safety precautions should be practiced to minimize worker exposure and drift.

For many pesticides, proper placement of the pesticide includes ensuring that the material adequately covers the plant surface. Horticultural oils mainly control pests by suffocating them, so thorough coverage of the plant surface and target insects is vital for optimal control. Many fungicides are applied before infection takes place, so susceptible tissue must be completely covered.

Smart sprayers use sonar or laser technology to detect plant branches or leaves and turn their nozzles on only when plant material is in the path of the nozzle. This reduces the amount of spray that goes offsite, saves growers money, prevents environmental contamination, and keeps pesticide where it should be—on the tree.

Timing. Proper timing of a pesticide application can also allow for the selective use of pesticides. Some pesticides are more effective at different life stages of the target pest. Understanding the biology of the pest will help determine its most susceptible life stages and allow for more effective application timing. The timing of an application can also avoid exposure to nontarget organisms and beneficial species. Some herbicides, for example, may be toxic to the crop or ornamental plant as well as to the weed species during a certain growth stage. Check the label for precautions on using pesticides during inappropriate life stages of the host.

Injury to nontarget organisms can often be avoided by timing the application to periods when nontarget organisms are not present in the treatment area. Honey bees, for instance, forage only during warm, daylight hours. Recommendations indicating that applications should take place during early morning, late afternoon, or on cold, cloudy days reduce bee exposure to pesticides. Likewise, timing insecticide and miticide applications while perennial plants are dormant limits toxicity to spring- and summer-active natural enemies.

PESTICIDE RESISTANCE

Pesticide resistance is a genetic trait a pest individual inherits that allows it to survive an application of a pesticide that kills most other individuals in the population. After surviving the pesticide application, the resistant individual then passes the gene(s) for resistance on to the next generation. The more the pesticide is used, the more susceptible individuals are eliminated and the larger the proportion of resistant individuals grows, until the pest population is no longer effectively controlled (see Figure 3-14, in Chapter 3).

Pesticide resistance is most common in arthropods, with over 500 species of resistant insects and mites reported worldwide. However, pesticide resistance is increasing with other types of pests as well, with certain populations of bacteria, fungi, vertebrates, and weeds resistant to more and more pesticides. Frequently, resistance develops in only certain local populations (biotypes) of a species. Elsewhere in the state, other populations of that species may still be susceptible to that pesticide.

Several mechanisms may be involved in the development of pesticide resistance. Resistant individuals may naturally possess enzymes that break down the pesticide rapidly or have behaviors that reduce their exposure, or the pesticide may not affect them in the same way as other individuals of their population.

Although the challenges presented by pesticide resistance are powerful, resistance can be managed. PCAs must be aware of the factors that influence resistance development and use tools such as monitoring, an understanding of population biology, and integrated control options to limit resistance development. IPM, which relies on a variety of management methods, offers one of the most powerful tools for managing and avoiding pesticide resistance.

Factors Influencing Selection for Resistance

Generally, the development of pesticide resistance is similar for insects, mites, fungi, bacteria, weeds, and vertebrates. It involves a combination of genetic, biological, and other operational factors.

Genetic Factors. Resistance is essentially a result of natural selection. When selection pressure is high, such as with the continuous use of one pesticide, individuals able to tolerate that pesticide will be able to survive and reproduce while others die. Genes that confer resistance may always be present in the pest population but are not expressed unless selection pressure is relatively high. Resistant genes can vary in frequency, dominance, and fitness. They can exist in multiples or in multiple sites; multiple alleles (forms of a gene) are known for many resistant genes. Genes can also mutate to initiate resistance. Mutations of a single gene in plant pathogenic fungi can reduce the attraction of the fungicide to the target site, alter the ability of the fungicide to be absorbed, or increase detoxification.

Mechanisms for resistance include biochemical responses that decrease the sensitivity of the target site (such as nerves), detoxification of the pesticide by enzymes, and reduced penetration of the pesticide through the cuticle or the plant epidermis. Once a pest exhibits resistance to one pesticide, resistance to others may follow more quickly. This phenomenon, called cross-resistance, occurs when the pest is resistant to two or more pesticides, and the same genes mediate the resistance. Multiple resistance occurs when pests have several distinct mechanisms to withstand pesticide chemicals, allowing them to tolerate several classes of pesticides that are unrelated to each other chemically (multiple genes typically mediate multiple resistance).

Populations of resistant individuals within a species are called resistant strains or resistant biotypes. A biotype is any population within a species that has a distinct genetic variation from other populations. Sometimes resistant biotypes may dominate in some areas of the state but not be present in other areas.

Biological Factors. The biology of the pest species influences the rate at which resistance will occur. Biological characteristics include the life span of the pest, its reproductive capabilities, and mobility. Typically, a short-lived, rapidly developing, immobile pest population that produces a large number of offspring will develop resistance rapidly. Resistance evolves more slowly when there are untreated refuges available or when the pest species (insect, pathogen, or vertebrate) is highly mobile. In weeds, resistance is favored by high rates of seed production and germination. High seed production increases the probability of a mutation that may lead to resistance through the process of natural selection.

Operational Factors. Operational factors can be controlled by people. Those that favor resistance include the pesticide's type, persistence, mode of action, and application method (the rate applied, frequency, whether it was mixed with other pesticides, and timing in relation to the dynamics of the pest population).

Management decisions involving these factors can promote or reduce resistance. Repeated use of a single pesticide increases the risk of resistance, especially when other control methods such as biological or cultural controls or pesticides with a different mode of action that would eliminate resistant individuals are not used. Table 5-20 lists the modes of action of different herbicide classes and the number of resistant weed species currently known.

Resistance Management Strategies

To slow the process of natural selection that leads to the development of resistance, implement an IPM program and apply pesticides only when necessary. Use control methods other than pesticides where feasible. Resistant cultivars, biological and cultural controls, and other nonchemical management tactics can be used to reduce the number of pesticide applications and to reduce selection for pesticide-resistant individuals. By using monitoring information and economic thresholds, pesticide applications can be more precisely timed, resulting in fewer applications with more appropriate rates. Preferred pesticides are selective and short lived. If repeat applications are required, rotate applications with pesticides that have different modes of action.

Consider the history of pest management practices, especially pesticide use at the site, to reduce selection for resistance. For instance, the risk of resistance is likely to increase in sites where broad-spectrum insecticides have been continually used. Repeat applications of these materials intensify selection pressures and eliminate the natural enemies that may control some insect pests. Using pesticides with different modes of action, on the other hand, may result in lower selection pressure and lower incidences of resistance. This can be very important in managing herbicide and fungicide resistance where mixtures of materials or alternating materials with different modes of action are used to prevent or overcome resistance.

To learn about individual pesticides' mode of action and specific information about managing pesticide resistance, consult the relevant website of the international Resistance Action Committees: the Insecticide Resistance Action Committee (IRAC), www.irac-online.org; the Herbicide Resistance Action Committee (HRAC), www.hracglobal.com; or the Fungicide Resistance Action Committee (FRAC), www.frac.info.

Modification of management practices to reduce the development of resistance has been demonstrated by the management of spider mite resistance in San Joaquin Valley cotton in California. Acaricide applications are made one or two times per season using a number of different materials to minimize selection pressure. In addition, broad-spectrum persistent insecticides are avoided to preserve and encourage natural enemy populations. The result is that high levels of *acaricide* resistance are not common in San Joaquin Valley cotton fields. However, to ensure that these materials remain effective, resistance management practices must be used in conjunction with other IPM strategies.

Other Factors That Influence Pesticide Efficacy

Pesticide resistance is not the only reason pesticide applications sometimes fail to control pests. Erratic pest control also results from improper spray coverage or applying an incorrect rate of pesticide. Also, sometimes large numbers of insects or mites may migrate in after a pesticide application, giving the appearance of a pesticide failure. These failures are sometimes mistaken for pesticide resistance.

Improper timing of application frequently leads to a lack of control. Whenever pesticides are recommended, applications applied at the most susceptible stage of the pest achieve the most effective control. Time of day, pH of the water, incompatibility with other pesticides or adjuvants in a tank mix, temperature, or other factors can impact pesticide efficacy. Be sure that the application equipment is properly calibrated and maintained, the spray rig is driven at the correct speed, the plant height is not above the spray zone, and that the right amounts of pesticide and water are being used to evenly distribute the pesticide over the application area. Nonsystemic sprays covering only the upper leaf surfaces will not kill pests on the underside.

To ensure that the pesticide is thoroughly covering the application area, commercially available water- or oil-sensitive cards can be used to determine how adequately the targeted area is being treated. The cards can be clipped or stapled to the specific plant parts targeted for treatment and retrieved after the application is completed. Monitoring spray coverage can be a useful tool to ensure that the right amount of pesticide is distributed evenly over the application site.

OTHER RELATED PEST MANAGEMENT/PRODUCTION SYSTEMS

Many growers, landscapers, and gardeners have become interested in using production or pest management practices that are more environmentally sound, reduce hazards to

TABLE 5-20.

Summary of number of weeds resistant to various herbicide classes in 2011. Go to www.weedscience.org for the most recent figures from the international Herbicide Resistance Action Committee (HRAC) and the Weed Science Society of America (WSSA).

Herbicide group	Site of action	HRAC group	WSSA group	Example herbicide	Total # resistant weeds
ALS inhibitors	inhibition of acetolactate synthase (ALS)	B	2	Chlorsulfuron	110
photosystem II inhibitors	inhibition of photosynthesis at photosystem II	C1	5	Atrazine	69
ACCase inhibitors	inhibition of acetyl CoA carboxylase (ACCase)	A	1	Diclofop-methyl	40
synthetic auxins	synthetic auxins	O	4	2,4-D	28
bipyridiliums	photosytem-I-electron diversion	D	22	Paraquat	25
ureas and amides	inhibition of photosynthesis at photosystem II	C2	7	Chlorotoluron	22
glycines	inhibition of EPSP synthase	G	9	Glyphosate	21
cinitroanilines and others	microtubule assembly inhibition	K1	3	Triflualin	10
thiocarbamates and others	inhibition of lipid synthesis—not ACCase inhibitions	N	8, 16, 26	Triallate	8
chloroacetamides and others	inhibition of cell division	K3	15	Butachlor	5
PPO inhibitors	inhibition of protoporphyringen oxidase	E	14	Oxyfluorfen	4
triazoles, ureas, isoxazolidiones	bleaching: inhibition of carotenoid biosynthesis (unknown target)	F3	11	Amitrole	4
nitriles and others	inhibition of photosynthesis at photosystem II	C3	6	Bromoxynil	3
caratenoid biosynthesis inhibitors	bleaching: inhibition of carotenoid biosynthesis at the phytoene desaturase step (PDS)	F1	12	Flurtamone	2
arylaminopropionic acids	unknown	Z	25	Flamprop-methyl	2
4-HPPD inhibitors	bleaching: inhibition of 4-hydroxyphenyl-pyruvate-dioxygenase (4-HPPD)	F2	28	Isoxaflutole	1
glutamine synthase inhibitors	inhibition of blutamine synthtase	H	28	Glufosinate-ammonium	1
mitosis inhibitors	inhibition of mitosis/microtubule polymerization inhibitor	K2	23	Propham	1
cellulose inhibitors	inhibition of cell wall (cellulose) synthesis	L	20, 21, 27	Dichlobenil	1
organoarsenicals	unknown	Z	17	MSMA	1
unknown	unknown	Z	25, 26	Chloro-flurenol	1
Total number of unique herbicide-resistant weed biotypes					359

people, and meet social goals. These practices include sustainable agriculture and organic production systems. Some product marketing programs or grocery store chains promise that environmentally friendly, or "green," production practices were followed. Because integrated pest management is a strategy for managing pests and is not reliant on specific types of products, these programs often incorporate IPM strategies.

Sustainable Agriculture

Sustainable agriculture refers to farming systems that can maintain their productivity indefinitely. They must be resource conserving, environmentally compatible, socially supportive, and commercially competitive. While sustainable systems generally use fewer off-farm inputs, they typically require higher levels of trained labor and management skill. There is substantial emphasis on practices that enhance and protect soil health, including the use of cover crops, composts and mulches, and reduced tillage. Efficient use of inputs with a maximum reliance on natural, renewable, and on-farm inputs is also emphasized. Biologically based IPM is a key element of sustainable agriculture systems, but because these systems consider production practices as well as marketing, labor, and other social factors, pest management is just one component. Stewardship of both natural and human resources is of prime importance, and programs generally take a systems perspective, which extends all the way from the individual farm and local ecosystem to community and global impacts.

Organic Farming

Organic farming relies on ecologically based practices such as cultural and biological pest management and excludes the use of synthetic fertilizers and pesticides. Organic farmers rely heavily on composts, green manures, cover crops, crop rotation, mechanical cultivation, naturally occurring pesticides such as copper fungicides and sulfurs, horticultural oils, biological pest controls, and botanical, microbial, and other "soft" pesticides. The USDA National Organic Program regulates the standards for organic farming, processing, and labeling. For instance, the use of synthetic pesticides and fertilizers must be discontinued for several years before planting any organic crop.

The USDA also accredits public and private organizations or persons as certifying agents for organic agriculture. Certifying agents ensure that organic production and handling practices meet the national standards. In California, the largest certifying organization is California Certified Organic Farmers (CCOF). CCOF develops organic farming standards, provides verification of those standards, recommends organic practices, and compiles a materials list for use in organic crop production.

The Organic Materials Review Institute (OMRI) is a national nonprofit organization that certifies products for use in organic production and processing according to the standards of the USDA National Organic Program. Manufacturers must pay a fee to get their product listed by OMRI, so some organically acceptable products are not OMRI certified.

Beyond the legally enforced limitations on materials that can be used, organic farmers are generally bound by a farming philosophy that is based on promoting long-term soil health. Organic growers commonly believe that organic practices feed the soil, which in turn feeds the plant. Many organic certification programs encourage or require the use of rotational systems, cover crops, and organic soil amendments. The IPM philosophy of relying on preventive methods, employing cultural and biological control methods, monitoring before applying treatments, and applying low-risk pesticides is very compatible with organic farming.

6 Monitoring and Decision-Making Guidelines

Monitoring includes any of a variety of procedures used to observe, measure, and record over time the activities, growth, development, and abundance of organisms or the factors affecting them. Monitoring is fundamental to IPM and is a prerequisite for effective decision making. Pest monitoring provides information on populations and infestation levels of pests. Additionally, it provides information on weather, crop development, the effectiveness of management practices, and populations of beneficial organisms. The records generated by monitoring establish a pest history for the field, orchard, or managed site.

This chapter provides the information necessary to develop a monitoring program and explores the variety of techniques available to the PCA. Many monitoring programs involve procedures to quantify or sample population numbers. However, not all monitoring techniques involve sampling. Being able to choose the appropriate technique for a specific pest problem is basic for successful decision making. Numerous IPM information resources are available, including various publications from university or Cooperative Extension offices. In California, the UC IPM Pest Management Guidelines, year-round IPM programs, and Pest Note series publications on the www.ipm.ucdavis.edu website, and IPM manuals provide monitoring techniques developed for pests in specific cropping and noncrop situations. It may be necessary to adapt them for use in other crops or for other pests since the research base is not complete for every situation. Learning what is known about the monitoring and biology of important pests is essential for choosing the best strategy for making informed decisions about pest status. Check the "Resources" section for additional information.

MONITORING INCENTIVES

The primary reason to monitor is to improve your ability to make good pest management decisions. Monitoring helps reduce the inherent risk of poor pest control by providing site-specific information about pest populations, natural enemies, crop growth, environmental conditions, and proper timing for control measures. Monitoring records can be invaluable in justifying pest management decisions to the client and in determining patterns of infestations over time. A good monitoring program will give a PCA or crop consultant a competitive edge, because clients appreciate a PCA who clearly understands the activities of the pests in their fields, orchards, or landscapes. Monitoring increases knowledge of the pests, which allows development of more efficient and effective sampling techniques. In addition, post-treatment monitoring provides information on effectiveness of pest control activities. Lastly, monitoring is a sound economic investment for both the PCA and client; if conscientiously applied, it will help avoid pest damage and give you the confidence to eliminate unnecessary pest control treatments (Table 6-1).

MONITORING OBJECTIVES

A major objective of monitoring is to obtain information to help make pest management decisions. Monitoring can provide immediate data for decision making, comprehensive data on the far-reaching effects of a pest organism over time or over an area, and a site-specific historical record that can measure the effectiveness of management actions. Monitoring can also be useful in predicting the location and abundance of a pest population.

TABLE 6-1.

Advantages of monitoring for pest control advisers.

- ☐ Develops a historical record of pests and beneficials for future decisions.
- ☐ Provides information to help time control measures properly.
- ☐ Provides information to help identify potential nontarget impacts.
- ☐ Helps justify pest management decision to client and provides confidence for those decisions.
- ☐ Provides competitive edge.
- ☐ Provides information to improve or simplify future sampling programs.
- ☐ Provides an economic justification for the recommendation to save client from an unnecessary treatment yet avoid crop loss.
- ☐ Provides earlier warning of potential pest problems.
- ☐ Provides feedback about whether pest control activities are working.

Using monitoring techniques specific to the pest and the site, a PCA can estimate the density of the pest population, evaluate potential impacts on crop yield and quality, and determine the need for a control action or treatment. A good monitoring program begins by assessing the status of the target pest with particular attention to the life stages present and the rate of population development of the stage that is most damaging to the host. Routine sampling helps provide an accurate estimate of the density of the pest species and the approximate areas infested, enabling the PCA to determine the total area that must be treated and whether spot-treating is possible. For many diseases, the actual pathogen may not be visible; monitoring the conditions that contribute to disease outbreaks is key. Later pathogen monitoring may involve recording the location and severity of disease symptoms. For weeds, identifying species and monitoring their density and stage of growth as well as cultural practices that may have contributed to weed invasions are important in making effective management decisions.

Monitoring also evaluates other factors that can impact crop damage and pest populations. Routine observations of the patterns of growth and development of the plant host and the presence of beneficial organisms can assist in developing better-timed and more diversified pest management strategies. Additionally, regular monitoring more accurately assesses the impact of weather and other environmental factors on the crop and pest populations. For example, cool, wet springs often reduce the abundance of web-spinning spider mites in tree crops. Monitoring the population of mites at this time can be a valuable cost-saving measure, because acaricide applications may not be necessary. It is also important to continue monitoring after management actions such as pesticide applications have been implemented. Post-treatment monitoring can help evaluate the effectiveness of the control on the pest and the impact of the control on the quality and yield of the crop. This is particularly true where broad-spectrum insecticides are used because post-treatment monitoring provides information on secondary pest outbreaks caused by disruption of biological control. For weeds, post-treatment monitoring may help assess the weed seed bank to plan for management in future crops.

ROLE OF SAMPLING IN MONITORING PROGRAMS

Monitoring includes a range of activities and procedures carried out to detect and document the presence, growth, population development, or population levels of organisms or to assess factors such as weather or physical conditions that can affect development of pest populations. Many, but not all, monitoring programs involve a quantitative evaluation of pest population numbers. These types of monitoring procedures are called sampling programs.

Some types of monitoring do not involve sampling. For instance, measurements of climatic factors may involve discrete measurements of rainfall or minimum and maximum temperatures in a field. Maintaining landscape or field maps for major weed infestations or vertebrate burrows records the location but does not involve counting the number of individual animals or weeds (Figure 6-1). Forecasting programs for disease outbreaks and disease monitoring generally do not include a sampling component; rather, they often rely more on weather data and field histories. These types of monitoring programs will be discussed later in this chapter.

FIGURE 6-1

Maintenance of field maps is a monitoring technique that does not involve sampling. Field maps provide information on the location of infestations but not on population numbers.

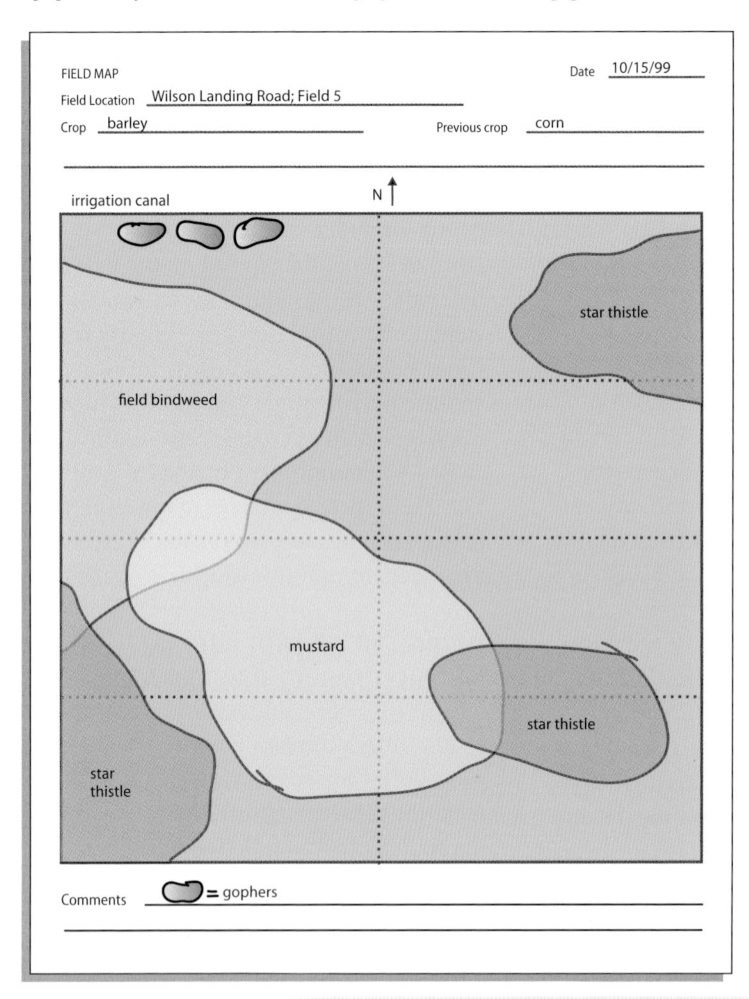

DEFINING THE SAMPLING UNIVERSE AND SAMPLING UNIT

Counting every weed seed or spider mite or measuring the length of every internode in a field is not feasible. However, a well-planned sampling program can provide the accuracy needed to evaluate a pest infestation. A sample is a set of measurements taken from part of a population or a subset of the physical features of an environment (Figure 6-2). The degree to which the sample represents the whole field or managed system will determine the accuracy of the sampling program.

The first step in setting up a sampling program is determining the boundaries of the area for which a specific pest management decision is being made. This will define the *sampling universe*, or the area that will be covered in a single sampling program. The sampling universe, or *block*, can consist of a single habitat, such as a whole field of corn; several smaller areas of a field if the areas can be managed independently of one another; or even certain plants or plant parts if spot-treating is feasible. For instance, if a grower treats different blocks of an orchard at different times, these blocks can be monitored and evaluated separately as different sampling universes. If blocks have different varieties, planting times, or environmental conditions,

FIGURE 6-2

To assess the population of aphids in a wheat field, take samples. Each sample involves counting all the aphids on an individual tiller of wheat. Research has determined that 40 to 50 samples taken throughout a field are needed to give a reliable estimate of potential pests.

Russian wheat aphid

Sampling in a seedling field

A tiller is a branch stem which develops from a crown bud in a grass plant.

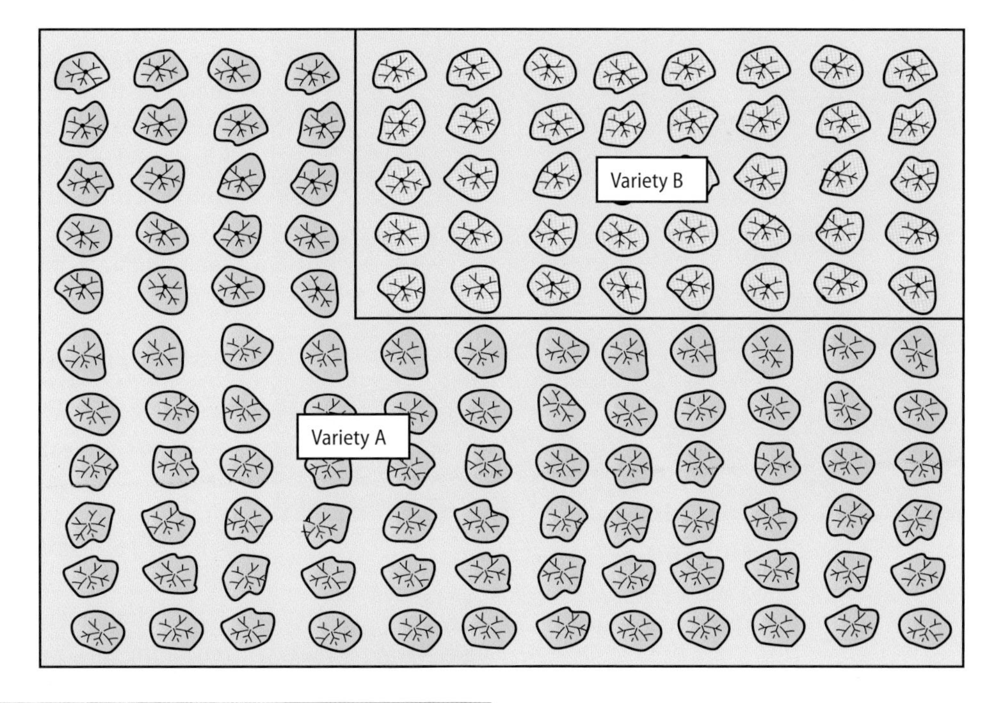

FIGURE 6-3

Define the sampling universe. An orchard with two varieties that are managed separately can be evaluated as different sampling universes. Separate groups of samples are collected from Variety A and Variety B blocks and are evaluated independently of each other.

FIGURE 6-4

When sampling for tomato fruitworm eggs on tomatoes, the sampling unit is the leaf just below the highest open flower.

it is best to sample them independently from one another (Figure 6-3).

The Sampling Unit

The plant part or entity sampled in the sampling universe is the *sampling unit*. It may be a leaf, a bud, the area of plant covered by one sweep net sample, a square foot of soil, or the soil contained in a soil core sample. The choice of the sampling unit should take into account the size, abundance, and distribution pattern and mobility of the pest organism, as well as balancing the cost of sampling against the accuracy of the information obtained. The sampling unit should be large enough to adequately assess the population of the target pest, taking into account the pest's biology, but small enough to economically permit an adequate number of samples in a reasonable amount of time. In general, choosing a smaller-sized sampling unit and collecting a larger number of samples provides better information than using a larger sampling unit and fewer samples. Where there are no research-based recommendations, choosing the sample unit and the number of samples requires the PCA to balance the reliability of the sample with the cost of obtaining it.

The sample unit varies with the pest species and the sampling universe but must be clearly defined. Choosing as the sample unit the place where the pest is most likely to be found can increase sampling efficiency and reduce search time. For example, the sampling unit for tomato fruitworm eggs is the leaf below the highest open flower because that is the preferred site for egg deposition (Figure 6-4).

The sampling unit also depends on the pest organism, its mobility, distribution, and method of dispersal. When sampling for weed species, the area within a Daubenmire square can be a sampling unit (Figure 6-5). The soil core pulled from an *auger* is a common sampling unit for various species of nematodes and soilborne pathogens or weed seeds in the seed bank.

FIGURE 6-5

A Daubenmire square or quadrat is a standard-sized square, rectangular, or circular sampling unit of known dimensions used to define the area (one sample) in which all weed species will be counted. The size of the unit will depend on site constraints.

The first step when deciding what sampling unit and method to use is to search the literature to determine procedures that have worked for others. Research-based sampling programs for many major pests are published in the online UC IPM Pest Management Guidelines. Table 6-2 lists examples of sampling units and methods for a variety of pests.

Sampling Accuracy and Precision

Sampling provides the PCA with an estimate of the size of the pest population or extent of damage. Accurate and precise sampling allows assessment of whether control action is needed or has been effective. *Accuracy* is how close the sample mean is to reality. When sampling is accurate, its results truly reflect the total population. *Precision* refers to the variability among samples, or how close measurements are to one another. The standard deviation and other statistics quantify this variability as described in Sidebar 7.1, "General Formula to Determine a Fixed Sample Size," in Chapter 7. When precision is

TABLE 6-2.

Examples of sample units for different pests in specified monitoring programs.

Pest (host)	Sample unit	Sampling technique	Number of samples	Reference[1]
cotton bollworm (cotton)	main stem terminal (5 per stop)	visual presence of bollworm larvae	5 terminals at each stop, 100 plants total/field	*UC IPM Pest Management Guidelines: Cotton*
lygus bug (cotton)	sweep net (50 sweeps in 1 row)	sweep net counts	1 row for each quarter of the field	*UC IPM Pest Management Guidelines: Cotton*
aphids (street trees)	3" × 2" water-sensitive card	honeydew on card	4 cards per tree	*Pests of Landscape Trees and Shrubs*
nematodes (cotton)	root system of whole plant	visual rating of damage	15–20 root systems per 10-acre block	*IPM for Cotton*
web-spinning spider mites (almond)	15 leaves per tree	visual (hand lens)	minimum of five 15-leaf samples per orchard	*IPM for Almonds*
acacia psyllid (landscape)	3 beats of a branch on a beating tray	branch beating	2–4 branches from 3 or 4 different plants at a location	*Pests of Landscape Trees and Shrubs*
citrus cutworm (citrus)	new growth flushes	timed searches	5 minutes in 4 locations in the orchard	*UC IPM Pest Management Guidelines: Citrus*
California red scale (citrus)	pheromone card	count male scale on 20% of card and multiply by 5	1 trap card per 2.5 acres	*IPM for Citrus*
tomato fruitworm (tomato)	leaf below highest open flower on plant	count number of eggs and parasitized (black) eggs per leaf	30 leaves per field selected at random	*IPM for Tomatoes*
weeds (landscape)	marked point on string	transect count (identify weed on marked points every 12 inches on string)	1 for every 2–4 feet of landscape bed length	*Pests of Landscape Trees and Shrubs*

[1]. Complete information is given in the "Resources" section.

poor (and therefore standard deviation is high), a larger number of samples are required to generate an accurate (or realistic) mean and a reliable estimate of the pest problem. Such a sampling procedure is relatively inefficient and therefore costly because of the required amount of effort.

Factors Affecting Sampling Accuracy

Many factors affect how easy it is to obtain an accurate sample. Characteristics of the pest such as its size, how easily it can be found, its mobility, how fast it develops, and its distribution patterns must be considered. Other factors include the choice of sampling unit and method and whether sampling is as consistently and carefully performed as intended. For instance, changing the location of the sample collection on the plant or in the field alters the accuracy of sampling. Sampling weed species in areas where the soil is more moist than the rest of the field could result in a higher population estimate than may actually be present in the field overall.

Many PCAs intentionally or unintentionally spend more time sampling areas they know to be prone to pest problems. Sampling from all sections of the field more accurately reflects the overall pest population. Consciously sampling hot spots is a good strategy to detect early infestation and more easily find beneficial insects, and it is also helpful when trying to detect when pest numbers are starting to build; however, these separately sampled areas should also be treated separately. In general, counts from areas known to have high pest abundance should not be used to make treatment decisions for the whole field or orchard. Rather, such information should serve as an early warning to PCAs that pest populations may be increasing.

Sampling accuracy and efficiency also varies according to how consistently the sampling method is performed. For example, sampling leaves that vary in size, age, and location may affect population estimates on leaf samples for mites. Most sampling methods provide a relative measure of whether pest populations, natural enemies, or damage are increasing, decreasing, or remaining about the same (see Sidebar 6.2.). Sampling must be done the same way every time to make the results comparable among sample dates throughout the year and from year to year.

Consistent sampling is especially important when scouts take the samples because accuracy can vary when different people sample the same site. For example, timed-search samples record the number of pests spotted in a certain period of time. A more experienced PCA using time searches may find more cutworm larvae or other pests than an inexperienced scout due to differences in skill or observational abilities. Differences in body height, visual acuity, and strength can also affect sample results. When using scouts, clearly define the sampling unit and methods for selecting samples to minimize variations in technique (for more information, see the section "Using Scouts Effectively," in Chapter 7). Also, consider differences in sampling skill, observation, care, and the size and strength of individual scouts when gauging the accuracy of the sample.

Factors Affecting Sampling Precision

Precision, or the variability among samples, is affected by the biology of the pest or characteristics of the area being sampled. This includes the pest's distribution pattern and the field size and shape. For example, due to their increased variability, small and irregularly shaped fields require more samples per unit area to provide accurate results than do larger, more uniform fields.

When precision is likely to be poor, accuracy can usually be improved by increasing the number of samples taken. The pest population is more precisely estimated and the risk of error is lower with more samples; conversely, taking fewer samples increases the potential for error. Also, as the pest population approaches the action threshold, take more samples to improve accuracy and help make the best decision. Otherwise, treatment may take place when it is not needed, or treatment may not be made when needed.

Taking more samples is an especially good idea when confronted with a new or unfamiliar pest or a new field or orchard. The less known about the pest, the more samples that may need to be taken until the optimal sample size can be established. When research-based recommendations are lacking, the number of samples needed to provide results of a desired precision can be estimated by calculating the mean and variance of preliminary samples as described in Sidebar 7-1, "General Formula to Determine a Fixed Sample Size," in Chapter 7. If pest distribution in the field is highly clumped, it is often more difficult to detect, and more samples will be required to accurately estimate its density compared with pests with more *uniform distributions*, as explained in Sidebar 6-1.

Improving Sampling Results

Using a backup technique can help evaluate the results of a new sampling technique or improve the accuracy of the overall monitoring program. For example, for western flower thrips in lettuce, you can sample thrips by using a beating sheet. To confirm the severity of an infestation, supplement that count by checking for feeding scars and thrips in plants. One sampling technique can also be a *trigger* for another sampling technique. For example, yellow sticky traps are often used as a detection tool for the presence of greenhouse whitefly (Figure 6-6); once the pest is detected, leaf counts are used to determine the density of the population. Also, to identify weed species, soil samples are sometimes taken in winter. The weed seeds are then germinated and the seedlings grown in greenhouses for identification. PCAs can use this information to choose preemergent herbicides or plan a monitoring program for weeds in the field in the spring.

FIGURE 6-6

Yellow sticky traps, like this one, are a good tool to detect adult whiteflies, but they are not useful for determining population density.

Handling samples carefully and taking accurate field notes can significantly increase sampling accuracy. Samples are often taken from the field and analyzed later or sent to a lab for a detailed analysis, as is done with soil samples for nematodes or plant pathogens. It is extremely important to keep these samples fresh and transport them with minimal disturbance. Keep field notes and sampling results in written form; analyze or graph the data to determine the size and distribution patterns of pest populations and to document their abundance throughout the season (see Sidebar 6-1).

Field notes should include information on the presence, population size, and stage of development of the pest organism; the field location of the pest; the degree of damage; and the stage of development of the host plant. Also include any cultural or other management practice that might affect pests, weather conditions, time and date, and the presence of other pests or beneficial organisms.

SAMPLING METHODS

Once the sampling universe, sampling unit, and pest distribution are defined, the PCA must determine the appropriate sample size, the most efficient *sampling pattern*, and the right sampling tools and techniques with which to gather the information. For sampling to yield reliable information, it is best to choose or develop a carefully defined sampling method, follow it at each visit, and record the method used so that accurate

SIDEBAR 6-1

Pest Distribution Patterns

To develop a sampling program, you must understand the target pest's distribution and ability to disperse within the sampling universe. Very mobile pests disperse rather rapidly through a field. Other pests move slowly out of initial infestation sites, perhaps at field edges or from individual plants in a field. Still others thrive in certain areas of the field where conditions are more favorable for them. For example, the distribution within a vineyard of a relatively nonmobile insect such as the grape mealybug is very different from that of a very mobile insect such as the grape leafhopper.

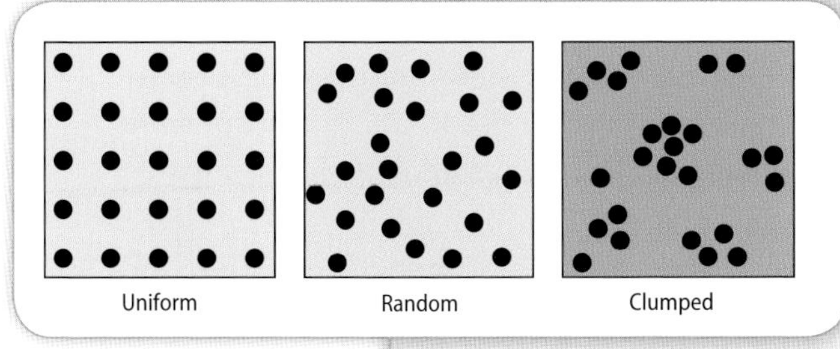

Examples of uniform, random, and clumped dispersion patterns.

As a result of these differences in dispersal ability, the distribution patterns of pest populations show varying degrees of clumping. If a pest is distributed evenly throughout the sampling universe—for example, every leaf on every plant has two aphids—the distribution is uniform. A *random distribution* is just the opposite: it occurs when the pests' distribution is not uniform and it cannot be predicted by any biological or environmental factors, such as mobility of pest or field edges; that is, the distribution is completely by chance. Pest populations are usually neither completely uniform nor completely random.

The distribution of pest individuals in natural populations is usually more or less clumped. Once one individual of a pest population in a sample unit is found, more individuals in the same sample unit will likely be found. Spider mites, aphids, mealybugs, and scale insects are examples of species that are usually clumped; eggs, nymphs, and adults of these species tend to aggregate closely near the site of a successful invasion, at least when populations are low to moderate. Cabbageworms and other Lepidoptera pests, on the other hand, are less clumped because the very mobile adult females can fly around the field, laying eggs on many plants. Plants infested with root pathogens are often found clumped around the initial infection site of the pathogen. For instance, Phymatotrichum root rot, or Texas root rot of alfalfa, often forms a ring-shaped pattern in the field as the fungus moves out from the center of the infection. In these cases, clumping occurs because pest populations tend to be offspring of the initial colonizer.

Clumping may also occur because environmental factors such as water, soil type, or shade favor establishment of the pest species in one area over other areas in the sampling universe. For instance, nutsedge prefers sunny areas with plenty of water. Once populations of many pests become high, they may spread throughout the field and appear more uniform or more random. When pest species are highly clumped, they are often difficult to detect and generally require more samples to accurately assess population density.

Once populations of many pests become high, their distribution appears more uniform, as shown here with this field bindweed infestation.

SIDEBAR 6-2

Absolute versus Relative Samples

Absolute samples measure the total pest population in a given area. They are an important research tool and are very accurate, but they are very time consuming. Counting all the individuals in a population in a given area is an absolute measurement that determines total population size. For example, counting all the ground squirrel burrows in an orchard or all the weeds infesting pots of chrysanthemums in a greenhouse are absolute measurements. In situations where a new or unfamiliar pest has invaded, absolute samples are often used to gather enough information to assess the problem and develop an appropriate sampling program. Absolute samples can also be used to compare or refine the accuracy of established sampling techniques or to evaluate the efficiency of a management action.

However, in most situations, it is not practical or economically feasible to count every individual pest present. A representative sample of the population can be used. The best procedure is one that gives a reliable and consistent estimate of pest infestation in a minimum amount of time. These types of samples, called *relative samples*, are commonly used in IPM programs. Relative samples estimate the total population or potential damage by sampling only a limited portion of the population. Examples of relative samples include counts of pests caught in beating sheets or sweep nets, weeds counted within a sampling frame, and visual counts per plant part. Relative samples typically require only simple equipment and allow for a large number of samples to be collected in a small amount of time.

comparisons of data can be made.

Sample Size (Number of Samples)

The number of samples (sample size) necessary to estimate damage or pest abundance depends on the pest species present, its biology and potential for damage, the sampling method and unit, the time of year, the area or crop being sampled, the crop stage, and the value of the crop. One of the major considerations in determining the number of samples to take is the level of precision required; the time it takes to sample is another important practical consideration. The more samples taken, the more accurately a population can be estimated, but there are limits to how much time can economically be spent in a single field (see Sidebar 6-2). It is always desirable to take as many samples as is practical.

The required sample size is correlated with the sampling technique. In sampling for nematodes, for instance, the sampling unit could be a core of soil; the sample size would equal the number of cores of soil that are taken in a field. A basic minimum sample size for one sample is typically 15 to 20 cores for every 5 acres (2 ha). On the other hand, if roots are being dug up to check for galling as a sampling technique, the required sample size may be different.

For many pests in major crops, university research has determined the number of samples required to get an accurate assessment of pest levels and potential damage. In California, these guidelines are available in the University of California's UC IPM Pest Management Guidelines and Pest Notes at www.ipm.ucdavis.edu. Similar guidelines are produced in other states. Unfortunately, research-based protocols and decision-making guidelines are not available for many pests, and PCAs must develop their own monitoring protocols. Experienced PCAs often modify methods used for similar pests in other crops and adapt them to the current pest situation.

For new PCAs, if a search of the literature reveals that useful guidelines are lacking, the best way to determine the appropriate sam-

TABLE 6-3.

Factors that can affect sampling reliability.

- ☐ Size of the sample.
- ☐ Changes in actual pest numbers due to population fluctuations; for example, migration of large numbers of pests from neighboring crops immediately after sampling takes place.
- ☐ Changes in the number of a particular pest developmental stage; for example, stink bug nymphs are occasionally found in cotton but control action guidelines apply to adult stages.
- ☐ Effect of time of day, temperature, or light on pest movement; for example, citrus thrips are most active and easiest to see as the day begins to warm up.
- ☐ Stage of crop growth; for example, tall corn is difficult to sample for mites.
- ☐ Changes in pest activity or distribution on the plant or part of the plant due to environmental stress or other factors; for example, spider mites increase greatly when plants are water stressed.
- ☐ Responsiveness of a particular sex or species of pest to the trap stimulus.
- ☐ Changes in trap efficiency; for example, pheromone effectiveness decreases with time.
- ☐ Change in the sampler's searching efficiency; for example, fatigue at the end of the day.
- ☐ Instability of the sampling unit; for example, the pest is mostly associated with one plant part at one time of year and less so at another time.
- ☐ Changes in pest distribution following treatment.

ple size is to try several sampling methods for accuracy and ease. Once a preferred sampling method is selected, take a large number of samples in the same field to determine variability of samples across the sampling universe. Then use a sample size formula (see Chapter 7) or compare the accuracy of decisions using fewer samples. New situations may initially require that more samples be taken more frequently than is ultimately necessary. Each sample will vary but by recording the counts and calculating basic statistics (such as the population mean or average and the standard deviation), pest abundance and distribution become evident and the cost of sampling and the number of samples required for making a pest management decision can be determined.

Sampling Efficiency. To make reliable pest management decisions, take enough samples in a cost-effective manner to provide an accurate estimate of the pest population; this is referred to as sampling efficiency. Field-applicable sampling methods do not usually provide an absolute estimate of pest density, and there are many factors that can impact the accuracy of a sampling procedure (Table 6-3). It is important to understand these factors, otherwise the significance of the pest species, its impact on the crop, and the impact of the natural enemies present is likely to be distorted.

Triggers. Different methods can be used to reduce the sampling time required to get an accurate estimate of a population while maintaining sampling efficiency. For instance, the sampler can use triggers to focus sampling during the most critical times in the life cycle of the crop or pest. Triggers rely on biological or physical events that have a direct relationship to host or pest development. They should occur early enough for focused sampling to be initiated so that vulnerable (treatable) stages or distinctive stages of the pest species can be tracked. For example, moths caught in *pheromone* traps signal when adult insects are flying and mating. This, in conjunction with monitoring temperatures and using degree-days, provides a trigger to initiate monitoring for eggs or larvae of the next generation.

For plant diseases, the presence of a susceptible crop and the occurrence of environmental conditions that favor pathogen infection and disease development are triggers to begin field monitoring for disease symptoms. Field personnel must usually assume that inoculum will be present, since inoculum usually cannot be visually confirmed unless diseased plants are present in the vicinity in other fields.

In an IPM program used in Colorado for early blight in potatoes, pathogen inoculum is monitored by trapping spores of the *Alternaria solani* fungus, which causes the disease (Figure 6-7). The critical time for managing early blight is just as tubers are forming, so tuber formation is the trigger to begin spore monitoring. Once the first tubers are detected (at about first flowering), greased microscope slides are placed in a weather vane spore trap just above the canopy level in the field. Slides are replaced and checked under the microscope twice a week to see if spores are present. *Alternaria solani* spores are fairly distinct in shape and size, but must be distinguished from other *Alternaria* species spores that may be trapped. The presence of spores indicates that secondary spread of the

FIGURE 6-7

Monitoring triggers for early blight of potatoes.

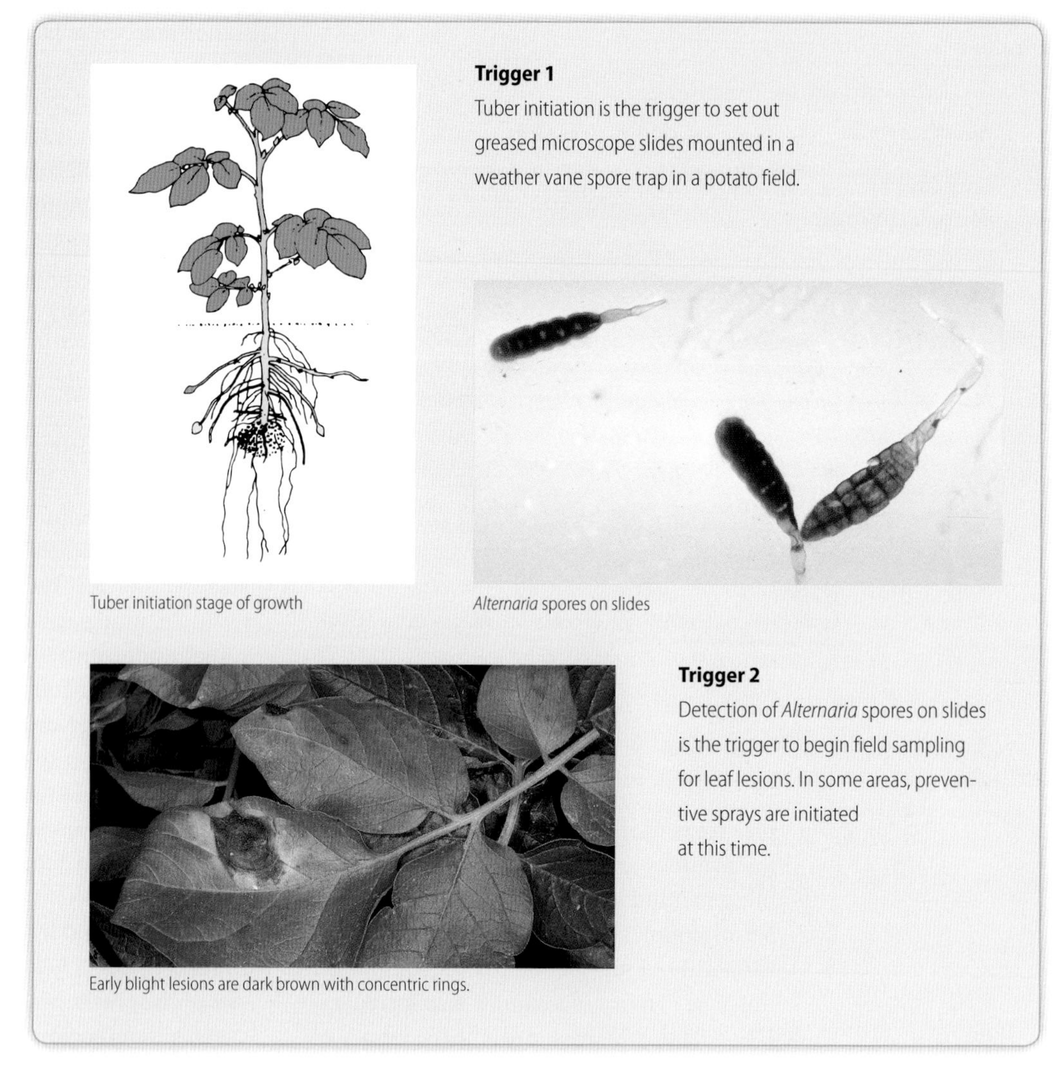

pathogen is occurring. In areas where early blight is a continual problem, spore trapping can be used to initiate fungicide applications or trigger more intensive monitoring including checking plants for blight symptoms and calculating degree-days.

Presence-Absence Sampling. Sampling programs that allow you to note whether a designated plant part has any pest individuals on it rather than counting each individual can greatly improve sampling efficiency. In these programs, called *presence-absence sampling*, each sample unit is recorded as either infested or uninfested. This technique is fast and simple to use. Presence-absence sampling can be used for all types of pests, but it is especially efficient for arthropod pests such as thrips, aphids, whiteflies, and spider mites that are tiny and abundant and take substantial time to count.

For presence-absence methods to be reliable, they must be based on research that correlates the proportion of infested sampling units (e.g., leaves) to the actual population density. Research must also assess the variability (precision) of presence-absence samples, an important determinant of required sample size. Because all individuals in the sample unit are not counted, in some cases more samples may need to be taken to achieve sufficient precision and ensure accuracy than if actual counts are made.

The reliability of presence-absence sampling is reduced when populations are very high and approach 100% of sampling units infested. Because absolute population densities can continue to increase after all leaves are infested, the utility of presence-absence sampling depends on the treatment threshold for the pest. For instance, in strawberries presence-absence sampling can be efficiently used when there are 10 or

FIGURE 6-8

Presence-absence sampling for elm leaf beetle. Clip sixteen 1-foot terminals (A) per tree, two in the inner canopy and two in the outer canopy of each side of each tree (B). Check each terminal for eggs on leaves. If one or more egg masses (C) are observed on a terminal, record it as infested (D). Treatment is warranted if 30% or more terminals are infested.

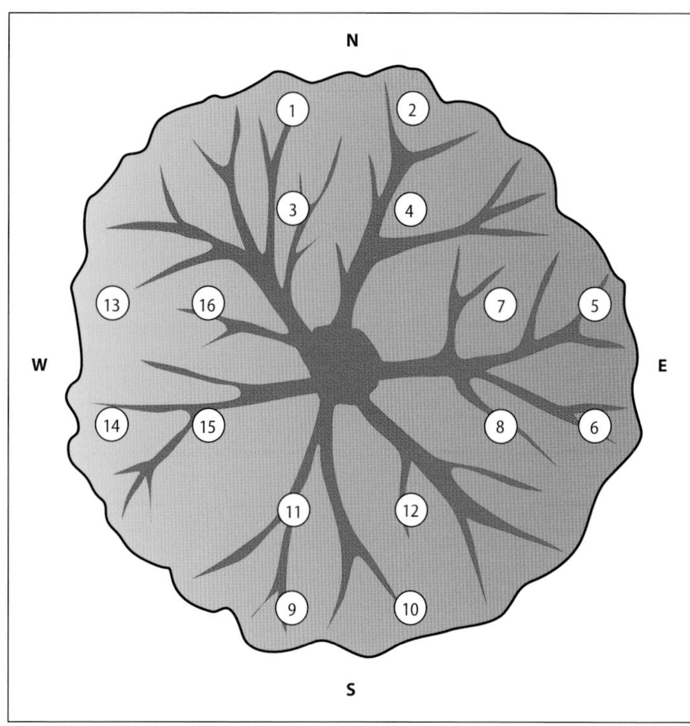

B. Aerial view of canopy showing sample sites.

A. Sample unit: 1-foot terminal.

C. Adult, first-instar larva and egg mass of elm leaf beetle.

TERMINAL	TREE #1	TREE #2
1	✗	
2		✗
3		✗
4		✗
5	✗	
6	✗	
7		
8		✗
9		
10	✗	✗
11		✗
12		
13		✗
14		✗
15	✗	✗
16		✗

D. Record sheet.

fewer spider mites per midtier leaflet. Above 10 mites per midtier leaflet, the number of mites per leaflet relative to the proportion of infested leaves increases rapidly, so that an unreasonably large number of presence-absence samples are needed to accurately estimate mite densities. This is why mite brushing methods are suggested when populations are greater than 10 mites per midtier leaflet. Also, since the treatment threshold is between 15 and 20 mites per leaf, depending on the strawberry variety, mite brushing is the recommended sampling method because populations are near the treatment threshold. See *Integrated Pest Management for Strawberries* for details about this program.

Presence-absence sampling works best in systems where the thresholds are well below 100% infestation levels and treatment need can be determined by taking a limited number of samples. An example is mite sampling in almonds. For brown mite and European red mite, the threshold is about 20% of dormant spurs infested with mite eggs. For web-spinning *Tetranychus* spp. mites, the threshold is about 50% of leaves infested. Actual thresholds vary depending on whether predatory mites are abundant or will be released, the number of samples collected, and the time of year. See the *UC IPM Pest Management Guidelines: Almond* at www.ipm.ucdavis.edu for details.

Presence-absence sampling can be used in landscapes to determine the need to treat for elm leaf beetle. This sampling program monitors eggs on leaves using degree-day calculations to predict when eggs are being laid. The degree-day calculations trigger foliage inspection and presence-absence sampling for eggs. Depending on the situation, samplers clip 16 terminals that are 1 foot (0.3 m) long from each tree and check all leaves for egg masses (Figure 6-8). If any eggs are found, the terminal is marked as infested. To reasonably predict defoliation, at least 25% of the trees (and at least 3 trees) must be sampled at a site with at least 120 samples taken. In some cases, presence-absence sampling may be less efficient when the sampling universe is small. Some examples of other pests with presence-absence sampling programs developed by the University of California are listed in Table 6-4.

Visual Injury Scales. Another way to increase sampling efficiency is to use visual injury scales that the sampler can readily compare with known levels of infestation. When using injury scales, it is helpful to have a standard for comparison to avoid overestimating damage. Injury scales are sometimes used to get a rapid estimate of the extent of foliar disease infestation. They can also be used in cotton to estimate the levels of root knot nematode infestation. This system, known as the weighted nematode rating (WNR) system, is convenient because samples can be taken from about mid-July until plowdown, when problem areas in the field are readily identified by the presence of unhealthy plants.

No special equipment or laboratory tests are required for WNR sampling. A sample is made up of 15 to 20 root systems taken in a random pattern from as broad of a representation of each block as possible. The degree of root galling for each plant is categorized by visual comparison to the root silhouettes in Figure 6-9 and is used to determine the percentage WNR. Fields should not be planted back to cotton if the WNR threshold is exceeded unless a treatment is made. More details are available in *UC IPM Pest Management Guidelines: Cotton* at www.ipm.ucdavis.edu.

Sequential Sampling. *Sequential sampling* is a method for efficiently concentrating sampling efforts where they will provide the most benefit. Sequential sampling can greatly increase sampling efficiency because it requires fewer samples to arrive at a decision with the same degree of certainty as standard (or fixed) sampling,

Fixed sampling involves taking the same predetermined number of samples at each visit to the field. For instance, in a fixed sampling program for cotton bollworms, 100 terminals are checked each time and the larvae present are counted throughout the damaging period. In sequential sampling, the number of samples can vary at each visit, depending on pest density; samples are taken only until the accumulated results show that the pest population is clearly above or below the threshold level.

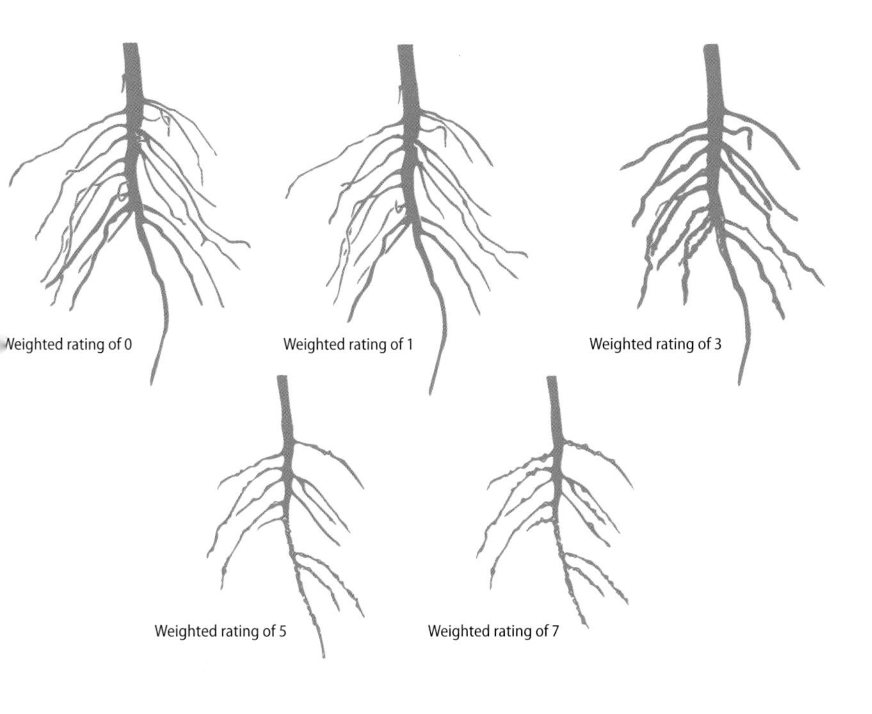

FIGURE 6-9

Example of a visual injury scale for root knot nematode in cotton. These diagrams of cotton root samples assist in determining the degree of root galling and decreased root mass when sampling for root knot nematode using the weighted nematode rating (WNR) method. A rating of 0 is given for healthy roots with no galls. Rating 1 is for root systems that are 1–25% galled; 3 is for 26–50% galled; 5 is for 51–75% galled; and 7 is for 76–100% galled. For current thresholds and how to calculate the WNR of root samples consult the online UC IPM Pest Management Guidelines: Cotton.

Sequential sampling concentrates efforts in fields where the pest population is close to the treatment threshold. When a population is either very high or very low relative to the action threshold, it takes only a few samples to show that it is above or below the threshold. For a population near the threshold, on the other hand, more samples are needed to reach the correct decision. However, extensive prior research is required to develop sequential sampling guidelines for pest managers. They cannot be quickly improvised by PCAs.

An example of a sequential sampling card and an explanation on how to use it is shown in Figure 6-10. The "Don't treat" and "treat" columns are boundaries set by the treatment threshold for each pest and by the error rate (the chance of making a mistake). Extensive preliminary research is required to establish sequential sampling guidelines.

Monitoring Multiple Species. Monitoring multiple pests and beneficial species at the same time can increase sampling efficiency. Many university guidelines, such as those in the UC IPM Pest Management Guidelines at www.ucipm.ucdavis.edu, provide for monitoring multiple species at the same time. In almonds and stone fruits, armored and soft scales and the overwintering eggs of aphids and mites can all be monitored on the same dormant spur samples. In floriculture and ornamental nurseries,

TABLE 6-4.

Examples of major California pests for which a presence-absence sampling program is established.

Pest and host	Sample unit	Reference[1]
spider mites on cotton	infested leaves	*IPM for Cotton*
green apple aphid on apples	infested shoots	*IPM for Apples and Pears*
twospotted mites on strawberries	oldest fully expanded leaves	*IPM for Strawberries*
citrus red mites on citrus	infested leaves	*IPM for Citrus*
elm leaf beetle on elms	1-foot-long terminal	*Pests of Landscape Trees and Shrubs*
spider mites on almond	infested leaves	*IPM for Almonds*
Pacific spider mites on grapes	infested leaves	*Grape Pest Management*
potato aphid on tomato	infested leaves	*IPM for Potatoes*
cabbage aphid on Brussels sprouts	infested leaves	*IPM for Cole Crops and Lettuce*

1. Complete information is given in the "Resources" section. These sampling programs are also described in the UC IPM Pest Management Guidelines for each crop at www.ipm.ucdavis.edu.

SEQUENTIAL SAMPLE CARDS

SAMPLE CARD A

threshold = 50% infested

sample no.	don't treat	tally	treat
1	–	0	–
2	–	1	–
3	–	2	–
4	–	3	–
5	–	3	–
6	3	4	7
7	4	5	7
8	4	5	8
9	4	6	9
10	5	6	9
11	5	7	10
12	6	8	10
13	6	9	11
14	7	10	11
15	7	10	12
16	8	11	12
17	8	12	13
18	9	13	13
19	9	___	14
20	9	___	15

Treat

SAMPLE CARD B

threshold = 50% infested

sample no.	don't treat	tally	treat
1	–	1	–
2	–	1	–
3	–	2	–
4	–	2	–
5	–	3	–
6	3	4	7
7	4	5	7
8	4	5	8
9	4	5	9
10	5	6	9
11	5	6	10
12	6	6	10
13	6	___	11
14	7	___	11
15	7	___	12
16	8	___	12
17	8	___	13
18	9	___	13
19	9	___	14
20	9	___	15

Don't Treat

SAMPLE CARD C

threshold = 50% infested

sample no.	don't treat	tally	treat
1	–	1	–
2	–	2	–
3	–	2	–
4	–	3	–
5	–	4	–
6	3	5	7
7	4	5	7
8	4	5	8
9	4	6	9
10	5	7	9
11	5	7	10
12	6	8	10
13	6	8	11
14	7	8	11
15	7	9	12
16	8	10	12
17	8	10	13
18	9	11	13
19	9	11	14
20	9	11	15

Recheck

FIGURE 6-10

To use a sequential sampling card, take a series of samples and keep a running tally of results in the center column. The kind of sample depends on the pest you are monitoring. As shown by dashes at the tops of the "treat" and "don't treat" columns, a certain minimum number of samples is needed before reaching a treatment decision; in this case, the minimum is 6. Beyond the minimum, you can reach a treatment decision whenever the number in your tally matches the corresponding number in one of the boundary columns. If the tally reaches the number in the "treat" column (Sample Card A), the pest population is above the treatment threshold. If the tally matches the "don't treat" column number (Sample Card B), the population is below the threshold. Continue sampling as long as the tally remains between the boundary column numbers. If you reach the bottom of the card and the tally is still between the boundaries (Sample Card C), the population is too close to the threshold for you to make a reliable decision; return and sample again in 1 to 3 days.

the adults of fungus gnats, leafminers, psyllids, shore flies, thrips, whiteflies, winged aphids, and various parasites can be monitored with yellow sticky traps.

Sometimes, however, it is not appropriate to sample two organisms at the same time. Citrus thrips cause damage when they feed on citrus fruit. They are most active and easiest to observe between 10 AM and 2 PM when it is sunny. Their predator mite, *Euseius tularensis*, on the other hand, avoids intense light and is more likely to be observed when weather is overcast or at dusk or early morning and in the shade. Therefore the pest and its predator must be sampled separately: thrips on exposed fruit and *Euseius* predators on the underside of leaves in the inner canopy.

Where university guidelines are lacking, pest managers can devise their own multiple species sampling strategies and use sampling tools and techniques to gather the most information possible in the shortest time span. In nearly all pest management situations, crop consultants must monitor insect, weed, disease, and vertebrate pests, as well as crop development; in other words, they must observe the big picture. Combining techniques, such as noting weed populations while checking pheromone traps or taking soil samples to assess nematodes and

FIGURE 6-11

Examples of pathways for random samples. For sampling to yield reliable results, sample in a pattern that covers the whole field. The best sampling pathways to use are the X, V, W, and U patterns. Pathways that sample only in one area or along field edges (below) often produce misleading results. When returning to a field to resample, vary your sampling pattern or enter from a different part of the field so new areas will be sampled at each visit.

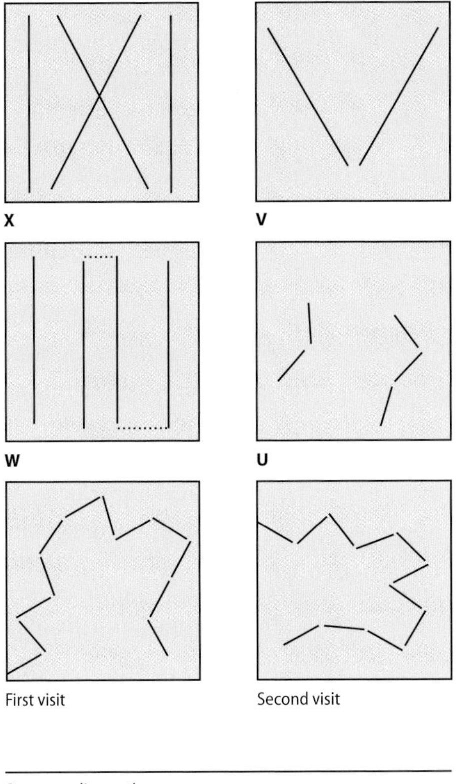

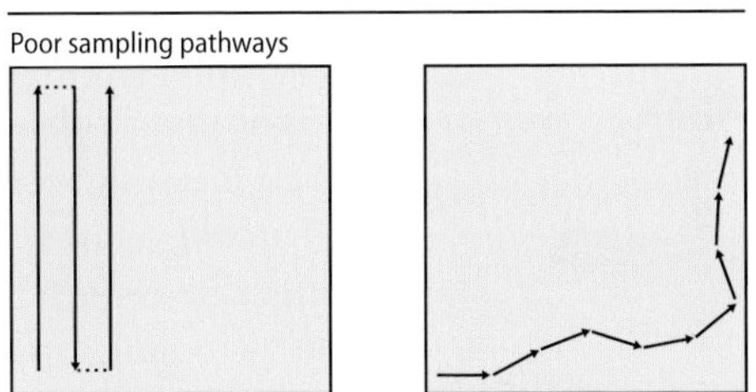

weed species, increases efficiency. Tools such as all-terrain vehicles and global positioning systems (GPS) help track locations and facilitate assessment of multiple pest species over larger areas. Computers and telecommunications devices such as personal digital assistants (PDAs) and smart phones are useful for electronically recording samples in the field, providing quick access to weather data, and otherwise assisting in efficiently sampling and managing multiple pests in a field, orchard, or landscape.

Sampling Pattern. The path followed when collecting samples (the sampling pattern) affects how representative and objective the assessment of the pest population will be. Different sampling patterns achieve different objectives; but with all sampling, it is important that the patterns be unbiased, otherwise the potential for error increases. Several different types of sampling patterns are used for pest monitoring, including random, stratified, and systematic sampling.

Random Sampling. The most common sampling pattern is random. With random sampling, every sample unit has an equal chance of selection, proving an unbiased estimate of the population. Conducting random sampling is not easy; the sampler has to concentrate to avoid favoring certain sample types, which would skew the results so the samples would not be representative of the sampling universe. Sampling the easiest plants to reach, damaged plants, or the damaged part of a plant is a natural tendency in sampling, but it will invalidate the objectivity of the *random sample.* In a tomato field, for instance, it is natural to focus on red tomatoes during sampling because they are seen first. The randomness of the sample is lost, as these fruit may be the most likely damaged.

To assist in keeping the sample random, predetermined systematic patterns, such as those shaped in a U, X, or V, are often used to walk a field (Figure 6-11); a different sample pattern can be used at each sampling interval. Samples can be taken at randomly selected intervals; for instance, walking 10 steps to take one sample, thirty to take the next, and so on based on a printed table or electronically generated series of random numbers. Another technique to help keep the sample selection random and prevent zeroing in on the specific pest or damage symptom is to throw rings or tennis balls along the sampling route and sample where the objects fall. This technique is often used when sampling for weeds.

Stratified Sampling. Another way to help keep the sampling random and allow for better interpretation of results, especially when there is great variability in the sampled area, is to stratify the sampling universe. *Stratified sampling* partitions the habitat into subunits that may vary in character. Each subunit (or stratum) is then independently sampled. Sampling efficiency is improved by isolating the different types of habitat within a field, such as high and low areas, parts of a field with low vigor, or the north and south sides of a tree. This

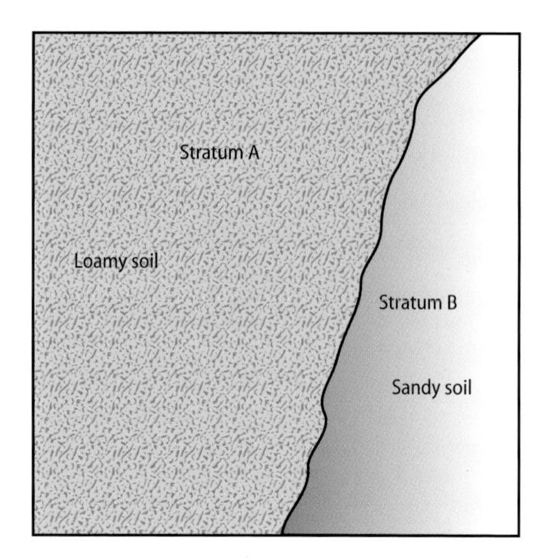

FIGURE 6-12

In stratified sampling, a field is divided according to physical or environmental factors that may make a big difference in pest damage or numbers—in this case, soil type. Each stratum (the loamy soil area and the sandy soil area) is sampled, evaluated, and treated separately. A field may be divided into several strata if necessary.

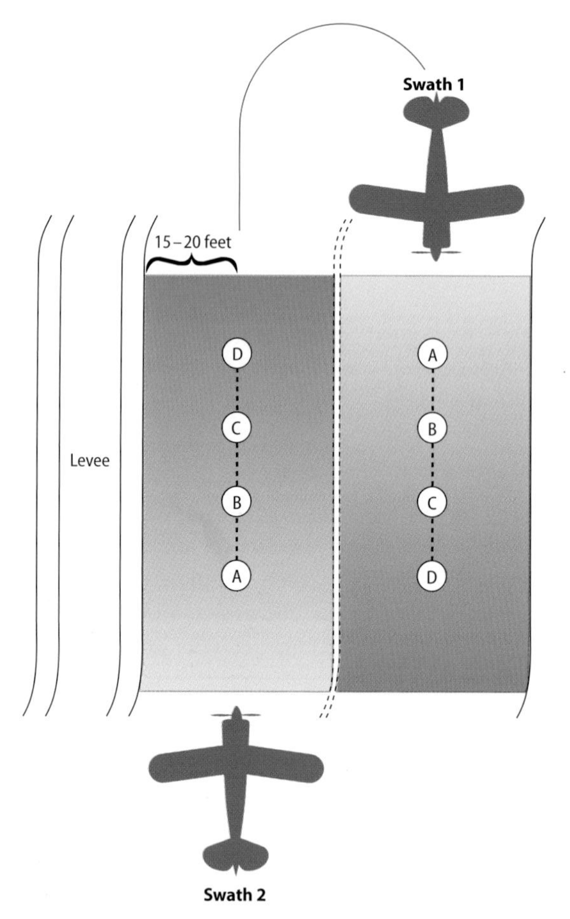

FIGURE 6-13

In systematic sampling, each sample unit is chosen systematically from a randomly selected starting point. For instance, in sampling for rice water weevil, samples are taken in rows about one airplane swath width apart.

helps ensure that samples are collected from all areas of the habitat.

In Figure 6-12, a field has been divided into areas with different soil types that may affect pest numbers. Each area type (stratum) is then treated as a separate universe and samples are randomly collected from each. Information resulting from stratified sampling can indicate the in-field distribution of a pest species as it relates to the factors you separated out in the stratum (in this case, soil type).

Stratified sampling is important in research and ecological studies and requires knowledge of the field before sampling. Biological knowledge of the pest is also important and helps to eliminate habitat where the pest is not found, increasing the efficiency of the sample. A stratified sampling program that treats field edges separately from the field center is helpful in many agricultural settings. A common sampling practice is to avoid the edge of the field and move in toward the center before beginning to sample. A consequence of this, particularly in small fields, is that a substantial portion of the habitat is not sampled. This can be especially problematic if the pest species is likely to move from the field edge to the center or is likely to build up to damaging levels along the field edge. Spider mites, for example, are likely to build up along field edges, particularly dusty ones. Sampling the edge can help ensure the population is not missed, detect an early population, monitor movement into the field, or lead to a recommendation to spot-control the population at that site. However, sampling the field edge or other part of the system separately makes economic sense only if the edge or part can be managed differently from the rest of the field or if sampling is used as an early warning for problems that may lie ahead.

Systematic Sampling. Systematic sampling is widely used, largely because it is easy to implement. Each sample unit is chosen systematically from a randomly selected starting point. Examples include sampling every 50th plant, taking a sample every 3 feet (0.9 m) in a row, sampling in a zigzag pattern, or pulling a sample every fifth time around (Figure 6-13). Systemic sampling can be appropriate when the sampling universe is relatively homogenous, such as a machine-planted row crop.

SAMPLING AND DETECTION TOOLS AND TECHNIQUES

The success of any monitoring program depends on the effectiveness and efficiency of the detection and sampling tools and techniques. Depending on the objective, tools include traps, nets, microscopes, and molecular test kits. Techniques include knockdown, netting, trapping, and visual inspection. Looking for characteristic damage symptoms or signs of pest presence (such as feces or fruiting bodies) can also be useful, depending on the pest. Various methods and tools for monitoring invertebrate species are listed in Table 6-5.

TABLE 6-5.

Selected invertebrate pest monitoring methods and tools. Beneficials may be monitored, too.

Method or tool	Pest monitored
visual inspection	Most exposed-feeding invertebrate species and their damage. Evidence of parasitism and predation. Monitoring may require a hand lens or other magnifier.
timed counts	Individuals of exposed beneficial and invertebrate pest species (e.g., lady beetles, caterpillars) or certain types of damage (rolled leaves, twig strikes) that are relatively large and obvious but occur at relatively low density so they are not observed faster than they can be counted.
knockdown techniques: branch beating, shaking plants, or tapping containers over a collecting surface such as drop cloth or clipboard with a white sheet of paper	Adults and larvae or nymphs of easily dislodged invertebrate species, including bugs, lacewings, lady beetles, leaf beetles, leafhoppers, mites, nonwebbing caterpillars, adult parasites, psyllids, thrips, and adult whiteflies.
suction techniques (D-VAC)	Relatively mobile adults and larvae or nymphs of invertebrate species including bugs, caterpillars, adult whiteflies; beneficial species such as big-eyed bugs, minute pirate bugs, lacewings, and spiders.
sweep nets	Adults and larvae or nymphs of invertebrate species, including weevils, caterpillars, and bugs that are free-living on foliage.
rotary traps	Flying and windborne invertebrates such as winged aphids.
bait traps	Invertebrate pests such as flies, ants, cockroaches, snails and slugs, and certain species of moths.
pheromone traps	Many moths, certain beetles, males of some scales, and other insects; certain parasite adults are attracted to their host's pheromone.
light traps	Night-flying adults of moths, some beetles (e.g., chafers, white grubs, other scarabs, and some leaf beetles), lacewings, and others.
sticky traps	Adults, including fungus gnats, leafminers, psyllids, shore flies, thrips, whiteflies, winged aphids, and parasites.
pitfall traps	Adult weevils, predaceous ground beetles, ground-dwelling spiders, Collembola, and possibly others such as squash bugs.
carbon dioxide exhalation and shaking	Thrips hidden in buds are stimulated to move by a long, gentle breath into terminals, and are dislodged and revealed by shaking plant tips over white or black paper on a clipboard.
indicator or key plants	Most exposed-feeding species and their damage. The same infested plants are inspected before and after treatment to determine whether pest is in the stage that is susceptible to the control action, to compare numbers to previous sampling, and to determine effectiveness of control.
degree-day phenology	Many pests and beneficial species for which temperature development thresholds and rates have been determined.
soil drench or flushes using pyrethrum, soap, or water	Relatively mobile species in soil or hidden places, including centipedes, millipedes, symphylans, and larvae of fungus gnats and shore flies. Thrips and other species in buds may be flushed.
trap boards	Adult weevils, snails.
potato traps	Root-feeding fungus gnat larvae and symphylans, which migrate to feed on underside of potato pieces. Push pieces into soil and pick up and examine the underside of each disk and the soil surface for larvae once or twice a week.
host collection and rearing	Immature stages of species such as parasitoids that feed inside host. Only the adult stage of many insects can be positively identified to species.

Visual Sampling

Visual sampling involves examining a specified number of plants, plant parts, or infested areas for pests or their damage. It is the most common sampling technique for most types of pests. An alternative to counting the numbers of pests per sample is to visually inspect for the presence or absence of the pest species or damage, as discussed earlier in this chapter (Figure 6-14).

Another method to visually sample is to count the number of individuals that can be seen during a predetermined time interval. These timed searches are used to monitor fruittree leafroller, western tussock moth, and citrus cutworm in citrus. Treatment decisions are based on the number of larvae counted during a fixed amount of time (e.g., larvae per hour of search). Considering the potential for damage (e.g., the crop or fruit development stage) together with the larval counts often gives a better estimate of the problem. This is true with other sampling techniques as well.

Weed Surveys. Weed management programs survey weed populations to identify the species present, record their abundance and stage of development, and monitor population changes from year to year. Weeds are usually sampled visually at least twice a year. Survey for winter annuals during fall and winter and for summer annuals during spring and summer. Check for perennial weeds during all surveys, but be aware that certain species die back and become dormant during some seasons. The first weed survey for a field crop, if possible, should occur while the previous crop is in the ground. Repeat weed surveys during the growing season to identify any new weed species and to check on the effectiveness of weed control.

Weed surveys commonly involve walking through a field in a random pattern and rating the degree of infestation for each weed species on a weed survey form (see Figure 6-35). It is helpful to sketch a map of the field, noting areas of high weed density, perennial species, and the location of fence rows, ditch banks, and any other features that may impact management.

In many tree crops, weeds in row middles can be beneficial as long as they do not include perennial species or interfere with activities such as harvest. In tree crops, weeds in and between rows are often recorded in separate columns on the same survey form. Examples of weed survey forms for various crops are available with the UC IPM Pest Management Guidelines at www.ucipm.ucdavis.edu.

Knockdown Techniques

Knockdown techniques sample arthropod pests that are easily dislodged from their habitat. Several methods are available for a variety of pests and cropping situations; all rely on physically removing pests and beneficials (if present) from a portion of a plant

FIGURE 6-14

Visual sampling includes counting the number of pests per leaf or area or inspecting for the presence or absence of the pest.

onto a tray, cloth, white-painted garbage can lid, or other collection device, then counting them. Sweep nets placed under the canopy can also be used. Knockdown techniques work better in warmer weather when pests and beneficials are more active.

A collection device is held beneath a branch or plant as a collecting surface to take a beating sample. Arthropod pests are dislodged from trees by beating two to four branches from different parts of each of several trees (Figure 6-15); in strawberries and tomatoes, pests are dislodged by vigorously shaking the foliage. Count the number of individual pests and record them separately for each species. Determine the average pest density for each species by dividing the number of pests by the number of samples:

Average pest density = number of pest individuals in all samples ÷ total samples inspected

Alternatively, for a quicker but less accurate estimate, instead of counting each individual pest, record whether or not a particular species is present in each beat sample (presence-absence). Then estimate the relative pest density for each species by dividing the total number of samples containing that species by the number of branches or plants beaten:

Relative pest density = number of samples with species present ÷ total samples inspected

The drop cloth is another common sampling tool, especially in vegetable and field crops. It is good for collecting species such as lepidopterous caterpillars, leaf beetles, and stink bugs that readily fall to the ground when disturbed. A sheet, roll of plastic or canvas, or tray is placed along a 12- to 18-inch (30- to 45-cm) section of the plant row; shaking the plant dislodges insects. For example, in the case of leafminers in tomatoes, as the maggots prepare to pupate, they drop onto the trays placed under the plants and can be monitored (Figure 6-16). A white or light-colored cafeteria tray works well, too. The drop cloth method is highly efficient and low in cost but can be time consuming.

Suction Techniques

Vacuum removal of arthropod species using suction equipment, such as the D-Vac, modified leaf blowers, or other suction devices, dislodges pest species that are somewhat mobile. Vacuum devices must have enough suction to capture target arthropods without destroying them or injuring the plant being sampled. For instance, lygus bugs in strawberries are sometimes captured with suction

FIGURE 6-15

Certain pest species such as adult acacia psyllids are monitored by branch beating or dislodging them onto a collection device that is held beneath a branch or plant.

FIGURE 6-16

Trays placed under tomato plants can be used to monitor leafminer maggots as they drop to the ground to pupate.

machines (Figure 6-17). Vacuum devices are more effective than beating sheets for detecting lygus bugs and other particularly mobile insects that are found on the plant canopy, but they are not appropriate for other kinds of insects, such as soft-bodied aphids, which would be crushed in the device. They can be effective when plants are too small to be sampled with a sweep net.

Netting Techniques

Many arthropod pests are sampled using netting techniques because of their ease of use and low cost. Sweep nets are used in many agronomic crops to dislodge arthropod species for collection. For example, sweep nets are used in alfalfa to sample Egyptian alfalfa weevil larvae, alfalfa caterpillars, and armyworms. A standard sweep net consists of a cone-shaped cloth bag 2 feet (60 cm) deep fitted into a wire loop 15 inches (38 cm) in diameter and attached to a handle 26 inches (65 cm) long. Using these measurements for all sweep nets helps to ensure that the area sampled will be consistent. In alfalfa, a single sweep consists of a 180° arc taken when stepping forward (Figure 6-18). Sweeps may be taken singularly or consecutively. Individual insect species are separated after a series of sweeps; counts of each species are based on an average of all samples.

Trapping Techniques

Traps are used to sample mobile insects. They attract insects either through visual or chemosensory attractants or randomly catch them. Bait traps provoke a response by using the pest's food source as a stimulus. For example, bait traps are used to monitor driedfruit beetles in figs. These traps, baited with culled fruit, water, and yeast, attract beetles before the fruit begins to ripen (Figure 6-19). When trap counts begin to drop off, it may mean that the beetles are being attracted to the fruit and should be controlled. Bait traps are also used for moth pests, such as oriental fruit moth, where mating disruption with pheromones is being

FIGURE 6-17

Vacuum devices can be used to monitor mobile insect pests such as lygus bugs in strawberries.

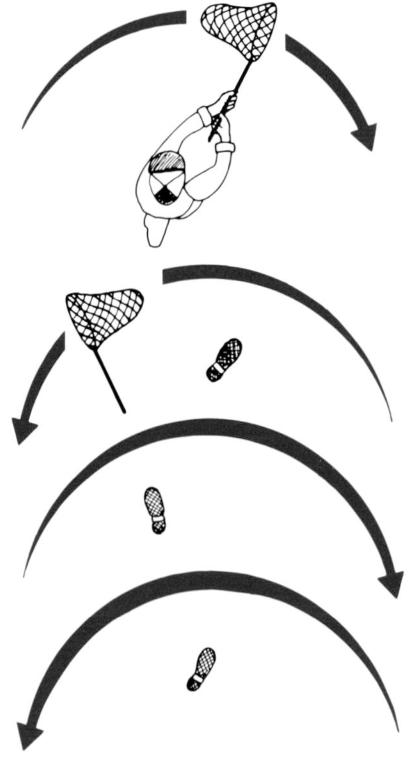

FIGURE 6-18

When taking sweep net samples, swing the net from side to side in a 180° arc, keeping the net below the tops of the plant until the end of the sweep.

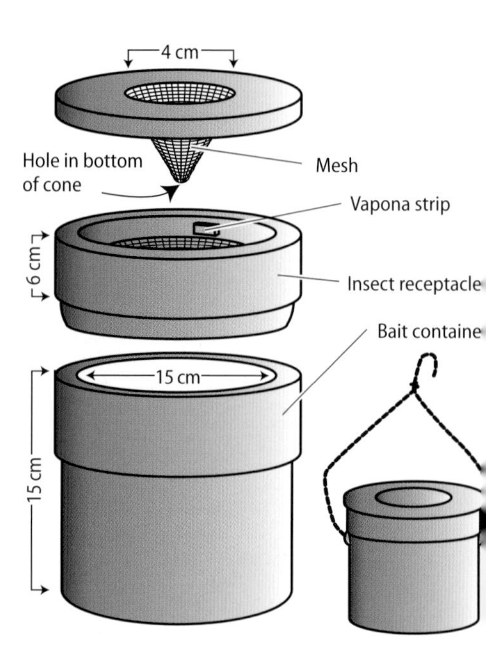

FIGURE 6-19

Bait traps for driedfruit beetle should be placed in fig tre before the fruit ripens and becomes attractive to the insec The bait trap consists of a large container that is baited with culled fruit, water, and yeast.

used as a management technique and pheromone traps are therefore not effective. Using bait traps for oriental fruit moth monitors population changes over time as with pheromone traps, except that both males and females are caught.

Pheromone traps. Pheromones are chemicals that allow insects to communicate with other individuals of the same species. Most pheromone traps mimic the sex pheromone that pests release to attract an individual for mating. These chemicals are added to various types of dispensers, or lures. Pheromone traps (Figure 6-20) containing a pheromone lure are available in different styles and have different efficiencies based on the behavior of the insect species being attracted.

Pheromone traps allow managers to determine when reproductive adults are in the field, aiding in the timing of management actions. These traps operate by attracting one sex, usually the male of the species; they are very economical and are specific to the target species. A few of the species for which pheromone traps are used in California include codling moth, oriental fruit moth, peach twig borer, pink bollworm, citrus cutworm, tomato pinworm, potato tuberworm, several clearwinged moths, California red scale, and San Jose scale. Bucket traps are used for variegated cutworm, black cutworm, western yellowstriped armyworm, cabbage looper, beet armyworm, and tobacco budworm. DA lure traps for codling moth contain a plant attractant (a pear ester) and are useful for monitoring codling moth in orchards where a mating disruption program using pheromones is being used. These lures are also sold in combination with a pheromone.

A few commercially available pheromones are not sex pheromones. An example is the consperse stink bug pheromone, which is an aggregation pheromone, attracting males, females, and some nymphs. Commercially available traps available for cockroaches feature an attractant and a sticky coating on the inside to which the lured insects adhere. These traps are more important as an indicator of cockroach infestations than as a control. Aggregation pheromones are also used in traps for detection of bark beetles as well as for management of certain species.

Light traps. Light traps are another type of trap that attracts and captures arthropod pests, including many species of night-flying insects, pest and nonpest alike. A typical light trap consists of an ultraviolet fluorescent light above a collection jar that contains a killing agent. Light traps have been used primarily to monitor moths, some beetles,

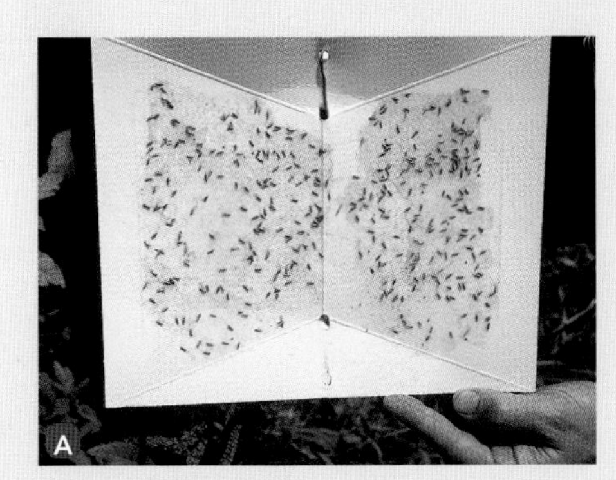

FIGURE 6-20

Examples of pheromone traps used to monitor various insect pests in California: tomato pinworm wing-type trap (A); stink bug jar trap (B); and California red scale card trap (C).

FIGURE 6-21

A sticky trap used to monitor small insect pests in chrysanthemums in a greenhouse is placed vertically so that the bottom of the trap is even with the top of the canopy. As the plants grow, move the trap up so that the trap bottom remains even with the top of the plant canopy.

FIGURE 6-22

Life cycle of a typical soft scale. Monitor crawlers with traps of double-sided sticky tape. Place two or more traps in each of two or more plants at each location in spring before the crawlers hatch. Wrap the traps around twigs near female scales.

and mosquitoes. They can also be useful for monitoring the emergence or migration dates of certain other pests. One disadvantage of light traps is that they attract so many different species that considerable time and skill is required to sort the samples. Because they may also attract insects from a considerable distance outside the field, they are best used as an indicator of relative abundance in a general area rather than in assessing population densities. Light traps are most commonly used in research, but they can also be used as a trigger to indicate the presence of certain species, indicating when other sampling techniques should be initiated.

Sticky traps. Adult whiteflies, thrips, aphids, and other pests and beneficial insect parasites are monitored with sticky traps. Sticky traps either randomly catch insects or attract them through the use of color. The ones used in most crops are bright yellow cards, each 3 by 5 inches (7.5 by 12.5 cm) or larger, covered with clear sticky material. Research has shown that certain wavelengths of yellow (about 550 to 600 nm) are more attractive to many insects than other wavelengths.

Light blue sticky traps can be used if the primary species of concern is western flower thrips. Blue traps can capture more thrips than yellow sticky traps, but they are more difficult to discern against the darker color. In most situations, the yellow traps are more efficient because the yellow attracts thrips as well as other insects, so the same trap can be used to monitor many different pests. To catch the most insects, hang the traps vertically, so that the bottom of the trap is even with the top of the plant canopy (Figure 6-21). Placement of traps depends on the crop and the pest. A general recommendation for most greenhouse insect pests is to use one trap per 10,000 square feet (930 sq m) of growing area.

The effectiveness of sticky traps is influenced by the stickiness of the adhesive; as it loses its ability to capture insects, by getting dirty or wet or after extensive use, trap efficiency declines. Inspect trap cards weekly. Replace traps when they become fouled or when they lose their adhesiveness.

Sticky tape traps (Figure 6-22) are used

to monitor crawlers of many scale insects, including California red scale in citrus, San Jose scale in fruit and nut trees, and various scale insects in ornamental plants. The tape, transparent cellophane tape that is sticky on both sides, is wrapped tightly around small branches before crawlers emerge in the spring. When crawlers hatch and begin searching for sites to settle, large numbers of them get stuck on the tape, signaling appropriate treatment time. By preserving tapes between white paper and clear plastic, a permanent record can be maintained for visual comparison of trap results among sample dates.

Pitfall traps. To randomly capture arthropod species, such as ground beetles, spiders, and Collembola found on the soil surface, pitfall traps are often used (Figure 6-23). These traps consist of a glass or plastic container sunk into the ground with the mouth even with the soil surface; they capture any arthropod that falls in. The traps can be baited or treated with a preservative to immobilize the trapped arthropod. Pitfall traps are cheap and easy to use, but catches are difficult to interpret. Be careful when locating traps with preservatives in them in areas where pets, wildlife, or children can access them. Population size, density, and movement influence trap efficiency.

FIGURE 6-23

Pitfall traps are used to monitor the population of arthropod species on the soil surface, for example, black vine weevils in landscapes. The trap consists of a funnel or smaller cup in a collecting jar buried in the soil so that beetles walking on the soil surface fall into it.

Damage Estimates

In many situations, the first indication of a pest problem is damage visible on the host. Damage can be a key indicator for certain pest species. Estimating the type, amount, and patterns of damage may provide the only indication of which pest is responsible. When using plant damage as a sampling tool, specifically note the plant part damaged (such as fruit or leaves) and the degree of damage. Observe the density of the stand in damaged areas versus nondamaged areas. Look for patterns.

Looking for damage symptoms is often a good technique for pests that are hard to detect. For example, katydids in citrus or nectarines take a single bite from a fruit and move to another feeding site; they eat holes in leaves and maturing fruit. A few katydids can damage a large quantity of fruit in a short time. Damage symptoms are used to initiate sampling for the presence of katydids, and if katydids are found, a control option is initiated.

When using damage estimates to monitor pest populations, severity indices are often used. A damage severity index rates the degree of damage to the sample unit (per leaf or per plant) to indicate whether a treatment is needed. For example, the weighted nematode rating (WNR) system (see Figure 6-9) assesses root gall symptoms in fields where cotton will follow cotton. The rating system is based on the relative intensities of root galling throughout the field or in stratified sections of the field.

Levels of pest damage in previous years' harvested crop are often used in decision making. For instance, the percentage of walnuts or apples infested at harvest in previous seasons gives an indication of the codling moth severity in an orchard and the need for treatment of the first generation the following spring.

Damage estimates must be used with caution, however. The level of damage inflicted on the crop or plant is also influenced by other factors, such as soil type, climate, water, age and health of the plant, in addition to the size of the pest population. These factors can combine to increase damage to the host or mimic pest damage. For example, heavy lygus bug infestations in Acala cotton can result in flower bud (square) loss, but abiotic factors such as cool weather can also cause the same damage. The incidence of disease, which also can cause square loss, increases during periods of cool, moist weather. Sweep net samples should be used in conjunction with square monitoring to confirm that lygus levels are high enough to cause observed damage and justify treatment.

Damage sampling is only an indicator of pest activity and should be combined with other sampling methods for accurate decision making. Also, make certain that susceptible stages of the pest are present before making treatment decisions. Often, once damage symptoms become apparent it is too late to effectively control a pest to prevent further damage. In some cases, the pest may be gone.

Clues That Indicate Presence of a Pest Population

Besides damage, pest populations often leave other visible clues indicating their presence. For example, the California oakworm, which can cause extensive defoliation to oak trees, leaves fecal pellets (frass) that are characteristic of the species (Figure 6-24). Several methods of sampling frass are used to estimate density and damage. Sticky cards, shallow trays, or cups are placed beneath the canopy; these frass traps are collected at regular intervals, and the volume of frass is recorded for comparison with other sampling dates or locations.

Several species of small mammals such as voles can damage crops, orchards, and landscapes. Often the only indication of their presence is their network of runways and burrows or damage to the crop or plant. Become familiar with the various vertebrate species that damage different crops and landscapes.

Damage caused by vertebrate pests is often mistaken for other pest damage, stress, or other environmental factors. Close inspection, however, will reveal chewed roots, gnawed trunks, or eaten fruit. Monitoring for the first signs of tracks, droppings, burrows, or mounds can help prevent some of this damage (Figure 6-25). Learn to recognize these signs early.

Honeydew and sooty molds provide clues that populations of aphids, whiteflies, or soft

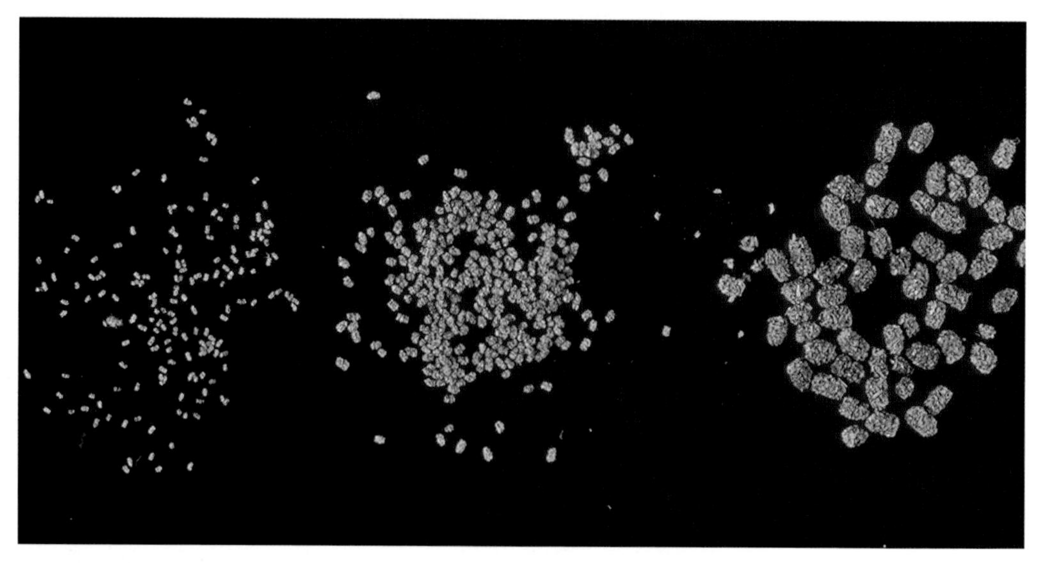

FIGURE 6-24

Frass collection can be used to monitor the California oakworm. The larval instar present can be determined by the size of the pellets. The three piles of frass (from left to right) were produced by the first instar, third instar, and fifth instar.

FIGURE 6-25

Clues that indicate the presence of vertebrate pests. The presence of vertebrate pests is often indicated by various clues such as tracks, droppings, burrows, or mounds. By monitoring early and knowing the vertebrate pests associated with your site or crop, you can use a combination of these clues to positively identify which pest is present.

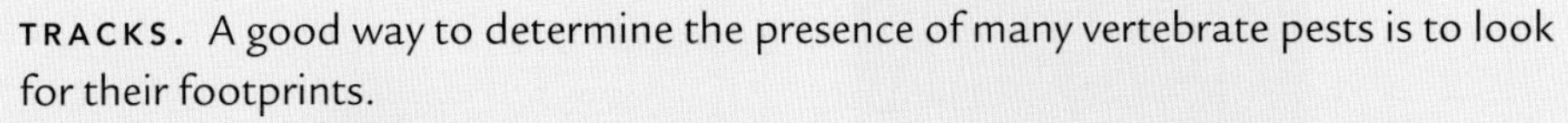

TRACKS. A good way to determine the presence of many vertebrate pests is to look for their footprints.

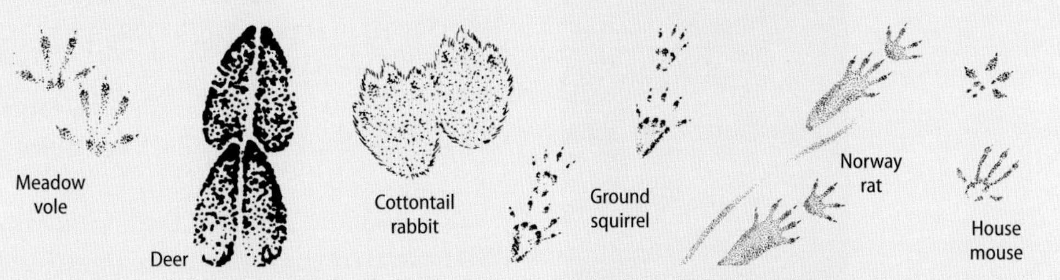

DROPPINGS. The size, shape, and grouping of vertebrate droppings often provide clues that help with identification.

rabbit droppings

deer pellets

MOUNDS AND BURROWS. The size and shape of burrows, where they are located, and whether they are shallow or deep or part of a network of runways and burrows can be used to identify vertebrate pests.

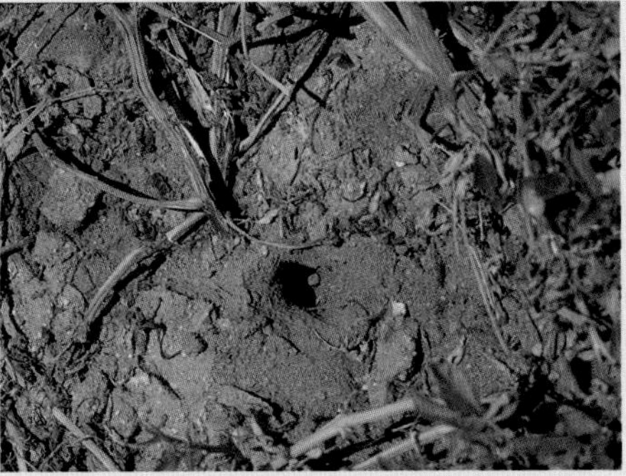

meadow vole burrows and runways

ground squirrel burrows

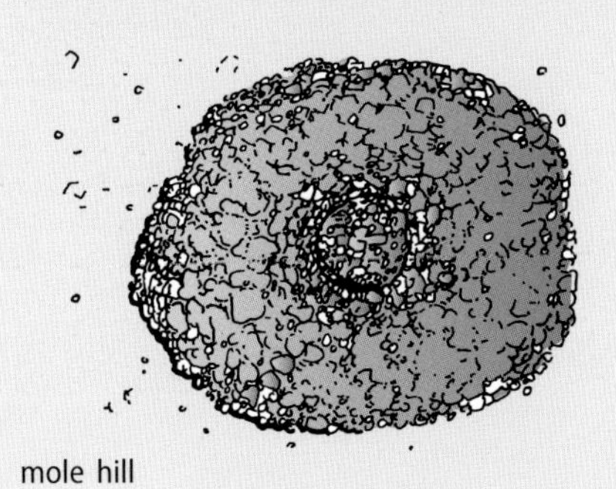

mole hill

gopher mounds

SIDEBAR 6-3

Honeydew Monitoring

Use water-sensitive cards to efficiently monitor honeydew from sucking insects such as aphids. Start monitoring in spring beneath two or more plants that have had pests in previous years. Attach cards to a stiff background. Use four cards per plant, placing one card beneath the lower outer canopy of each quadrant of the plant. Place the cards beneath the plants for the same period once each week, such as from 11 AM to 3 PM on a day when no rain or irrigation is expected.

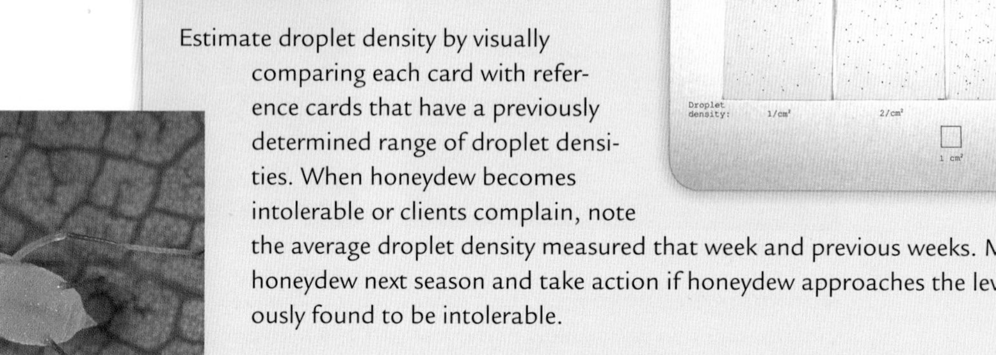

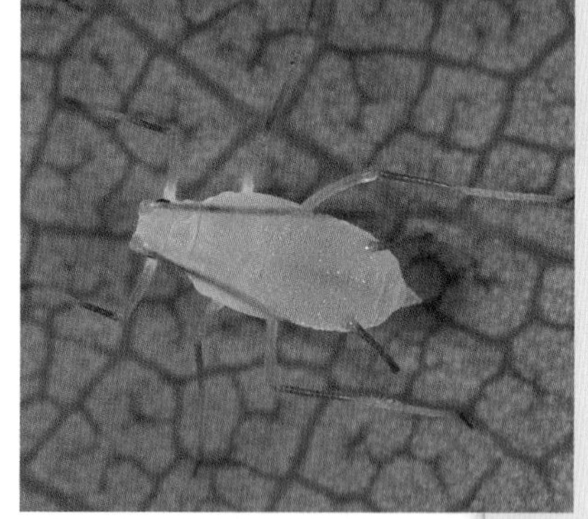

Estimate droplet density by visually comparing each card with reference cards that have a previously determined range of droplet densities. When honeydew becomes intolerable or clients complain, note the average droplet density measured that week and previous weeks. Monitor honeydew next season and take action if honeydew approaches the level previously found to be intolerable.

Above, examples of honeydew droplet density on cards placed under tulip trees to directly measure pest damage from tuliptree aphid.

To calculate average aphid honeydew density, count the total drops on each card and divide by the area of each monitoring card. A sample monitoring form is shown below.

PLANT SPECIES Tulip Tree MONITORING DATE 6 June 93
MONITORING TIME 11am-3pm PERSON MONITORING Jan Doe

Area of each monitoring card = length x width = 3 x 2 = 6 sq. inches

Tree Location	Card Quadrant (or No.)	Total Drops	Card Area	Total Drops / Card Area =	Average for Each Card
4th & Main	E	68	6		11.3
	N	49	6		8.2
	W	71	6		11.8
	S	92	6		15.3
				Sum of Averages	46.6

Number of cards 4

Sum of averages / Number of cards = Overall average = 11.7 drops/sq. in.

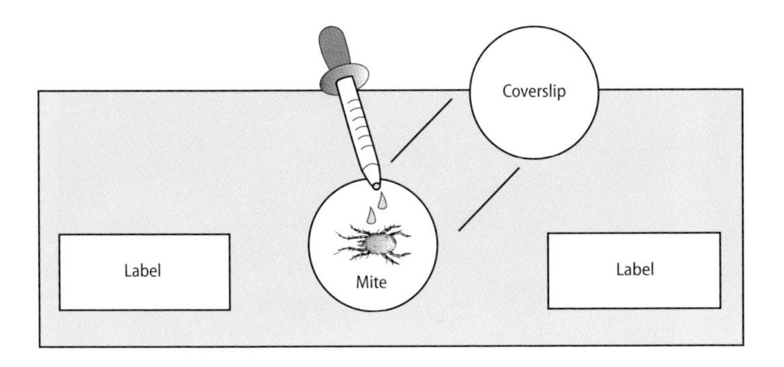

FIGURE 6-26

To positively identify small arthropod species such as tetranychid spider mites, place a male spider mite onto a microscope slide. Cover the specimen with two or three drops of Hoyer's mounting solution. Cover with a microscope cover glass. The species of spider mite is identified using a phase contrasting compound microscope to examine features such as the arrangement of hairs on the body and legs and from the shape of its aedeagus (male sexual organ).

scales are present. Honeydew is sugary water excreted by many homopteran species that ingest sap. It is generally harmless to plants, except when it becomes so abundant that black sooty mold, a dark fungus, grows, affecting plant growth or contaminating harvestable commodities. Columns of ants on trunks can also indicate that plant-juice-sucking insects are present, because many ants feed on honeydew.

Honeydew can be efficiently monitored using yellow water-sensitive paper or cards. Place cards beneath the plant or suspend cards several inches beneath the branches of trees. Estimate the average honeydew droplet density by visually comparing each card with reference cards that have a range of droplet densities previously determined (see Sidebar 6-3).

Laboratory Tests

When sampling for pests, it is often necessary to take field samples back to the lab or office for further examination. Depending on the monitoring procedure, a simple laboratory or office outfitted with a microscope and other basic equipment may be sufficient. In other cases, it is necessary to send samples to laboratories equipped with specialized equipment and staffed by trained specialists. Laboratory analysis or observation is almost always needed to confirm the identification of certain pests, such as plant pathogens or nematodes. However, diagnostic test kits for some pathogens are available, which makes diagnosis of certain pathogens possible in the field.

Using Microscopes for Arthropod Identification. Many tiny pest and beneficial species must be viewed under a microscope for positive identification. Tetranychid spider mites, such as the twospotted mite, can be positively identified to species only by fixing them on slides and viewing them with a compound microscope. Mites are placed on a slide with a droplet of Hoyer's mounting solution and pressed with a coverslip (Figure 6-26). The pressure causes the male sexual organ (the aedeagus) to be extruded; each species of mite has a differently shaped aedeagus.

In situations where parasites are important in managing an insect pest population, determine the percentage of parasitism by collecting the pest and confirming the presence of the parasite. To confirm parasitism in California red scales, for instance, flip over the third instar female scale cover and check under a microscope for parasites in or on the scale body. Figure 6-27 shows the egg, larval, pupal, and adult stages of the parasite as they appear under the scale cover.

Nematode Soil Samples. Soil samples are regularly taken for nematodes (Figure 6-28) and submitted to a qualified laboratory for extraction and, if necessary, identification of the nematode species present. It is important to sample when the soil is moist (less than 60 centibars) but not saturated. Contact the laboratory in advance and schedule sampling so that the lab can process the samples as soon as possible. Samples should be handled carefully and transported quickly to avoid overheating or drying out. Farm advisors, Extension agents, or other authorities can help find a lab equipped for extracting and identifying nematodes from soil samples.

Soil Tests for Weeds. Soil samples can also be collected and the weed seeds germinated to identify the presence of certain nondormant weed species. These tests are especially valuable when a weed survey

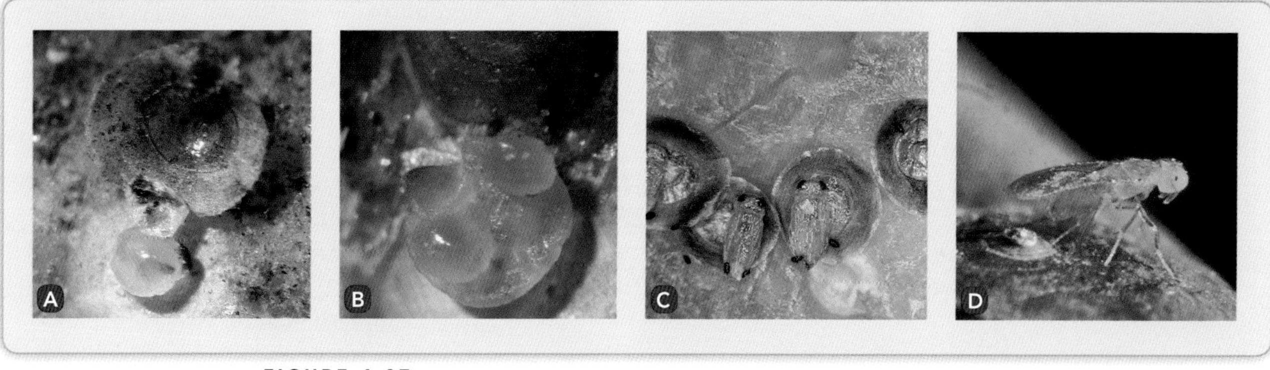

FIGURE 6-27

Confirming parasite presence in California red scale. (A) To confirm parasitism by Aphytis melinus *Debach, turn over the scale cover of third-instar California red scale.* Aphytis *prefer to oviposit on the third-instar scale larvae because of the instar's large size. (B) To recognize the larval stages of* Aphytis, *look for an elongate sac with body segments evident; individuals in this stage will vary in size. Two parasite larvae are feeding on this scale. (C) The pupal stage of* Aphytis *is recognizable from the development of eye pigmentation, which is initially clear but over 4 to 5 days changes from red to reddish brown and finally turns green. (D)The adult* Aphytis *emerges 1 day after the eyes turn green. Photos by M. Badgley, L. Forester, and E. Grafton-Cardwell.*

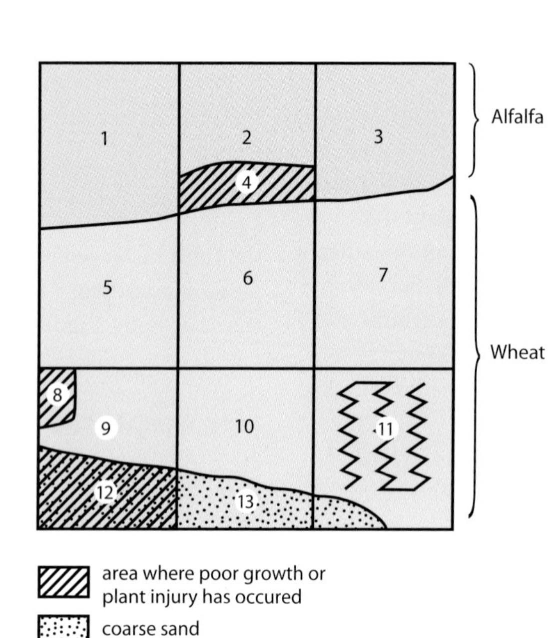

FIGURE 6-28

Collection and care of samples for identification of a soil nematode pest. Collect samples from areas that show symptoms (if present) and from healthy areas. Keep samples from different areas separate and label them. Unless other methods are recommended, collect at least one sample from each area; each sample should consist of 15 to 20 cores. Sample from the root zone of each plant and include the roots. Collect cores at random, using a zigzag pattern as shown in block 11. Keep samples in an insulated cooler at 54°F (12°C). Pack in a sturdy cardboard box or coffee can and immediately send to a processing lab. The field shown is stratified into separate sampling universes showing a recommended sampling pattern.

cannot be done or prior weed management information for a field is not available. They are also useful in testing the effectiveness of an herbicide application. Germination tests are not good indicators for all weeds—for example, the parasitic weed dodder would be missed—but they nonetheless provide a starting point for building a weed management strategy.

Collect soil samples from the top 2 to 3 inches (5 to 7.5 cm) of soil at random from at least 10 to 20 locations in the field. Mix samples from each area, keep them moist in a greenhouse flat, and place them indoors in a warm, sunny location. Using weed identification handbooks, websites, and other materials listed in the "Resources" section, identify weed seedlings that emerge.

Plant Disease Diagnosis. Field diagnosis of plant disease is difficult and can be complicated by the presence of several pathogens and other interacting causes of disease. Samples must usually be sent to a laboratory for positive identification. Plant diseases have many symptoms similar to those caused by other pathogens, pests, and disorders. In some cases, characteristic fruiting bodies or other features of the pathogen may not be evident, or a previously unknown pathogen may be present. In any case, positive identification of many plant diseases can occur only by preparing

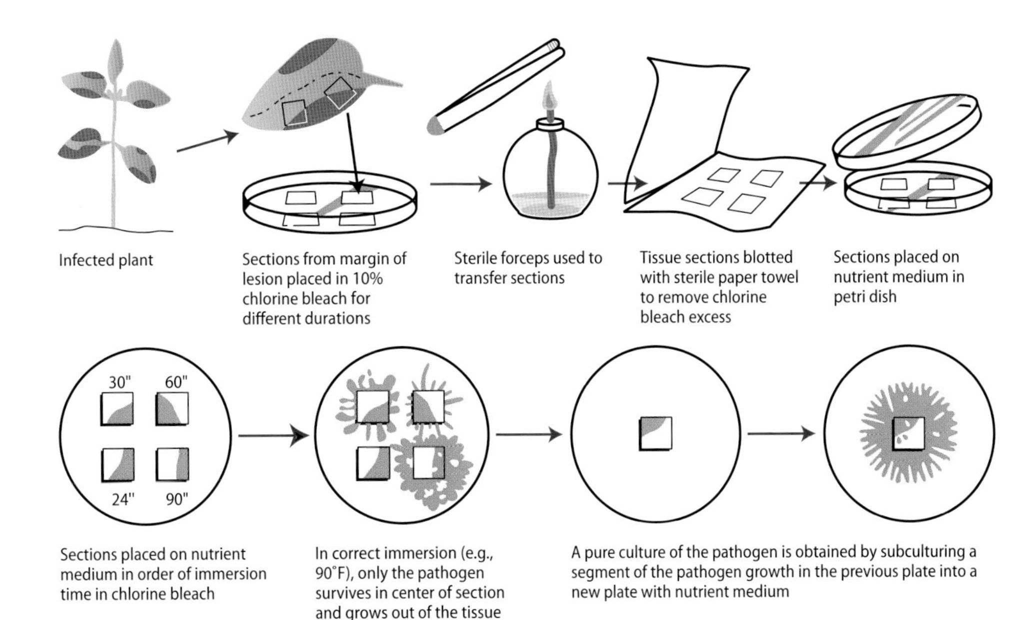

FIGURE 6-29

Isolation of fungal pathogens from infected plant tissue. After Agrios 2005.

cultures of the pathogen. If fruiting structures are present, they can be viewed under a microscope for identification. Often, a trained plant pathologist is required to confirm pathogen identity.

For some plant diseases, accurate identification involves isolating the pathogen from the diseased plant tissue. The most common method for isolating a pathogen is to cut several small sections of infected and healthy leaf, stems, fruit, or roots and place them in an agar nutrient medium. The steps involved in preparing an agar plate culture are shown in Figure 6-29. Infected tissue from a root or any plant tissue in contact with the soil must be thoroughly washed to remove all soil so that any saprophytic organisms present are removed. Fleshy roots or tissues are placed in a standard 10% bleach solution and soaked for 5 to 10 minutes to thoroughly wash adhering soil. These procedures are normally carried out in a plant pathology diagnostic lab that is specially equipped to perform these tests. Check with your Cooperative Extension office for lists of commercial labs.

Diagnostic Aids. Instead of sending samples out for laboratory testing, easy-to-use pathogen detection kits can be used in certain situations to identify common bacterial, fungal, oomycete, and viral plant pathogens. Field-usable test kits employ molecular methods such as immunochromatography, ELISA (enzyme-linked immunosorbent assay), and PCR (polymerase chain reaction).

The molecular methods used to detect pathogens in these test kits differ, but sample preparation and how they work are similar. Immunochromatography combines an immune response (antibodies that attack specific antigens such as pathogens) and chromatography (separating substances based on their different rates of movement through viscous gels or absorbent papers). ELISA detects pathogen-specific genes. PCR detects the species-specific protein coat of viruses. Diagnostic kits using molecular methods commonly involve macerating samples from diseased plants and using extraction and reaction solutions and coated detection containers or dip strips that change colors if specific pathogens are present.

Test kits are available for many plant pathogens including the oomycetes *Phytophthora* and *Pythium*, the fungus *Rhizoctonia*, pathovars of *Xanthomonas campestris*, which causes bacterial canker and blight, and viruses including impatiens necrotic spot and tomato spotted wilt viruses. Test kits offer the potential to detect pathogen presence in as little as 10

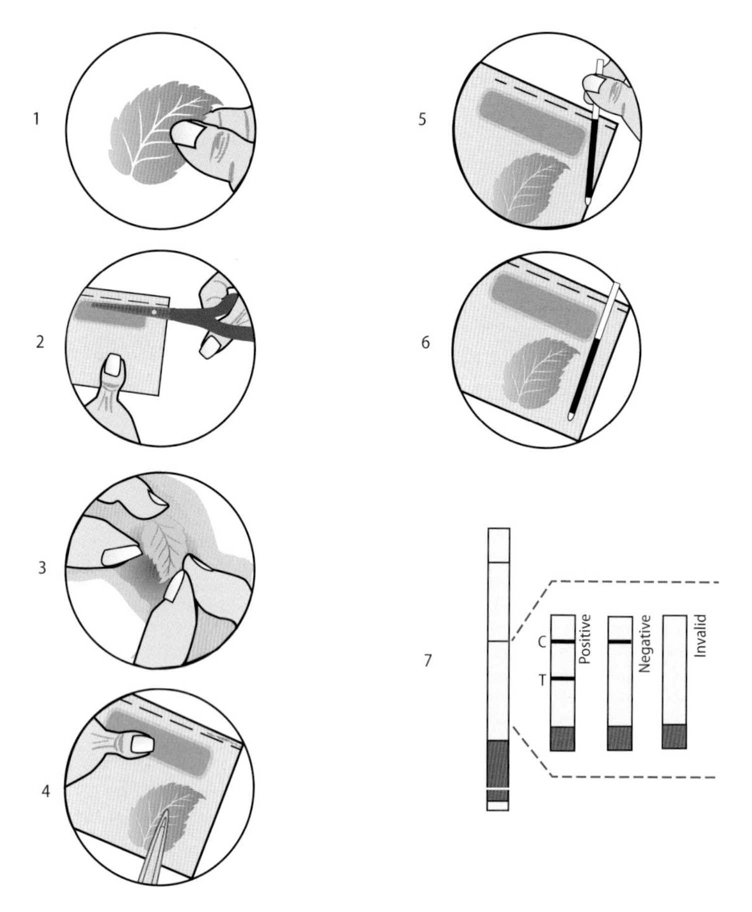

FIGURE 6-30

On-site test kits are available that can confirm the presence of many fungal, bacterial, oomycete,and viral pathogens. The Agdia ImmunoStrip Test kit is easy to use. (1) collect plant tissue suspected of being infected; (2) cut open sample extraction bag from kit; (3) insert sample into bag and seal; (4) rub the surface with a blunt object such as a permanent marker to extract the sample; (5) insert ImmunoStrip into the buffer-filled bag so that only ¼ inch is in the liquid sample; (6) allow the strip to remain in the sample for up to 30 minutes and (7) follow manufacturer's guidelines to interpret results.

minutes, for instance with the ImmunoStrip test kit illustrated in Figure 6-30.

Knowledge of whether specific pathogen inoculum is present allows for disease treatment decisions that maximize the effectiveness of control measures. Test kits require proper collection and careful processing of samples. Just because a pathogen is present does not mean that disease will occur or that the pathogen detected is the cause of any observed plant disease.

Pathogen test kits are currently being used primarily in floriculture and nursery crops to ensure that seeds or new planting stock is disease-free. They are also being used in research and to track the movement of newly established diseases that are spreading to new areas, such as tomato spotted wilt. New kits are continually being developed; contact a Cooperative Extension farm advisor or product suppliers for the latest information on pathogen test kits.

METEOROLOGICAL MONITORING SYSTEMS

Weather has a major influence on the development of plants and the pests that affect them. Temperature influences the rate at which insect and mite pests develop, and along with the availability of water and light, it also influences weed and other plant growth. Rainfall and humidity, together with temperature, are primary factors favoring the development of many foliar diseases, while soil temperature influences the development of certain soilborne pathogens, nematodes, and insects.

Having a reliable source of weather information is critical for making many pest management decisions. Keeping track of daily high and low temperatures is fundamental for using research-based biological models to predict pest and crop development. Disease forecast models incorporate moisture and temperature data. Weather influences the effectiveness of pesticide applications and can increase phytotoxicity or runoff.

Up-to-date weather information is also important for planning other management activities for the crop or landscape. Evapotranspiration data are useful for scheduling irrigation and calculating water requirements. Weather forecasts are essential for scheduling frost protection measures and irrigation activities for many crops and also to plan for protective treatments against foliar diseases. Accumulation of chilling hours helps predict future crops of stone fruits. Weather information must be incorporated into decisions about when to plant, when to harvest, and when to pollinate.

Weather data are also important for disease forecasting. Variations in weather can occur within short distances, owing to differences in elevation, storm paths, nearby large bodies of water, and the type of ground cover. Local variations in temperature, precipitation, relative humidity, and leaf

FIGURE 6-31

A computer-assisted weather station installed in a field can provide the most accurate weather data for calculating water budgets and using pest and crop models.

wetness therefore must be measured. To obtain the most accurate data, in-field weather stations are recommended.

Setting Up a Weather Station

In-field weather stations placed in key areas may be part of a network of weather stations that use radio telemetry to automatically transmit weather data to a computer that analyzes the data. A computer-assisted weather station generally consists of an electronic *data logger*, sensing devices, power supply, environmental enclosure, and a support structure (Figure 6-31).

Data Loggers

The data logger is the storage unit of the computerized weather station and consists of a microprocessor, storage memory, internal battery backup, transceiver, and power supply control unit. Data loggers collect data at specified intervals, store intermediate measurements in memory, process summary values at a specified reporting interval, and store the summary values. Data telemetry provides access to or transfer of summary values, typically on a near-real-time basis, to a central location for more general processing, long-term storage, and dissemination. The stored data can be downloaded into a computer to view, analyze, or graph the information.

Data loggers record signals generated by sensors. The sensors may be located in a crop canopy or may be part of a standardized reporting station at the edge of the field or other nearby location. Sensors typically monitor air temperature, precipitation, relative humidity (or other humidity requirements), and leaf wetness. Leaf wetness is a very site-specific variable; to improve accuracy, the sensor must be placed in the appropriate place relative to the crop canopy, yet be protected from direct exposure to sunlight. Other variables required for certain data loggers include solar radiation, soil temperature and moisture, wind speed and direction, net radiation, and light quality.

Stand-alone data loggers are also available for monitoring degree-days, chilling hours, or frost degree-hours and for tracking the powdery mildew index. These hand-held battery-operated loggers work without having to be attached to a microcomputer; they also measure and accumulate minimum-maximum temperatures.

Sources of Temperature Information

Temperature information can be obtained locally through radio stations or through a number of different sources on the Internet. The National Weather Service broadcasts local and regional weather information on National Oceanic Atmospheric Administration (NOAA) Weather Radio and VHF channels, and it also posts pinpoint local forecasts on the National Weather Service website. Evapotranspiration information is available from the California Department of Water Resources CIMIS program website, www.cimis.water.ca.gov. Current weather information is also available online by subscription from private weather services that can provide localized forecasts for a fee. The type of information available includes daily agricultural weather forecasts, 7-day temperature and rain forecasts, short-term forecasts, hourly weather observations, and daily weather summaries.

In California, temperature and other weather information is available through the UC IPM California Weather Database (see the UC IPM website, www.ipm.ucdavis.edu). The weather database provides current and historical daily weather data for approximately 400 weather stations throughout California. Choose a station and obtain daily data over a range of dates and use the data with the degree-day utility (DDU) or various phenological models to help make management decisions.

PREDICTIVE TOOLS

Increased knowledge about how weather affects pest and crop development has allowed the creation of research-based models that help pest managers forecast or predict events later in the season. These models, which include phenological models, disease forecasting, *expert systems*, and plant mapping, are not foolproof because they are based on a simplified conception of the factors that influence an organism's development. However, when backed up with monitoring of actual events in the field, they can be extremely helpful and greatly improve decision making and monitoring efficiency.

Phenology Models

Phenology is the study of the relationship between weather and cyclic or periodic events, such as flowering, growth, or reproduction, in the life of an organism. *Phenology models* predict when such events will occur. The development of organisms that cannot regulate their own internal temperature depends on external temperature. Invertebrates, including insects and nematodes, require a certain amount of heat to develop from one point in their life cycle to another, for example, from egg to egg hatch or from a newly hatched larva to an emerged adult. Because of yearly variations in weather, calendar dates are not a reliable basis for predicting when these events will occur. Measuring the amount of heat accumulated over time, therefore, provides a physiological time scale that is biologically more accurate than calendar days.

All organisms have temperature thresholds above and below which growth does not occur. These thresholds vary from species to species. The lower *developmental threshold* for a species is the temperature below which development stops. The upper developmental threshold can be defined either as the temperature at which the rate of growth or development is at its maximum or as the temperature at which development ceases. Upper thresholds are not used in some areas of the country where daily maximum temperature extremes do not commonly exceed an organism's upper developmental threshold. For many organisms the upper thresholds are not used because such data are lacking. Both lower and upper thresholds are determined through carefully controlled research and are specific for different organisms. The first step in the development of a phenological model is to determine the lower threshold for development and, if possible, the upper threshold.

The average amount of heat above the lower developmental threshold required to complete a given organism's development—a combination of temperature (between thresholds) and time—can be determined through research. This measurement, called physiological time, is often expressed in units called degree-days (DD) (Figure 6-32). A degree-day is the amount of heat that accumulates over a 24-hour period when the average temperature is 1° above the lower developmental threshold of an organism. For instance, if a species has a lower developmental threshold of 52°F, and the temperature remains at 53°F for 24 hours, 1 degree-day (°F) is accumulated. Similarly, at 60°F constant temperature (8°F above the lower developmental threshold), it takes only 3 hours to accumulate 1 degree-day; 8 degree-days are accumulated in the 24-hour period. Degree-days can also be calculated using the Celsius scale.

Each stage of an organism's development has its own total heat requirement. Development can be estimated by accumulating degree-days on a daily basis between the organism's developmental temperature thresholds throughout the season. The accumulation of degree-days from a starting point can help predict when a developmental stage will be reached. The date to begin accumulating degree-days, known as the *biofix date*, varies with the species. Biofix dates are usually based on specific biological events such as planting dates, first trap catch, first occurrence of a pest, or a specific developmental stage of a pest or crop. Sometimes January 1 is used as a default biofix point, but in more temperate areas where different life stages of a given pest overwinter, the biofix must be accurately linked to some observation. Ideally, biofix points are established from research-based information. Phenological models are of little value without a reliable biofix date.

Once the biofix has passed, degree-days are calculated and added up for each day. The simplest way to estimate the number of degree-days accumulated in a day is to get the average temperature for that day by adding the highest and lowest temperatures and dividing by 2. Then subtract the lower developmental threshold. This manual method (Table 6-6) provides an approximate guide but can be less accurate if the minimum temperature is far below the lower threshold or if there is an upper developmental threshold. More precise formulas are available for use with programmable hand calculators and computers. Degree-days are added up on a daily basis until some important biological event is forecast by the total accumulation.

Most computerized environmental monitoring and control equipment has a built-in ability to calculate degree-days and can continuously record temperatures and calculate degree-day accumulations. Degree-day calculators linked to the UC IPM Program's Weather database can be accessed online at the UC Statewide IPM Program's website. Degree-day software that uses local weather data is also available for use on personal computers.

TABLE 6-6.

Approximating degree-days (DD) manually.

Add the daily minimum and maximum temperature and divide by 2 to get the average daily temperature.	(78°F + 54°F) ÷ 2 = 66°F
Subtract the lower threshold temperature (for example, 52°F for codling moth) from the average daily temperature. The result is the approximate number of degree-days accumulated that day.	66°F – 52°F = 14 DD
Add the degree-days accumulated for each day until you reach the sum when specific actions, such as an insecticide application, are recommended.	see below

Day	Minimum	Maximum	Average	Degree-days
1	54	78	66	14
2	56	76	66	14
3	56	78	67	15
4	54	74	64	12
5	55	79	67	15
6	56	82	69	17
ACCUMULATED TOTAL				87

This manual method of degree-day estimation becomes significantly inaccurate when temperatures are far below the lower developmental threshold. Also, it ignores the upper developmental threshold. Other estimation methods are more accurate and are recommended. Users can obtain degree-days online at the UC IPM website, www.ipm.ucdavis.edu. You can also purchase environmental computer monitoring systems or compact electronic temperature recorders that calculate and display degree-days automatically.

Disease Forecasting

Disease forecasting systems use weather, host, and pathogen data to predict times when disease outbreak is likely. Growers in a number of crop areas have used simple disease forecasting models for many years. For instance, W. D. Mills at Cornell University devised a chart in 1954 that correlates the length of time it takes spores of the apple scab fungus, *Venturia inaequalis*, to infect apples at different temperatures (Figure 6-33). Leaves must remain continuously wet at these temperatures for infection to occur. By measuring average temperatures in the orchard and how long leaves remain wet, growers or PCAs can determine whether a preventive fungicide treatment is needed. A similar model for development of powdery mildew in grapes, based on temperatures over a 3-day period, is widely used in California.

In many areas and for specific crops, networks of weather stations allow growers and PCAs to receive the latest temperature, rainfall, relative humidity, and leaf wetness readings on which to base pest management decisions. In-field weather stations automatically transmit weather data to a central computer that gathers the data. The potential for an outbreak of specific diseases is calculated based on past weather patterns and disease incidence.

A plant disease model is a mathematical description of the interaction between environmental, host, and pathogen variables that can result in disease. Host variables can include crop growth stage and cultivar; pathogen variables include inoculum levels and maturity of spores; and environmental factors include temperature and moisture. The interaction that occurs among these different variables is specific to the site.

Expert Systems

Expert systems are computer programs designed to simulate the problem-solving behavior of a team of well-trained professionals. In agriculture, integrated expert systems combine the expertise of various disciplines such as plant pathology, entomology, weed science, nematology, and crop production into a framework that manages and coordinates information to address specific on-site concerns. Integrated expert systems assist decision making by offering alternative control options, calculating the impact of multiple interacting or competing factors, and exploring how certain actions would

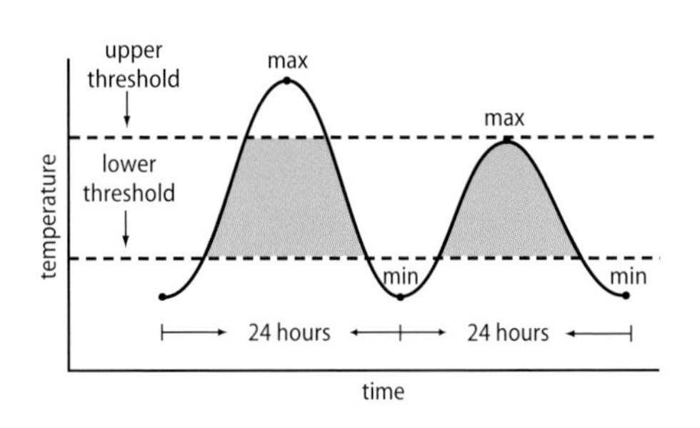

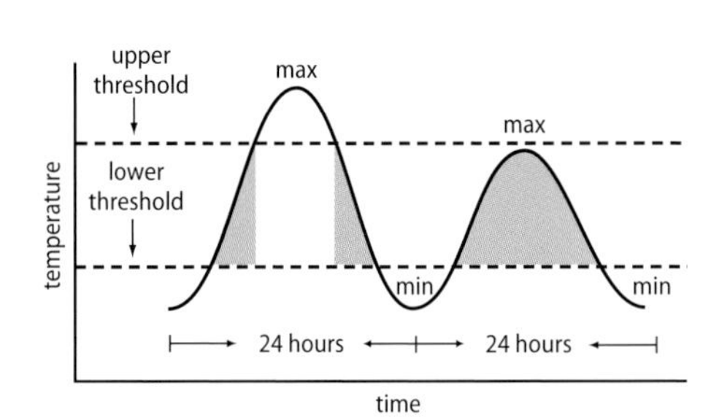

FIGURE 6-32

Degree-days (DD) are the accumulated product of time and temperature between the developmental thresholds for each day. The lower developmental threshold for an organism is the temperature below which development stops. The lower threshold is determined by the organism's physiology and is independent of the method used to calculate degree-days. The upper developmental threshold is the temperature above which the rate of growth or development begins to decrease or stop as determined by the cutoff method. The physiological interpretation of the upper threshold depends on the cutoff method. The horizontal cutoff method for calculating degree-days (shown here) is used when the organism's development rate is assumed to remain constant above the upper threshold. One degree-day is 1 day (24 hours) with the temperature above the lower developmental threshold by 1°. The colored area in the figure represents total degree-days accumulated. The vertical cutoff method assumes that no development occurs when a temperature is above the upper threshold and no degree-days are accumulated for that organism.

affect crop development, yields, or pest management. For instance, The CALEX/Cotton Expert System developed in the 1990s has about 20 agronomic and pest management rule bases that interact with one another (Figure 6-34). While expert systems have been useful for research and demonstration, currently available expert systems have required more field data collection and input than PCAs or growers have been willing to undertake for day-to-day decision making.

Plant Mapping

Plant mapping is a useful tool for certain crops, enabling growers or PCAs to monitor the developmental status of plants in their fields. Plant mapping systems are computer programs that use plant data from field samples to provide an indication of crop health and development. Cumulative field measurements track plant growth throughout the growing season and alert managers to signs of stress. In cotton, for instance, plant mapping measurements include plant height, number of vegetative branch nodes, number of fruiting branch nodes, and first-position boll retention in the top and bottom five nodes. These measurements are recorded and calculated on a plant monitoring form or entered into a computer containing a plant-mapping program. By knowing precisely the stage of crop development, accurate assessments can be made to determine such events as when certain types of pests will be damaging (e.g., bud or seed feeders) or when the crop will be ready for harvest.

Precision Farming

Precision farming is a technology that uses computers and electronics to produce a sophisticated site-specific approach to crop management. It provides a set of tools based on the measurement and consideration of variability between fields and within fields, supplying more information to make better decisions. Precision farming systems use global positioning systems (GPS) to collect data and *geographic information systems* (GIS) to manage it.

The first step in implementing precision

Average required temperature (°F)	Hours of wetting for infection
78	13
77	11
76	9.5
61–75	9
60	9.5
57–59	10
55–56	11
54	11.5
52–53	12
51	13
50	14
49	14.5
48	15
47	17
46	19
45	20
44	22
43	25
42	30
33–41	*

* Not known

FIGURE 6-33

The Mills and La Plante Table: temperature and hours of wetting requirements for apple scab infection. The need for preventive fungicide treatment for apple scab can be determined by keeping track of daily average temperatures and hours that leaves are wet each day. The Mills Chart indicates how many hours leaves must remain wet at various temperatures for infection to occur. Treatment is warranted if wetting hours are exceeded. Shown here are apples damaged by scab.

farming is the production of yield maps and soil samples. Yield maps are data files that depict the variability of the crop yield across a field. They are produced through the use of GPS to precisely identify locations in a field; the maps are linked to yield data. GPS relies on strategically placed satellites that orbit the Earth; signals are emitted that can be picked up by a receiver and converted to determine a precise geographical position in the field. The GPS receiver is typically fitted on top of a tractor or harvester that has a vehicle positioning system integrated with a recording system. This type of precision can be used to plan more accurate and finely tuned pest management programs. For instance, using GPS, the application rate of a pesticide can be varied in different parts of the field according to known differences in soil type or known pest populations preprogrammed into the system.

The data produced through the use of GPS are managed through the GIS, a data management program designed to manipulate and display data on computerized maps. A GIS organizes, statistically analyzes, and displays data; links each set of data to precise GPS locations; and produces new sets of data by combining overlays. Examples of overlays include soil type, topography, and cover crop. Other features recorded include crop management inputs such as seed, pesticides, and fertilizer. The advantage of this system is that by combining these maps, more precise information on the interactions between yield, pest damage, and treatment history can be obtained. The record-keeping system maintained with GIS systems can be especially valuable for IPM programs.

PESTICIDE RESISTANCE MONITORING

Pesticide resistance monitoring programs detect tolerance to pesticides in pest populations before resistance becomes widespread. Early detection provides the opportunity to integrate other pest management options to prevent or reduce selection for resistance. For example, combining different weed control practices such as increased cultivation and crop rotation with herbicides can reduce the occurrence of resistant weed biotypes. By alternating pesticide treatments with different modes of action, it may be possible to use some pesticides for a longer period of time because a single pesticide mode of action is no longer the principal source of pest mortality.

Detection and Monitoring

The key to effective management of pesticide resistance is monitoring and early detection. Detection and monitoring techniques document that pesticide resistance has occurred and confirm that lack of pest control is not due to other factors such as poor application, changes in management practices, or environmental changes. Detection and monitoring can also provide early warning that resistance is developing, help identify the presence and frequency of resistant individuals in the target population, and determine whether strategies introduced to avoid or delay resistance have been effective.

Once pesticide resistance has been confirmed, detection and management strategies aid in pesticide selection and in the detection of cross resistance and multiple resistance. There is considerable interest in using resistance monitoring to predict a pest's predisposition to resistance based on the history of pesticide use against a target pest, site-specific ecological and biological factors, and the genetic attributes and defense mechanisms of the pest species. At this point in time, however, complete information is lacking for most major pest species.

At present, most monitoring for pesticide resistance is being carried out by researchers in Cooperative Extension offices, universities, and pesticide companies. If you suspect that resistance is developing to one of the pesticides you are using, contact your local Cooperative Extension office for help.

Resistance management strategies for insects, weeds, and fungal pathogens all include rotating classes of pesticides (i.e., pesticides with the same mode of action, such as pyrethroids, organophosphates, carbamates, etc.). Pesticides with the same mode of action have been assigned the same group number by respective pesticide resistance action committees: IRAC (Insecticide Resistance Action Committee), FRAC (Fungicide Resistance Action Committee), and HRAC (Herbicide Resistance Action Committee). These group numbers have been included in the treatment tables to help clarify which new class of pesticides to rotate to. However, the strategies used in rotations differ. For example, with fungicides, it is suggested that classes be rotated every application. With insecticides, a single chemical class should be used for a single generation of the target pest followed by a rotation to a new class of insecticide that will affect the next generation and any survivors from the first generation. Longer use of a single chemical class will enhance the chance of resistance since the survivors of the first generation and the next will most likely be tolerant to that class. Rotating through many chemical classes in successive generations will help maintain efficacy. The websites for IRAC, HRAC, and FRAC include more information about managing pesticide resistance and reports of pesticide resistance development in common pests.

HOW TO KEEP MONITORING RECORDS

Pest management decisions are complex. An integrated approach to pest management demands organized and dependable records. Complete monitoring records provide information on weeds, insects, diseases, and all pest species associated with the crop at the particular site. They also include general field information and history and information on crop health, cultural and management practices, and weather patterns. Good written records include the date, specific location, host plant, sampling method, and

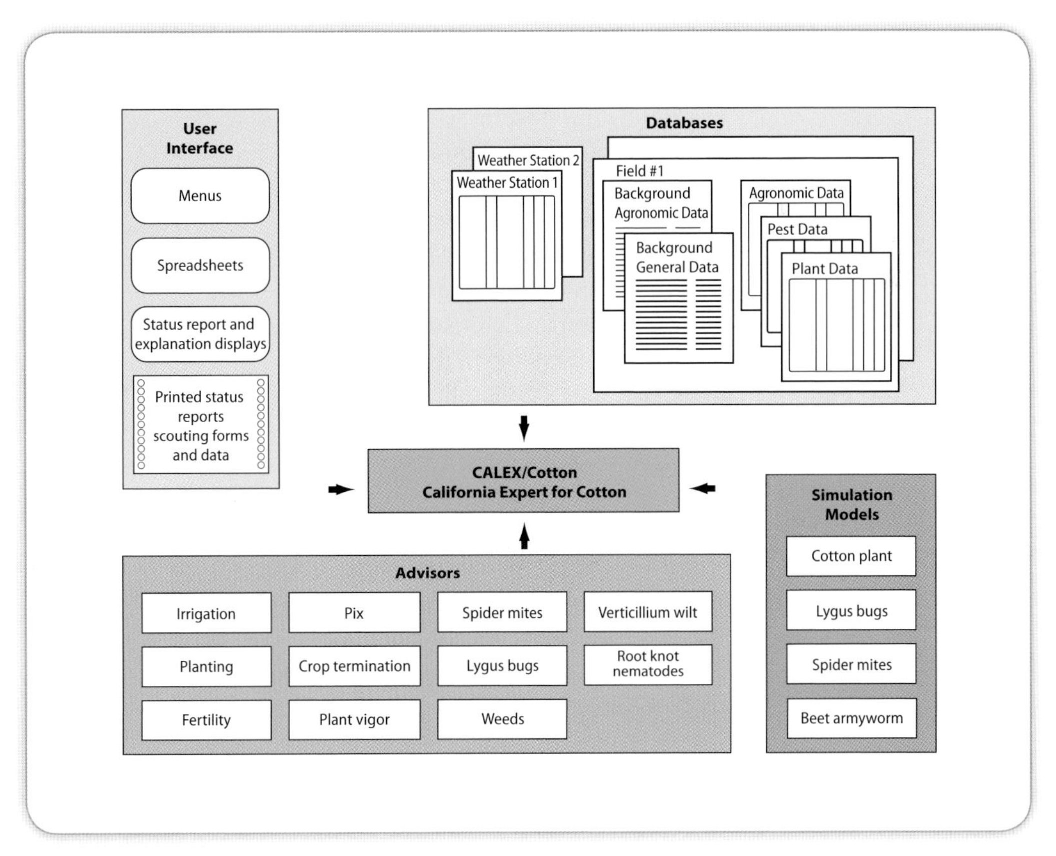

FIGURE 6-34

Schematic representation of the components of the CALEX/Cotton Expert System. Source: Goodell et al. 1990.

pests sampled. Records should also include the management action taken, when it was taken, and the result of that action.

Sampling Records

After determining the sampling procedure appropriate for the pest and the location, recording the data collected in a consistent and useful format is one of the most important steps in an effective monitoring program. Keep separate field sampling records for each pest species and for each location where the samples are taken. Use forms that are easy to complete and from which information can be easily accessed. Include the number of plants inspected and an assessment of the measure of damage or pest abundance.

Sample weeds by rating the infestation or using a numerical scale (Figure 6-35). List the common weeds for the area on a weed infestation record and rate the degree of infestation for each species. Conduct weed sampling or weed surveys twice a year, in spring to see which winter annuals are present and during the growing season to check the effectiveness of control measures and to look for summer annuals.

A fruit sampling form for monitoring armyworm and fruitworm on fruit in processing tomatoes is shown in Figure 6-36. This is a sequential sampling program, and treatment thresholds are noted in the "Treat" columns. Damaged fruit is tallied in the center columns. The number of damaged fruit in the sample determines the number of samples that will ultimately be taken, as well as whether or not the field needs to be treated. Using the 2% damage threshold, if no more than 1 fruit in 100 is damaged, no treatment is required. If the number of damaged fruit is from 2 to 6, sampling continues. A decision to treat is made only if there are 7 or more damaged fruit in a sample of 100. More samples are required if fewer than 7 are found.

Data sheets should also contain blank areas to record observations in the field. Note obvious differences in growth patterns and subtle variations in crop growth and development. Field notes are a handy place to record such information as budbreak, crop emergence dates, and other specific events that correspond to crop stages or trigger specific pest monitoring practices. Sampling records also provide the best record of field visits and are a log of the information relayed to the client. See the UC IPM Pest Management Guidelines at www.ipm.ucdavis.edu for monitoring forms for many pests.

Permanent Samples

Certain sampling methods provide a permanent record of pest or damage levels without the need for detailed written counts. These include honeydew droplet monitoring cards, scale crawler sticky tape traps, and possibly other methods such as reagent papers on which insecticide-resistance squash tests are conducted. These samples can be appropriately labeled and stored for easy visual comparison of samples from various dates or years.

Graphs

Graphs are used with sampling records and degree-day models to give a visual presentation of the data being gathered. When combined with a sampling form as in Figure 6-36, the recorded counts over time provide valuable information about the relative density and distribution of pest populations and the effectiveness of control measures. Graphs are also a good tool to use when explaining pest problems to the client.

Data Sheets and Files

PCAs should maintain a complete description of each site they manage. A good way to maintain this description is by keeping a comprehensive data sheet or computer file for each field for each season. The file should include management information such as variety selection, planting and harvest date, fertilizer and irrigation practices, and past pesticide applications that may influence pest activity and pest management decisions. Data sheets should also include general permanent field information such as field size, soil characteristics, and cropping history.

Field Maps

Field maps provide a site-specific visual representation of each field and depict in greater detail information that can impact pest infestations and pest management decisions. Maps should record information such as the location of waterways, roads, buildings, fences, and field position. Differences in soil type and texture can also be recorded for future reference.

Use field maps to record specific pest management information such as hot spots, location of traps, and areas likely to become infested first. This information is useful to pinpoint likely sources of infestation and to spot-treat specific areas. Record weekly sampling counts on field maps at the sampling site; also count and record pheromone traps at the location. Many online sources are available to help with mapping, including Google Maps, soil type databases (such as Soils-to-Go in California), and the USDA Natural Resources Conservation Service (NRCS).

Electronic Databases

Advances in electronic technology have brought about the development of computer databases and software packages that help to efficiently record information relevant to crop production and landscape management. Data collected in a pest management program can be entered and analyzed in data management programs. This software can be used to record sampling data, other field information, and weather data, and they also incorporate mapping capabilities, multiseason record keeping, and field activity tracking. To assist with the required paperwork, they can produce recommendations, state Notices of Intent (NOI), reports, and label warnings. Many commercial sources offer these products, including SureHarvest, AgCode, CDMS Advisor, Agrian, and Tiger Jill.

Most data management programs have been designed for field use and can, in many instances, replace written records. Smart phones and other hand-held electronic devices can access spreadsheets laid out like a sampling form, so that data can be entered into the computer in the field.

INTERPRETING AND USING MONITORING RESULTS

Monitoring provides information on the daily and seasonal status of field conditions, pests, the crop, and the weather. However, to be useful the information must be properly interpreted.

To efficiently use monitoring information, the sampling technique must accurately reflect fluctuations in the pest population and must be used with guidelines that effectively relate those populations to economic damage and treatment decisions. For example, in the 1970s, treatment for tomato fruitworm, *Helicoverpa zea*, was based on visual estimates of damage; the reliability of the estimates was limited and the risk of error was high. In addition, this approach resulted in treatment applications after damage had already occurred. A new sampling technique was developed that based the treatment threshold on the number of eggs found in leaf samples and levels of parasitization, resulting in more reliable fruitworm estimates, treatment of the more susceptible early-instar larvae, and a reduction of insecticide applications.

To gain confidence in a sampling method and accurately interpret the results, follow treatment thresholds when available and be attentive to other factors in the system that, combined with sampling results, influence pest management decisions.

Relating Monitoring Results to Treatment Thresholds

The estimates of pest population densities obtained from sampling are of little value if they cannot be meaningfully related to potential pest damage. The purpose of treatment thresholds is to relate monitoring results to the need for control action.

Established decision-making guidelines are available for many pests of agricultural crops, commercial turf, and pests of home and landscape. For California, the UC IPM

Grape—Early Season Weed Survey

Supplement to UC IPM Pest Management Guidelines: Grape

Grower/Vineyard: ________________________ Date: ________________

Comments: ____

Mechanical Control/Herbicide/Application Date: ____________________

Record weeds on the form below; use the map to record the location of problematic weeds.

Directions:

1. Survey wine or raisin vineyards in late spring or summer. For table grapes, survey in March (San Joaquin Valley) and January to February (Coachella Valley), after summer annuals have germinated.
2. Pay particular attention to perennials. Check for regrowth of perennials a few weeks after cultivation.
3. Pay attention to wet spots, as these may be problem areas in terms of weed growth.
4. Survey areas around the vineyard as these areas could be a potential source for wind-disseminated seeds, such as marestail and fleabane seeds.
5. Rate infestation either using a numeric scale from 1 to 5 (1 being the lightest), or use "light," "medium, or "heavy."

Summer Annual and Perennial Weeds

Weed	Row middles	Rows
Annual broadleaves		
hairy fleabane (flax-leaf)		
horseweed		
spurge (prostrate/spotted)		
puncturevine		
cudweed, purple		
knotweed, prostrate		
nightshade, black		
pigweed, prostrate		
nettle, burning		
lambsquarters, common		
willowherb, tall annual (panicle willowherb)		
annual morningglories		

Weed	Row middles	Rows
Annual grasses		
junglerice		
barnyardgrass		
crabgrass, large		
foxtail, yellow		
Perennial broadleaves		
bindweed, field		
Perennial grasses		
bermudagrass		
Other perennials		
nutsedge		
Other weeds		

FIGURE 6.35.

Weed infestation records should be prepared for each location that is surveyed. On each record, list the common weeds for the areas, such as is done in this example for vineyards, and rate the level of infestation, Include a field map so that infestations can be noted here as well. Source: UC IPM Grape Pest Management Guidelines.

Tomato—Sequential Sampling for Armyworms on Tomato Fruit
Supplement to UC IPM Pest Management Guidelines: Tomato

Grower: ______________________ Date: ______________

Field location ______________________________

Comments ______________________________

Fruit sampling directions

1. Pick fruit that are 1 inch or more in diameter to check for damage due to caterpillars.
2. Start with a sample of 100 fruit, picking them at random one at a time.
   - Count any fruit that contains a larva or feces, has a hole deeper than $^1/_{10}$ inch, or has a shallow injury that will not scar over by harvest.
   - Don't count shallow injuries that will scar over by harvest.
3. Tally damaged fruit in the columns below using either the 2% or 1% threshold.
4. To reach a treatment decision:
   - Treat if the total number of damaged fruit reaches or exceeds the number in the "treat" column.
   - If the total matches or drops below the "don't' treat" level for the number of fruit checked, stop and sample again next week.
   - If the total remains between the two columns after you have checked 300 fruit, stop and sample again in 2 or 3 days.

2% insect damage threshold

Number of fruit sampled	Number of fruit damaged by: Armyworms	Fruitworms	Don't treat	Total damaged fruit	Treat
100			1		7
125			2		8
150			2		9
175			3		10
200			3		11
225			4		12
250			4		13
275			5		13
300			5		14

1% insect damage threshold

Number of fruit sampled	Number of fruit damaged by: Armyworms	Fruitworms	Don't treat	Total damaged fruit	Treat
100			1		5
125			2		5
150			2		6
175			3		6
200			3		7
225			4		7
250			4		8
275			5		8
300			5		8

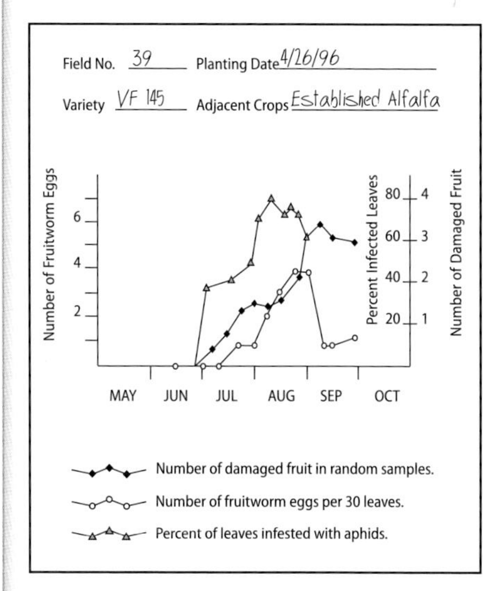

FIGURE 6-36

Sampling forms for monitoring caterpillar pests of fruit in processing tomatoes at the 1% and 2% "insect damage" thresholds at left. Charting sampling counts weekly will help in monitoring infestations in following seasons. Counts for different pests can be combined as in the graph above. Source: UC IPM Pest Management Guidelines: Tomatoes.

Pest Management Guidelines and the IPM manuals describe pest management actions suggested by University of California experts. Land grant universities in other states provide similar guidelines appropriate to their cropping systems. Use these guidelines to obtain descriptions of pests at a particular site and to assess the risk of damage. Follow the suggested monitoring or sampling procedures. Treatment timing and options using available biological, cultural, physical, or chemical controls are provided.

When adapting guidelines to new conditions or developing new guidelines for pests, define a treatment threshold appropriate to the situation, that is, determine at which point a pest management action needs to be taken to prevent economic damage. To make this determination in situations where guidelines do not exist, increase the sampling size and number. Note the stage of pest development and consistently rate the size of the pest population and the observed damage. Communicate with other PCAs and Cooperative Extension personnel to compare observations, sampling tactics, and pest management strategies.

Treatment thresholds are not static; they are subject to revision based on new pests, new varieties, environmental constraints, new management practices including new pesticide products, new marketing standards, and variations in commodity prices. Be prepared to be flexible. In new situations, set treatment thresholds low, to offset unknown risks, and increase monitoring; with experience and experimentation will come the opportunity to revise them.

The use of treatment thresholds presents more of a challenge for disease management because numerical thresholds are generally not established for pathogens. Disease management is based primarily on taking preventive management strategies when conditions are conducive to disease development and using methods that reduce the amount of primary inoculum or the rate at which the inoculum increases. Rather than measuring pest populations, as in insect treatment thresholds, disease treatment guidelines must predict disease outbreak before it occurs. Therefore, monitoring focuses on quantifying conditions likely to promote pathogen infection, development, or spread.

The concept of treatment thresholds is more complex for weed management. Unlike insects and diseases, where just a few species are key pests, weed infestations in a given area usually involve numerous species. Also, weed seeds and vegetative propagules can persist in soils for long periods, making new infestations likely whenever conclusive conditions occur. Nonetheless, the concept of thresholds has applications in weed management.

The most common thresholds being developed in weed management are damage, period, economic, and action thresholds. Damage thresholds identify the weed population at which a negative impact on the crop is detected. Period thresholds define the period during crop development when crop losses from weed interference are most likely to occur; for instance, certain weeds may seriously reduce the growth of crop seedlings but present much less threat once a crop has developed a substantial canopy. Economic thresholds are the weed population densities at which control measures must be taken to prevent economic loss. Action thresholds represent the weed population level at which some action is needed to avoid crop loss.

Other Factors That Influence Decision Making

Sampling provides an estimate of the number of individuals of a pest species at a particular site. Other monitoring techniques detect pests or conditions that favor outbreaks; reviewing monitoring records kept over time helps define trends in the pest population.

Other factors also influence decision making. To make good pest management decisions, become familiar with the cropping history at the site being monitored. Past crops, pests, and the success of pest management activities on those crops can give an indication of future pests. Likewise, future crops planned for the site can be a factor in the choice of control practices or the need to manage specific pest species. Be familiar with the crops and uncultivated areas surrounding the site, as pests often migrate into the field from surrounding areas.

Monitoring the crop is as important as monitoring the pests. The growth stage of the crop plays an important role in treatment decisions. At certain stages of crop development, the crop may outgrow the pest damage; at other stages, the pest may be present but the crop may not be susceptible. Also, depending on the health of the crop, treatment thresholds may change; a stressed crop is often more susceptible to pest damage, requiring treatment that would not be needed on a healthy crop. Field size and topography can also influence the pattern of certain pest infestations; keep good maps that indicate field size and any irregularities that contribute to pests.

The preferences and concerns of the client are a primary consideration in the decision-making process. The client may have treatment preferences or certain perceptions regarding the presence of pest species. The size of the crop may be a limiting factor affecting the cost of monitoring, control, or management options. Depending on the value and the market for the crop, the cost of monitoring may not be warranted. Likewise, depending on the profitability of the crop, the cost of a control action may not be economical. Ultimately, it is the grower or other client, not the PCA, who makes the final decision to treat.

The market for the crop can affect pest management decisions. Become familiar with any restrictions the processor may have regarding pesticide use and know the marketing specifications for the commodity. Produce grown to be exported out of California or out of the United States may have quarantine or trade restrictions to prevent movement of pests into new areas. These restrictions (see the USDA's Maximum Residue Database for more information, www.fas.usda.gov/htp/MRL.asp) often require pesticide sprays that can disrupt IPM programs. In landscape situations, the aesthetic preferences of clientele may vary; the level of tolerance toward various pest species will dictate many pest management decisions.

Be aware of all management practices that affect pest populations. For example, the pesticide use history of a site can provide information regarding pest species of concern and the effort to control them. It can also be key if secondary pests are present and for managing pesticide resistance.

Follow-Up Monitoring After Treatment

Monitor after every treatment to learn whether the control activity was successful. This includes monitoring when no treatment has been recommended; nontreatment decisions should include follow-up monitoring until the pest is no longer a threat to the crop. After treatment, monitor to look for signs indicating that the level of control achieved is adequate.

After a pesticide application, check for damage to natural enemy populations and nontarget species in the area. Look for other problems that may occur, such as phytotoxicity, pest resurgence, and secondary pest outbreaks. After a fungicide application follow-up monitoring is important to verify that coverage was thorough and the pathogen has been suppressed. Also, if the conditions for pathogen development continue, ongoing monitoring will be necessary to determine whether further management actions are required as unprotected foliage becomes susceptible.

When herbicides have been applied, follow up to see which weed species were not controlled or only partially controlled. Also look for damage to nontarget plants. Record observations for comparison with records before the treatment actions were enacted. This information will also be useful for future decision making.

Evaluating the Efficiency of the Monitoring Technique

The best indication that the treatment decision was correct is reduction of the pest population and a profitable crop for the grower or an aesthetically pleasing landscape for the landscape manager. Efficiency in regards to time spent monitoring will improve with experience and familiarity with the site. But never take for granted the reliability of the monitoring technique. Pest management systems are constantly changing; differences in pest populations, weather patterns, and management practices can impact monitoring results.

Evaluate the efficiency of the monitoring program on a regular basis. Is enough time spent sampling? Is the site visited often enough to adequately evaluate the pest? Is the sampling technique the best one for the pest species? These and other questions should be a regular part of monitoring.

Also be aware of what is going on in other locations. Communicate with other PCAs and Cooperative Extension advisors and compare sampling results. Be aware of changes within the system, such as changes in weather patterns and management practices and how they impact monitoring. Counts in traps, for example, can vary when the pheromone lures are old or when the traps get wet. To avoid the risk of error, use more than one sampling technique if a backup method is available. For instance, when monitoring with egg traps for navel orangeworms, periodic nut samples for navel orangeworm eggs or larvae serve as an indicator of whether the egg traps are providing accurate information. Sometimes efficiency can be improved by changing the sampling pattern or the placement of traps. The best tactic available to improve the efficiency of any sampling method is to be keenly alert and become aware of the pulse of the ecosystem; monitor the total system.

7 Setting Up Monitoring Programs and Field Trials

The previous chapter introduced the principles of monitoring, the various tools and techniques practiced in sampling programs, and guidelines for effective decision making. This chapter gives step-by-step instructions for setting up monitoring programs and field trials.

HOW TO DESIGN A MONITORING PLAN

Effective monitoring programs require forethought, knowledge, and flexibility. The keys to a successful monitoring program is knowing which pests are likely to cause damage at each site, the best time to monitor for each pest, and the appropriate sampling strategy for the pest species. No single program works for every pest; fine-tune the monitoring program to fit each situation. Decide what to sample for, how many samples to take, and how often to take samples. Remember that there is a trade-off between the number of samples and how accurately the pest density and distribution is assessed.

Efficient monitoring programs consider the phenology of the pest species, how the environment affects the pest, and the interaction of the pest with its host, other pests, and beneficial organisms. To increase efficiency, try to devise programs that allow you to monitor more than one pest where possible. In landscape situations, consider public tolerance for the pest species. Every situation should be monitored.

Specialized sampling programs, however, may not be warranted for all pests, especially those that rarely occur or do not have a high damage potential. For low-yielding or low-value crops, the cost of intensive sampling programs can cut deeply into profits, so monitoring programs must be streamlined. In some landscape situations, the pest may be tolerable; just keeping an informal eye on the situation is all the monitoring that is necessary.

Begin by evaluating the crop or host and the pest species present. Recognize the major pest species and the secondary and minor pests for each crop or landscape. If you are not familiar with common pests on a crop, consult the UC Pest Management Guidelines or other resources that can help you identify them. After working in a specific crop and location, the key pests that require constant monitoring will become evident. Those pests that require more defined sampling will also become apparent.

Research pest and crop associations and recommended monitoring strategies. When none are suggested elsewhere, adapt those that exist for similar pests or situations (check the "Resources" section). In addition, the ability to observe and ask questions is the most important resource for diagnosing a pest problem and designing a monitoring program that is specific for the pest and the site. Talk with the grower or other client about pest species of major concern, past management practices, and past control actions.

To design a monitoring program, follow the eight basic steps outlined in Table 7-1. The discussion below details what is involved in carrying out each step in Table 7-1.

Step 1. Identify the Pests

Design monitoring plans specifically for the pest species of concern. Many potential pest species are likely to reside within or close to the site, but not all will cause significant

TABLE 7-1.

Designing a monitoring plan.

- ☐ Step 1. Identify the potential pests.
- ☐ Step 2. Establish monitoring guidelines for each pest species:
  - ☐ select the sample unit.
  - ☐ define the sampling universe.
  - ☐ determine the number and size of samples needed.
  - ☐ determine how often to sample.
  - ☐ determine when sampling begins and when it ends.
- ☐ Step 3. Establish injury levels and action thresholds for each pest species.
- ☐ Step 4. Determine what host or crop developmental stages must be monitored to assess normal growth, predict timing of pest activity, or evaluate damage.
- ☐ Step 5. Determine the environmental factors that must be monitored.
- ☐ Step 6. Determine the production practices that can impact the development of the pest species.
- ☐ Step 7. Streamline the monitoring program to develop efficiencies.
- ☐ Step 8. Keep good written records.

FIGURE 7-1

Identification of hairy nightshade in the seedling stage is essential for choosing effective management methods.

enough economic damage to warrant a monitoring program. Make a list of the key pests, potential pests, and disorders likely to occur. Be aware of exotic invasive pests that threaten the crops and plants you manage.

For each pest, know the damaging life stage and the best time to monitor so that management decisions can be made. Weed management efforts, for example, improve when the weed species is identified in the seedling stage (Figure 7-1). Seedling identification allows time to develop a weed management strategy and implement a control option prior to seed set, thereby reducing the seed bank. Citrus cutworms are most easily controlled with microbial insecticides when they are in the youngest instars. Thus, detecting damaging levels just after they hatch is critical for control.

Resources are available for most crops and landscapes that outline major pest species and offer monitoring and management suggestions (Figure 7-2). Review the University of California IPM publications and the UC IPM Pest Management Guidelines for pest management information on various pests and crops; other helpful publications are listed in the "Resources" section. The grower, other local PCAs, Cooperative Extension farm advisors or agents, and local agricultural commissioners are all potential resources for identifying pests that cause the most damage in specific areas.

Step 2. Establish Monitoring Guidelines for Each Pest Species

Once the pest species of economic importance are known, using the resources mentioned above identify monitoring guidelines for each of these major pests. Become familiar with the seasonal development of each pest. Develop a regular monitoring schedule for use throughout the growing season. Most pests must be monitored only at certain times of the year. When and how often to monitor pests vary with the individual pest species, crop development stage or season, growing location, and factors such as weather conditions.

During the growing seasons visit each field or location at least once a week; during periods of crop development or when a population is

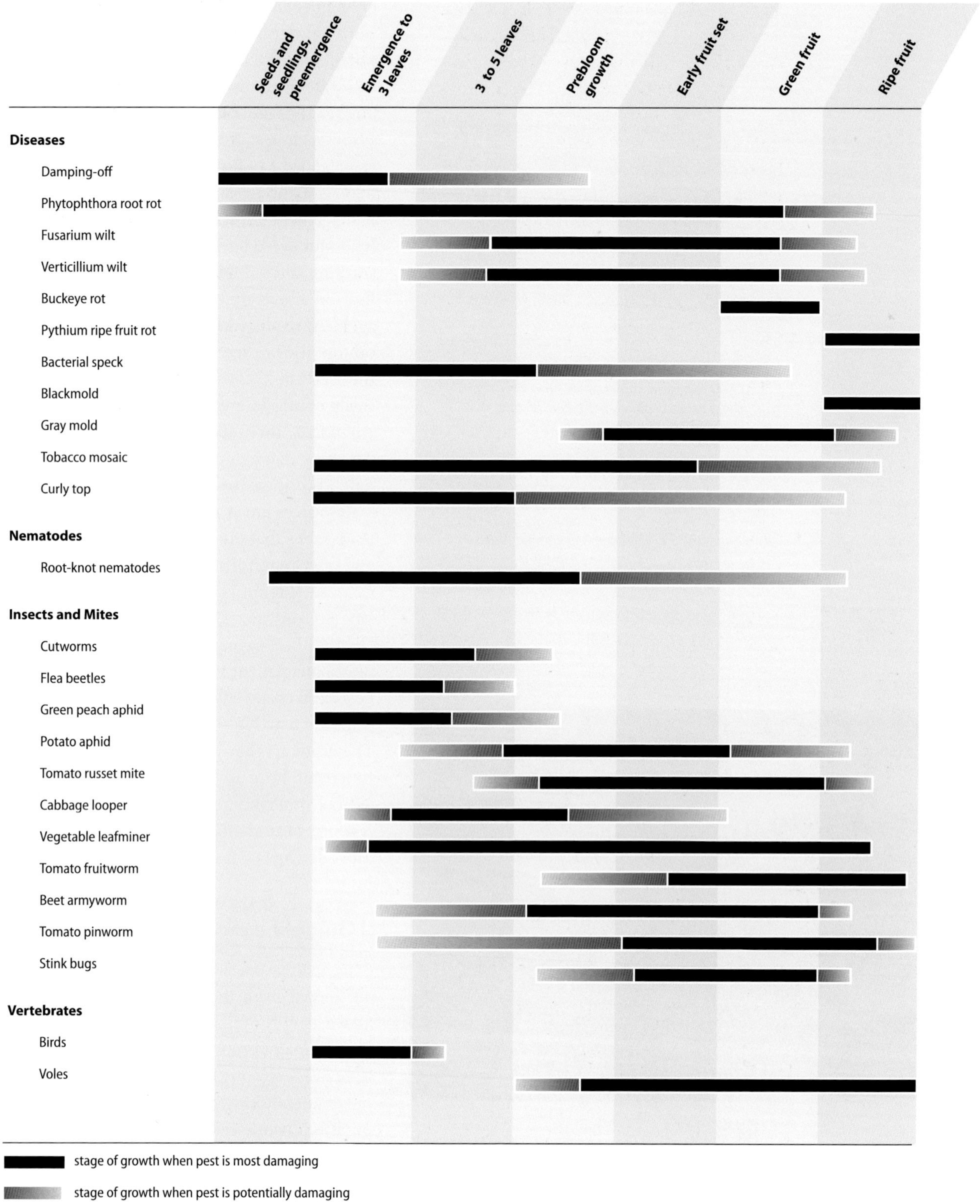

Seeds and seedlings, preemergence
Emergence to 3 leaves
3 to 5 leaves
Prebloom growth
Early fruit set
Green fruit
Ripe fruit

Diseases

Damping-off
Phytophthora root rot
Fusarium wilt
Verticillium wilt
Buckeye rot
Pythium ripe fruit rot
Bacterial speck
Blackmold
Gray mold
Tobacco mosaic
Curly top

Nematodes

Root-knot nematodes

Insects and Mites

Cutworms
Flea beetles
Green peach aphid
Potato aphid
Tomato russet mite
Cabbage looper
Vegetable leafminer
Tomato fruitworm
Beet armyworm
Tomato pinworm
Stink bugs

Vertebrates

Birds
Voles

stage of growth when pest is most damaging

stage of growth when pest is potentially damaging

FIGURE 7-2

Consult university resources for information about pests on your crop. This chart, from the University of California IPM book for tomatoes, provides information on the stage of growth when various pests should be of concern. After Rude and Strand 1998.

approaching a treatment threshold, visit every 2 or 3 days. In most crops and landscapes, the emphasis is on two or three species that occur regularly and threaten yields or quality. For some of these species, intensive sampling is needed only during a specific period of the host's development. Work sampling methods for all pertinent species into a single routine; a special pass through the field for each species is not necessary. For example, in grapes, leafhoppers, spider mites and mealybugs can be monitored at the same time on 20 randomly selected vines throughout a vineyard block (Figure 7-3).

Unless other formal methods are recommended, it is usually desirable to keep sampling patterns as random as possible when designing a monitoring plan. However, when the distribution of a pest is known to be nonrandom, concentrating efforts in the areas where the pest is most likely to be found can save time. Predetermined systematic sampling patterns, such as the U, X, and Z patterns discussed in Chapter 6, can assist in designing an unbiased monitoring plan. Use a sampling pattern adaptable to field conditions. Develop a system for making sure that the same sampling method is used each time and that the sample selection is random at each site.

Select the Sample Unit. Identify the sample unit (leaf, plant, fruit, twig) most appropriate to the pest species. The sample unit (Figure 7-4) varies according to each monitoring method and amount of time available to take the sample. Choose a

www.ipm.ucdavis.edu

Grape—Insect and Spider Mite Monitoring Form

Supplement to UC IPM Pest Management Guidelines: Table Grape

Directions:

1. Start monitoring weekly for leafhopper nymphs one month after budbreak or when nymphs first appear, and for spider mites after first leaves emerge.
2. Randomly select 20 vines in each block of the vineyard, each at least a few vines in from the end of the row.
3. Sample leafhoppers, spider mites, and mealybugs as outlined below.

Leafhoppers	Spider mites	Mealybugs
On each of the 20 vines: First generation nymphs • Choose one leaf at the 3rd or 4th node up from the basal node. • Count and record the number of nymphs on each leaf. Second and third generation nymphs • Choose young, fully expanded leaves in middle of cane. • Note whether you see grape leafhopper nymphs (G), variegated leafhopper nymphs (V), or both (B). All generations • Check the leaves for red, parasitized eggs (red or exit holes) • Note their presence (+) or absence (-) on each leaf.	On each of the 20 vines: Early in the season • Choose one leaf between the 2nd and 4th nodes. • Use a 10X or 14X hand lens and look for mites and mite predators. • Note if mites and mite predators are present (+) or absent (-). Later in the season • Choose the fourth expanded leaf back from the growing tip. • Use a 10X or 14X hand lens and look for mites and mite predators. • Note if mites and mite predators are present (+) or absent (-) on the monitoring form.	On each of the 20 vines: Early in the season • Inspect basal leaves for grape, obscure, and longtail mealybugs. • Inspect under the bark of trunks for vine mealybug. Later in the season (in table grape) • Inspect all plant parts for mealybugs. • Record with a check any vine that is infested.

Record your results on the table on page 2 of this form.

FIGURE 7-3

To increase sampling efficiency, try to monitor more than one pest at a time. For example, in grapes, monitor leafhoppers, spider mites and mealybugs at the same time on 20 randomly selected vines throughout a vineyard block, as shown in these procedures from the UC IPM Pest Management Guidelines: Grape.

FIGURE 7-4

A single healthy green citrus fruit is the sample unit for monitoring for citrus thrips in California oranges. It is essential to carefully inspect areas under the sepals of young fruit where thrips may hide. Each fruit with one or more thrips larvae is counted as infested. Take leaf samples for predatory mites that are the natural enemies of thrips at the same time.

sample unit and monitoring technique that allows for obtaining the most reliable sample in the most efficient time.

The best sample unit for the same pest may be different, depending on the time of year or stage of pest or crop; sometimes different sample units may be combined that complement each other. For San Jose scale in stone fruits, the dormant shoot is the sample unit used in the winter to assess the need to treat. However, fruit samples are the sample unit taken just prior to harvest and/or at harvest to identify San Jose scale infestations that must be treated the following year. Pheromone traps are used in the spring to help time treatments. In this case, the sampling unit is the mobile male stage of the scale.

Define the Sampling Universe. Determine the boundaries of the area being sampled. Each field can make up the sampling universe, or the universe can more accurately be defined if the pest's preferred habitat (e.g., soil, leaves, or fruit) is known. In orchard crops, the sampling universe is usually the orchard block that is easily identifiable by roads, readily monitored as a unit, and can be treated separately from the rest of the orchard if a control action becomes necessary. In large orchards, divide the orchard into 20- to 80-acre (8- to 32-ha) sampling blocks.

Any decision based on sampling applies only to the field sampled and only to that part where conditions are the same as in the area sampled. In a field with areas that differ in plant growth because of variable soil types, drainage, or other factors, sample each area separately. For example, if one section with poor soil has smaller plants than elsewhere, take a separate set of samples there (Figure 7-5). In orchards that have different varieties with marked differences in susceptibility to the pest, sample these varieties separately and be prepared to recommend treating them separately as well. There is no rule on the size of the sampling universe, but the more uniform the field and the more uniform the pest distribution in the field, the larger the sampling universe can be.

Determine the Number and Size of Samples Needed. The sample size and number should adequately reflect the pest population. Determining the number of samples or observations to take is a critical decision. For many pest species, fairly specific sampling guidelines can be found in University of California publications. Ideally, take the smallest number of samples necessary to acquire the level of precision acceptable for accurate decision making. There are many factors that will help determine the number of samples to take; some important factors to consider are listed in Table 7-2.

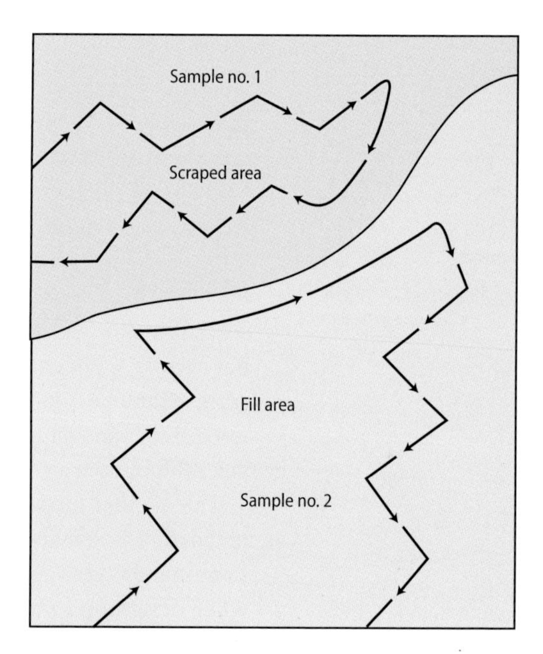

FIGURE 7-5

Each field can be defined as a sampling universe, but where there are differences within the field, such as soil type, separate samples should be taken for each area.

The sample provides an estimate of the pest population. Take enough samples to accurately estimate whether the pest population is increasing, decreasing, or being adequately controlled by natural enemies or pesticide treatment. The pest density will probably be different for each sample. By increasing the number of samples taken, the average density of the pest population and fluctuations in the population become known. If you are uncomfortable with or unsure of sampling results, take more samples.

Take advantage of field conditions, whenever possible, that make sampling more efficient. For example, knowing the species distribution in a field and on the host plants can assist in taking more informative and therefore fewer samples. Also, take advan-

TABLE 7-2.

Factors to consider when determining the number of samples to take for the pest species present.

- ☐ The pest species—its distribution patterns, mobility, and population density.
- ☐ The biology of the pest species—its phenological development.
- ☐ The potential for damage—host susceptibility and how seriously can the life stages of the pest species now present injure the crop at its current stage of development.
- ☐ The area and crop being sampled—differences in terrain, neighboring crops, or alternate hosts will make it more difficult to assess pests using just a few samples.
- ☐ The value of the crop—from an economic standpoint, the higher value the crop, the greater the number of samples that should be taken. Exported crops often have higher value and more rigorous pest control requirements.
- ☐ The sampling technique—how effective is the sampling technique in reflecting the population dynamics of the pest?
- ☐ Variability in samples—if you are finding a great variability in the pest numbers in your samples, you may want to take more.
- ☐ Experience—as you become more experienced in reading your sampling results and extrapolating them to the bigger picture, you may be able to reduce the number of samples.
- ☐ The level of precision required—the greater the number of samples taken, the more reliable the estimate of the pest population.
- ☐ The action threshold—the closer the pest population is to the action threshold, the greater the number of samples that must be taken; this is how sequential sampling works.
- ☐ Sampling time required to get an accurate estimate of a population—devote enough time for the required number of samples so that the information is reliable.
- ☐ Cost of obtaining a sample—if traps or test kits are used, how expensive are they relative to the value of the information returned? Remember that sampling time is also a cost.

TABLE 7-3.

Factors that determine the frequency of sampling.

Factor	Sampling frequency
rapid pest physiological development	+
rapidly increasing pest populations	+
close to action threshold	+
damage potential of the pest species:	
high damage potential	+
susceptibility of host growth stage	+ –
potential effect of weather and host stage on pest population	+ –
potential effect of natural enemies on pest population	+ –
cost of sampling:	
high sampling costs	–
low sampling costs	+
cost of control method:	
high cost of control	+
low cost of control	–
follow-up sampling:	
control effective	–
control ineffective	+

\+ increases sampling frequency
– decreases sampling frequency

tage of hot spots, where pests may be more of a problem. Sample hot spots separately to gather needed information about pest and natural enemy density and distribution and to get an early warning of when a pest population might be increasing. As populations increase or begin moving, adjust sampling patterns and techniques to reflect the pest population dynamics of the entire field. However, don't base population estimates on samples of hot spots alone; consider them in relationship to the total population when contemplating a treatment action.

Researchers use a series of equations called sample size formulas to determine the number of samples to take. As the example in Figure 7-6 illustrates, the greater the degree of precision needed, the more samples required. Use sample size formulas to estimate the number of samples to take and to increase the precision of the sampling technique (see Sidebar 7-1). Check the "References" section and "Basic Statistical Terminology and Concepts," Sidebar 7-2, for more statistical definitions and information.

Determine How Often to Sample. For sampling results to yield reliable information, the frequency of sampling needs to be established. Generally, insects are sampled at least once a week; but this varies with the pest species, time of year, and host. Insects that develop rapidly may need to be sampled every 3 days, while pests with populations that grow slowly, like soft scale, may need to be sampled only a couple of times per year. Other types of pests, such as weeds or nematodes, may require monitoring only once or twice a year at appropriate times. References are available that recommend sampling frequency for various pest species. Use these as a basis for a sampling program, but don't be afraid to adapt them to specific situations.

You will also encounter pests for which there are no guidelines. In these situations, consult with Cooperative Extension and other consultants for their recommendation or pattern the sampling program on pests with similar behavior or life cycles on similar crops. Adapt any available sampling methodology. Table 7-3 lists many of the

SIDEBAR 7-1

General Formula to Determine a Fixed Sample Size

Within a homogeneous habitat, the number of samples required to get an accurate assessment of a population is given by:

$$n = \left(\frac{ts}{D\bar{x}}\right)^2$$

where
n = the number of samples
t = values from 'student t' distribution tables
s = the standard deviation of a sample
D = the predetermined standard error of the mean
$\bar{x}$ = the mean of a sample

Definitions to interpret sample size formulas (see also page 215):

Mean: the average number of what is being counted in all samples.

Standard deviation: the square root of the variance (the degree of difference among the individual samples).

Predetermined standard error: the level of accuracy acceptable to you; you designate how close you want to be able to estimate the actual mean number of what you are sampling.

Student's t *test:* a statistical significance test to assess the compatibility of a set of data with the hypothesis that the treatment effects are equal (null hypothesis of no treatment differences). It is usually set to a probability level of 0.05 (or 5%).

Example:

Estimating the mean ($\bar{x}$) and standard deviation (s): To estimate $\bar{x}$ and s, define the sampling universe and unit; begin by taking more samples than you would normally hope to be sampling. For example, count the number of aphids on stem terminals.

10 15 18 25 20 17 22 25 19 11 21 16 15 22 14

Then calculate the mean ($\bar{x}$) and standard deviation *(s)* of the sample. This can be done on a simple statistics program on a computer or laboriously by hand. The long method for calculating the standard deviation of a sample is shown below.

First, calculate the **mean** by adding all the samples together and dividing by the total number of samples taken:

$$\Sigma x = 10+15+18+25+20+17+22+25+19+11+21+16+15+22+14 = 270$$

$$\bar{x} = \frac{270}{15}$$

$$\bar{x} = 18$$

mean ($\bar{x}$) = 18

Next, find the **variance** and **standard deviation.**

X (number of aphids/sample)	X-$\overline{X}$ (number minus mean)	$(X-\overline{X})^2$
10	−8	64
15	−3	9
18	0	0
25	7	49
20	2	4
17	−1	1
22	4	16
25	7	49
19	1	1
11	−7	49
21	3	9
16	−2	4
15	−3	9
22	4	16
14	−4	16
270	0	296

Subtract the mean from each individual count $(x-\overline{x})$; and square each $(x-\overline{x})^2$. Add them all together $S(x-\overline{x})^2$.

Divide this total by the number of samples minus one $(n-1)$ to find the variance of the samples

$$\frac{\Sigma(x2\overline{x})^2}{n-1} = \frac{296}{14}$$

The variance of the sample = 21.14

The standard deviation is obtained by finding the square root of the variance

$$\sqrt{21.14}$$

The standard deviation, s, = 4.6

Defining t and D:

t = a value from *Student's* t *tables:* determined by the level of accuracy required and the number of samples taken. For most cases where more than 10 samples are taken and a 5 to 10% error is acceptable, it is approximately 2.

D, or the predetermined error, is the level of accuracy required. For instance, to be correct 90% of the time, the error will be 10% and D = 0.1; for a 95% confidence, D = 0.05; for a 99% confidence, D = 0.01.

We can now determine the number of samples required to estimate aphid density with various confidence levels.

For a confidence level of 90%, the sample size formula would be as follows:

$$n = \left(\frac{ts}{D\overline{x}}\right)^2 = \left(\frac{2 \times 4.6}{0.1 \times 18}\right)^2 = 26$$ samples would need to be taken.

To increase the confidence level to 95%, you decrease D to 0.05. This adjustment greatly increases the number of samples required.

$$n = \left(\frac{ts}{D\overline{x}}\right)^2 = \left(\frac{2 \times 4.6}{0.05 \times 18}\right)^2 = 105$$ samples need to be taken.

To increase the confidence level to 99%, you must decrease D to 0.01. In this case, t also increases to 3.

$$n = \left(\frac{ts}{D\overline{x}}\right)^2 = \left(\frac{3 \times 4.6}{0.01 \times 18}\right)^2 =$$ 5,878 samples would need to be taken to ensure this level of accuracy—an almost impossible task!

factors to consider when deciding how often to sample.

Determine When Sampling Begins and When It Ends. From the standpoint of cost and efficiency, when to start sampling is probably the most important question a PCA needs to answer. When sampling begins relates directly to when potential damage may occur and when a management action must be taken to prevent it. In cases where controls must be applied early (e.g., preemergent herbicides, protectant fungicides, or applications for overwintering pests) or cases where preventive actions can be taken (e.g., sanitation, selective pruning, etc.), monitoring may need to begin in the previous season. For some pests, a biofix such as the first pheromone trap capture or a critical temperature reading may determine when to begin monitoring.

The time of day to take a sample is also an important consideration in the design of a monitoring program. Many insects, for example, are found in different parts of the plant or are not on the plant at different times of day, thereby affecting sampling results. Depending on the pest, sampling in the morning can bias counts because, for example, low temperatures reduce arthropod activity. In other cases, morning is the best time to sample because pests retreat to hidden areas or become flighty during the heat of the day. Other factors to note are light intensity and wind.

The most efficient monitoring programs are designed to take samples at the ideal sampling time; unfortunately, all fields cannot be sampled at the ideal time. Initially, take samples at different times on the same day; a comparison of the results should reveal any sampling bias. If the data show consistent patterns, use this information to adjust sampling numbers accordingly. A good approach is to monitor a specific field at approximately the same time of day at each visit.

Step 3. Establish Injury Levels and Action Thresholds for Each Pest Species

Injury levels (sometimes called economic injury levels) establish the amount of pest damage or yield reduction that occurs from given pest densities. The treatment or action threshold indicates when management actions are needed to avoid significant economic losses from pests. Guidelines for insect pests are generally numerical thresholds based on specific sampling techniques; they are intended to reflect the population level that will cause economic damage if no action is taken. For example, treatment thresholds for whiteflies in cotton in Arizona are based on sampling the fifth main stem node leaf on 30 plants at least 10 to 15 feet (3 to 4.5 m) apart at two sites in the field (Figure 7-6). Leaves are sampled for presence or absence of whiteflies. The established treatment threshold is 57% or more leaves infested.

Guidelines for other pests, including most pathogens and weeds, are usually based on the history of a field or region, the stage of crop development, weather conditions, and other observations. Action thresholds for many pathogen and nematode pests are not well documented, although this is changing as risk assessment models for some diseases are being developed. Injury levels and action guidelines for certain pests in California are listed in the UC IPM Pest Management Guidelines; other guidelines are available in other Cooperative Extension publications. When available, follow guidelines based on university research to establish a monitoring program. Unfortunately, for many pests, research-based guidelines have not been developed. In these cases, rely on experience or use the information presented in this chapter to develop site- and pest-specific guidelines.

Action or treatment thresholds are not rigid. Conditions such as weather or harvest schedule may influence the decision to treat, even though the pest population may be approaching the treatment threshold. Populations of beneficial species, for

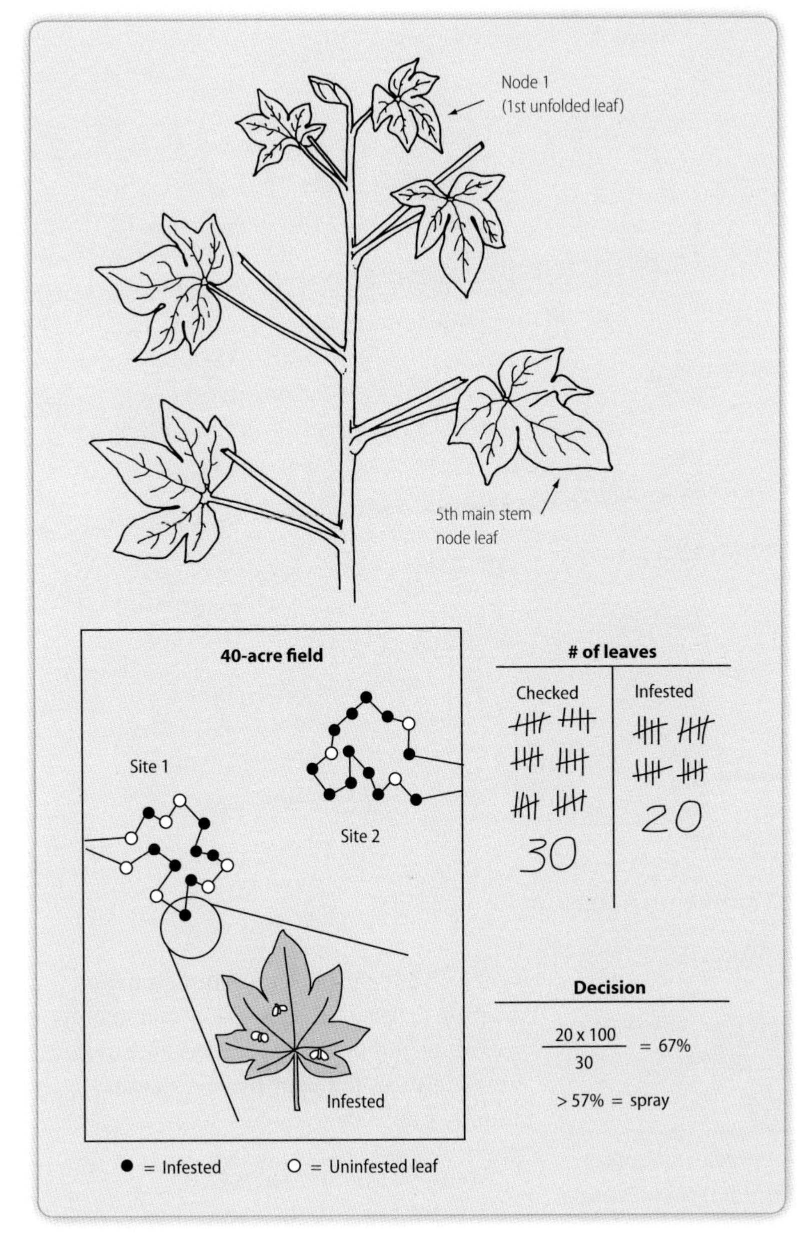

FIGURE 7-6

Treatment guidelines for insect pests are generally numerical thresholds based on specific monitoring procedures. For example, an action threshold for whiteflies in cotton in Arizona was established as 57% or more of fifth main stem node leaves infested. Infestation levels are determined by sampling leaves on 30 plants in two areas of the field using presence-absence sampling. Source: *Ohlendorf et al. 1996.*

instance, should also be monitored, and, if their populations are high, a treatment decision may be delayed. However, continue monitoring the pest population to determine whether it increases despite the presence of natural enemies. Other factors to consider before making a treatment decision include the life cycle of the pest, the susceptibility of the developmental stage present, weather conditions that may cause the pest population to decline or prevent the control action from being effective, the relative value of the crop, the cost of taking a control action, and the known or presumed efficacy of the control procedure.

The threshold usually assumes a given value of the crop and cost of a control action. These monetary considerations may change due to unforeseen conditions, raising or lowering the threshold. Control action guidelines often change as new varieties, cultural practices, or pest control procedures are introduced and as new information on pests becomes available. Thresholds on exported crops such as fruit may be much lower than on crops sold to domestic markets. In some cases where other countries have quarantines on pests, the threshold may be anything above zero.

Step 4. Determine What Host or Crop Developmental Stages Must Be Monitored to Assess Normal Growth, Predict Timing of Pest Activity, or Evaluate Damage

The ultimate goal of pest management is to produce a crop or landscape that is healthy, vigorous, and profitable or aesthetically pleasing. The plants should be monitored at some level throughout the season from planting to harvest or from budbreak through dormancy. Track key developmental stages to be sure the plant is on schedule. Stress from pest damage or other factors may make plants more susceptible to yield loss or injury from pests.

Crops and ornamental plants are most susceptible to certain pests during specific developmental stages. Monitor plant development and intensify pest sampling when the host nears these critical stages. For example, Figure 7-7 charts monitoring times for insect pests and diseases in potatoes according to potato developmental stages.

When setting up a monitoring program, identify these critical plant stages and predict when they may occur. Sometimes models for plant development are available for anticipating crop growth. Field monitoring of plants is essential for verifying that

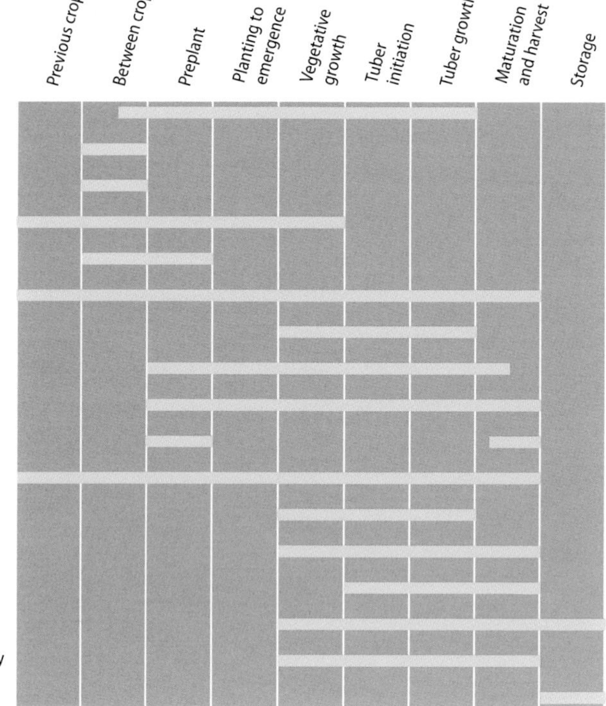

FIGURE 7-7

Monitoring activities important for potato pest management. Source: *Strand 2006.*

the predictions hold true.

Step 5. Determine the Environmental Factors That Must Be Monitored

Environmental factors can impact sampling counts; they can also determine monitoring strategy or timing. Because weather influences host and pest development, a reliable source of weather data improves pest management decisions. At a minimum, daily high and low temperatures are needed if degree-day accumulations are used to predict pest development. Free moisture for specific time periods is required for development of certain diseases. Soil temperatures affect seedling emergence of weeds and crop plants. Slower crop seedling growth can make plants more vulnerable to a number of seedling diseases. Soil moisture content and nutrients should also be noted. Local variations in terrain, vegetation, elevation, and other conditions affect pest development. Wind can reduce the flights of certain insects while helping to spread other pests. When monitoring, record the impact of specific environmental factors that may be necessary to interpret the data.

Step 6. Determine the Production Practices That Can Impact Development of the Pest Species

Production practices, including planting time, irrigation, fertilizer use, crop rotation, and field sanitation, can have a significant impact on many pest populations. It is important to understand the impact of these activities on pest management to anticipate problems that might occur and be able to react appropriately. Planting when the soil is too cool, for example, can increase the susceptibility of many seedlings to pests and disease or favor weed germination. Excessively high nitrogen levels can increase pests such as aphids or brown rot of stone fruits. The type, timing, or duration of irrigation can increase a plant's susceptibility to disease and induce germination of weeds (Figure 7-8). Water-stressed trees can promote spider mites. Certain corn pests are reduced if the field is rotated to another crop.

Step 7. Streamline the Monitoring Program to Develop Efficiencies

Combine monitoring or sampling activities to reduce the time needed in the field. Evaluate the efficiency of the current monitoring program and look for ways to improve it. Every year is different; be prepared to adjust timing and methodology to fit the pest situation at hand, not a convenient schedule.

Incorporate a system of checks into the monitoring program. Use multiple monitoring techniques: for example, check for egg laying as well as taking counts of adult males caught in pheromone traps. Look for damage symptoms. Be prepared to adjust control thresholds as changes in management practices or marketing requirements dictate. As experience is gained, the knowledge necessary to adjust sampling techniques to be more time efficient will also be gained.

Whenever possible, use sequential sampling, stratified sampling, or other techniques described in Chapter 6 for reducing numbers of samples.

FIGURE 7-8

Border, or furrow, irrigation can spread weeds and pathogens through orchards as the water moves through the area.

Step 8. Keep Good Written Records

Knowing what happened last year or last month can make many pest management decisions easier. Good written or electronically archived records become part of the field record, so it is usually best to take field notes on a standardized form that reduces the amount of writing or data entry required, provides a reminder of what to look for, and makes it easier to keep uniform records. A sample sticky trap form is shown in Figure 7-9. Keep a file, notebook, or electronic record of the following information for each area being monitored:

- weekly monitoring records, including counts of pests, natural enemies, damage, and crop growth and development
- weed surveys
- records of pesticide and fertilizer applications, including materials, rates, dates, and methods of application as well as the target pest and efficacy of applications as indicated by observations and subsequent sampling
- laboratory reports and records of diagnostic tests
- releases of natural enemies or other pest management activities
- irrigation, cultivation, and other cultural practices
- aerial photographs with reports attached
- agronomic information, including crops and cultivars planted, planting and harvest dates, yields, and grades
- disease reports

In addition, keep a file of local weather data, including charts of accumulated degree-days if they are being used for pest management decisions. Detailed field records have become even more important as geographic information systems and precision agriculture become more prevalent. In California, agricultural commissioners require reports of pesticide applications and records of PCA

Yellow Sticky Trap Record Form

Trap location ______________ Scout's name ______________

D a t e traps were set out or last checked ______ Date traps were collected or checked ________

Card no.	Trapping period (days)	Number of insects* Leaf-miners	Thrips	White-flies	Fungus gnats	Shore flies	Aphids	Other/Comments
Total								
Average								
Previous average								
Trend								

Comments/map:

*Insects per entire trap or insects per 1-inch-wide column (note which).

Be sure to record when traps were set out or last checked. Unless traps are changed each time they are checked, numbers in traps become culmulative. If traps are reused, subtract all previous catches in order to determine the number of insects caught during the most recent period.

FIGURE 7-9

A standardized form, such as the yellow sticky trap record form shown here, makes it easier to keep uniform records. Be sure to record the location of each trap using a map of the area. Source: *Dreistadt et al. 1998.*

written recommendations. These should become part of the file for each field or location.

Using Scouts Effectively

Scouts can be an effective tool in a monitoring program. For scouts to collect accurate information, however, they must be carefully trained and supervised. They must be precise and methodical when gathering data so that sampling data from multiple scouts will be consistent. If not, the sampling information will be misleading and the result will be poor decisions.

Explain precisely what the scout is to sample and how to take the sample. The field scout's job is to warn of the presence and concentration of pests for future decision making. Scouts should be able to identify the pests they are sampling and be able to recognize damage symptoms. They should recognize a pest not previously present or an unusual occurrence of a pest or damage. Provide precise information regarding sampling patterns and technique. Establish regular sampling intervals.

Provide sampling forms that are easy to use and summarize the results of the samples taken from each visit. Information recorded on sampling forms should include the scout's name, the date and time of sampling, field location or identifier, and observations on plant health, growth stage, and weather conditions.

Incorporating New Monitoring Techniques and Flexibility into the Monitoring Program

The agroecosystem or landscape being monitored is always changing. What occurred at a site one year might not happen during the next. Agroecosystems are exposed to constant pressures not only from pest species but also from neighboring fields, weather, and the management practices used at the site. Pest species are always present, evolving, and a threat to the viability of the resource. The design of the monitoring program has to account for this variability; otherwise, the risk of error in pest management decision making increases. Lastly, keep up with the latest research on monitoring techniques. Pest management is not a static science; because of changes in pest status, the introduction of new pest species, failures in control methods, new control tools, and research breakthroughs, new monitoring strategies are regularly being developed.

Investigate new strategies thoroughly. If a new strategy seems like it will fit the situation, try it in one location and compare it with the previous sampling method. Just because a sampling strategy exists or is new doesn't mean that it is the correct method for the situation. Shape and mold all monitoring plans with the intent of making them fit the circumstances.

When pest control options change, new monitoring practices may be required. For instance, new pesticides may need to be applied to different life stages than previous products, and monitoring programs will have to be adjusted. If releases of biological control agents are being substituted for pesticide applications, different monitoring programs may be required to determine application times.

FIELD TRIALS

Field trials are in-field comparisons of different treatments or practices. For example, they may compare different pesticides, rates, volumes of water in a spray tank, release rates of biological control agents, irrigation methods, or pruning practices. In some cases, they are designed using replicated plots and controls. In other cases they are purely observational. For example, comparing the rate of a pesticide application on one section of an orchard to that of a different rate on another section may give an indication of which rate provides better control. This conclusion, however, is based on an observation and can't be validated because it is impossible to know whether both sides of the orchard had the same pest population levels or whether additional factors might have been affecting pest levels differentially in each side. Because such field trials are not statistically valid, be careful about the conclusions drawn from them. Also, information gathered from a field trial at one location might not be relevant to another site. Although replicated field trials (see below) are best, observational field trials can be helpful tools for gathering additional knowledge about pest management options as long as their limitations are understood.

Purpose of Field Trials

There are many reasons to conduct field trials. Field trials are an important tool for testing the effectiveness of a new control method. They provide a means by which treatment timing can be understood and improved. Field trials also allow the establishment of control action thresholds and sampling methods.

A number of factors, however, must be taken into consideration before beginning a field trial. First, clearly define the goal of the trial. Scientists call this stating a hypothesis. Have a clear understanding of what you expect to achieve. Be sure that the idea that is being tested is amenable to the testing process that will be used. For example, when comparing two sampling techniques, use them in the same area and compare the results with absolute samples.

Field trials are conducted in a complex and dynamic environment. Factors that can influence the results of a field trial are called *variables*. The outcome of the trial is subject to environmental variables that cannot be controlled; management decisions can also interfere with the field trial results. Enlist the client's cooperation to minimize variability in irrigation scheduling and other management practices that might influence the outcome. To get clear results, it is important to minimize as many confounding variables as possible and to control variables that can be compared. Whenever possible, also incorporate an untreated control into the field trial to ensure that a pest population is present and to assess the magnitude of the population. If a control is not possible, make comparisons with the standard practice.

REPLICATED, STATISTICALLY DESIGNED FIELD TRIALS

The most reliable results will be obtained from field trials if they are conducted as scientifically designed, replicated experiments with controls. An experiment consists of units (a single leaf, a whole plant, a row of plants, or several acres) that receive different treatments. In an experiment, the treatments are replicated (i.e., repeated several times in the same field trial), and the experiment is designed so that the data can be statistically analyzed. Replicating the treatment allows for the measurement of variability across experimental units. It is the measurement of this variability that is the key to the results of a statistical test. If variability among replicates is high, it is very unlikely that a trial will yield statistically significant results unless the differences among treatments tested is also very high. Likewise, if variability among replicates is low, statistical significance can result among different treatments even though they may not differ by a large amount. Experiments are conducted to provide specific facts from which general conclusions or principles can be established.

Experimental Design

The design of an experiment is critical to its success. A well-thought-out experimental design will be efficient and suited to the conditions under which the experiment is conducted. The number of treatments, the number of controlled and measured variables, and the number of *replications* to be used in the experiment are major factors in the experimental design.

Before an experiment is begun, a pretreatment evaluation is carried out. Pretreatment evaluation establishes whether the information sought can be obtained. For example, it can establish that the population of the pest species is great enough that the effects of different treatments can be measured. For instance, pretreatment sampling may provide a baseline for future comparison following treatment. It also can provide useful information to improve the success of the experiment, such as information on the variability of a population across the area to be treated as well as an indication of the number of samples that must be taken.

A successful experiment defines the problem clearly and precisely. Objectives are stated in precise terms either in the form of a hypothesis or in the form of questions to be answered. Various treatments (different rates, materials, timing, etc.) are used to test the hypothesis. Treatments are typically compared to a control, either an untreated plot or a conventional standard treatment acting as a control.

Of the several important principles in experimental design, a key one is replication. Replication is the repetition of treatments. Typical agricultural experiments, for example, use from four to eight replications, with a minimum of three. The number of replications is determined by the degree of precision required. The type of pest, abundance of the pest being measured, and the magnitude of expected differences between treatments influence the degree of precision required. Typically, the higher the degree of precision required, the larger the number of replications. Similarly, the more variability among blocks, the more replication is required. Replication provides the basis for deciding whether an observed difference is real or just due to chance, and whether it is

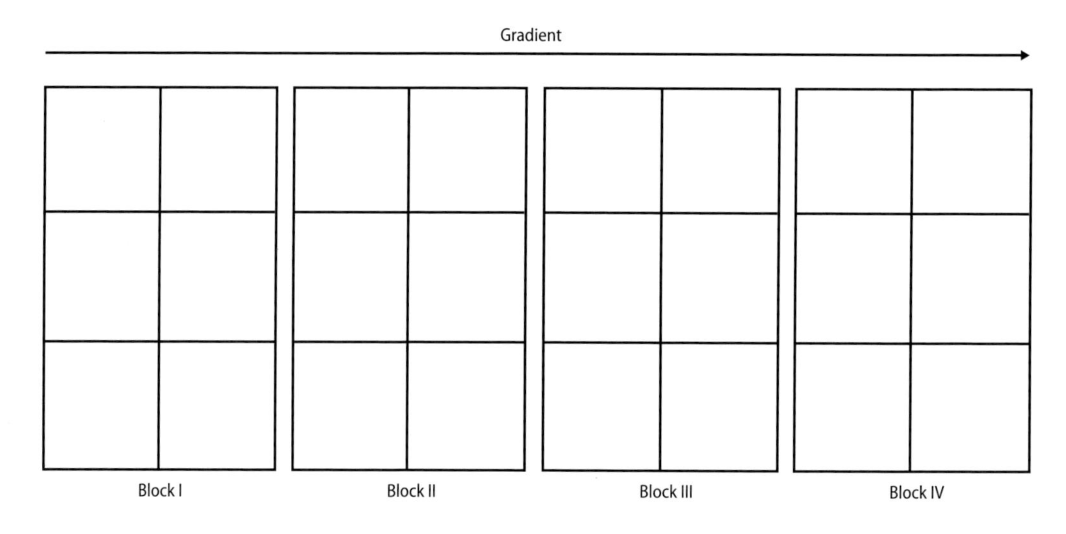

FIGURE 7-10

When conducting an agricultural experiment, the experimental area is typically divided into blocks. A randomized complete block with six treatments and four replications per block is shown here. The experiment should be designed so that each treatment is similarly exposed to environmental or biological factors that may influence the results.

important in validating the conclusions from an experiment. The results from an experiment or field trial can be influenced by many factors such as soil fertility, moisture, damage from other pest species, pest distribution, and crop management practices. Replication helps minimize the effects of these confounding factors, which can influence the results of the experiment. Replication also helps minimize experimental error.

Another important technique for controlling experimental error is blocking. A block is an area that includes one plot or replication of each experimental treatment (Figure 7-10). Treatments within a block should be arranged so that each treatment is similarly exposed to any environmental or biological factors that you believe might influence results. Because so much variation exists in a typical field or orchard, independently and separately assigning the location for all treatments in each block reduces the possibility that an environmental or physical factor will affect results. The blocks should be set up in the direction of the pest, soil, irrigation, or other gradient in the field. Each block represents a replication of the experiment.

Another important principle is randomization. Each treatment should be randomly placed in the block so that each experimental plot has an equal chance of receiving a particular treatment (Figure 7-11). Random placement of treatments guarantees unbiased estimates and reduces experimental error. A table of random numbers such as found in a statistics textbook (see the "Resources" section) is a convenient way to assign treatments randomly to each experimental block.

The final critical element in an experiment is the use of a control. The control is the treatment to which other treatments will be compared. When testing management techniques or products, the ideal control is a replicated untreated area within the experiment itself. Alternatively, a well-known pesticide or standard grower practice can be applied and used as the control, but this will not let you know what the level of the natural population is or account for natural pest population crashes or decreases in the pest population due to biological control. All treatments may appear to be effective when in reality the pests were in low numbers in all plots in the experiment. Only an untreated control would identify this situation.

Nonexperimental Field Trials

It is not always possible for PCAs or growers to carry out replicated field trials that meet the standards of scientific rigor. However, field testing of new materials or methods can be useful, and growers should be encouraged to try out new methods in small areas first. While results of nonreplicated field trials can provide experience, practitioners must be careful about the conclusions drawn from them. Without replication, it is not safe to assume that the same results would be obtained at another site or in another year or even in the same field if it were repeated another time. Without controls, there is no certainty that the results obtained are due to

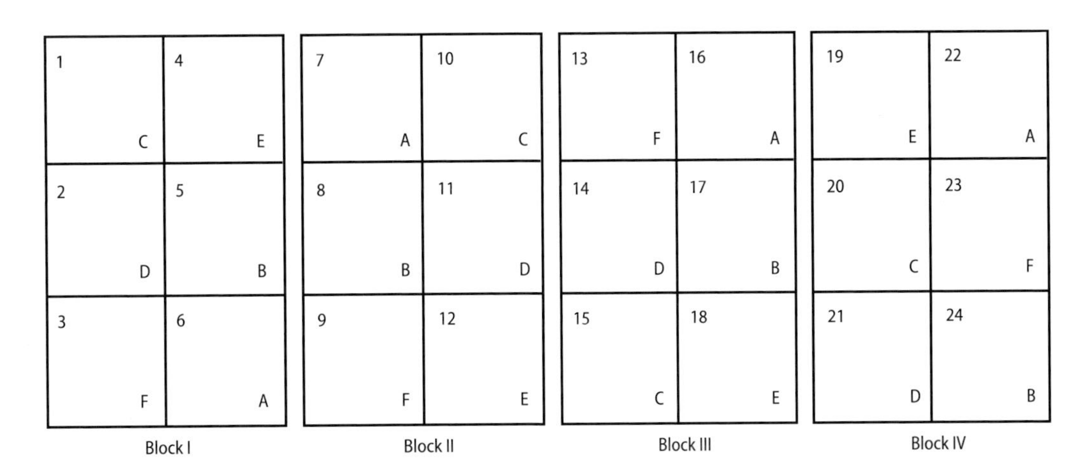

FIGURE 7-11

An example of a randomized complete block design with six treatments (A, B, C, D, E, and F) and four replications (I, II, III, and IV).

treatments and not some other factor in the environment.

However, several things can be done to enhance the value of nonreplicated field trials. Before beginning a field trial, sample across the site for variability. Pretreatment sampling will give an indication of the size of the pest population, the developmental stages present, and the variability within the test location. Assign the treatments to areas of the field with similar pest densities.

Site selection is an important consideration in field trials. When comparing treatments, choose sites that are as similar as possible; consider soil type, drainage, microclimate, habitat, and pest pressure, as well as current management practices. In every field trial, a control (an untreated area or area treated with an established method) must be established against which new treatments can be compared.

Evaluating Data from Field Trials

Most PCAs and growers normally have little time to conduct experiments. Experiments, properly done, can be very time consuming, and the data collected must be evaluated using accepted statistical analysis. Although treatments may tend to show different degrees of control, statistical analysis is necessary to show whether these differences are truly significant or can be explained by chance.

Even if experiments are not statistically designed, the results from research trials can still be useful. To evaluate the validity and the relevance of the data, however, requires a good understanding of the experimental process and the ability to evaluate data. Understanding the basics of experimental design will give the PCA confidence to try a new approach.

Evaluating the data from a field trial and determining whether the outcome of the trial is useful at a particular site require an understanding of the basic concepts and terminology of statistics. Some of these are discussed in Sidebar 7-2, below. For further information, refer to a basic statistics textbook (see the "Resources" section for some recommendations).

It is important to remember, however, that just because a trial was conducted and the results of a new treatment were a "success," it does not mean that current practices should be thrown out and past experience ignored. Review all research with a critical eye. Some common problems with research results are

- absence of a control
- no replication
- large standard error or standard deviation (variability)
- no indication of the statistical significance of results
- comparisons of different treatments tested at separate locations
- pseudoreplication, that is, observations are not statistically independent but treated as though they are

If a new practice is adopted based on someone else's research, conduct additional field evaluations on a small area of one or more fields to be sure that the technique works under the prevailing field conditions.

SIDEBAR 7-2

Basic Statistical Terminology and Concepts

Mean. When understanding data, it is often convenient to begin by finding average values. For example, individual counts from five codling moth traps are shown below. The mean represents the sample average. In this case, the mean is obtained by adding up the individual counts from the traps and dividing the sum by the number (n) of traps. The formula to calculate the mean is

$$\bar{x} = \frac{\Sigma x_i}{n}$$

The mean of these five samples is

$$\bar{x} = \frac{15 + 12 + 25 + 36 + 21}{5}$$

$$\bar{x} = \frac{109}{5} = 21.8$$

The mean should not be confused with the median, which is the middle value in a list of numbers, above and below which lie an equal number of values. In the list above, the median would be 21.

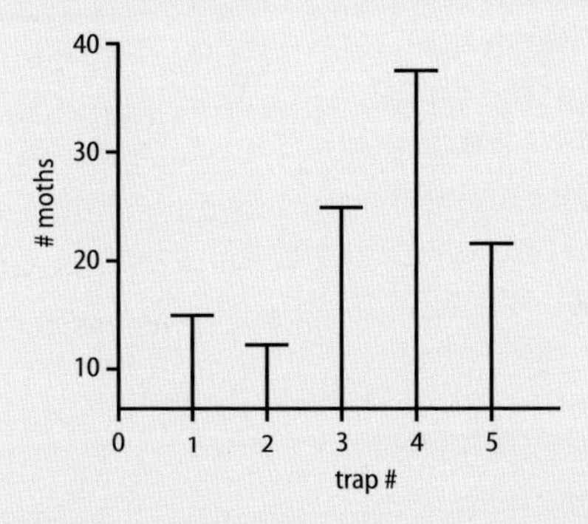

Standard Deviation. The most commonly used measure of variation is called standard deviation. This measure tells you whether the counts from the samples are close to the mean in a set of data by measuring the amount of variation in the samples. Standard deviation is easily illustrated using a bell-shaped curve. The bell curve is steep when the standard deviation is small, that is, when the counts are close together because there is little variation in the samples. When the second bell curve is rather flat, the standard deviation is large, and there is greater difference among the counts of the individual samples. A greater standard deviation means that there is a greater chance a given sample will be considerably different from the mean.

The formula for standard deviation (σ) in a population is

$$\sigma = \sqrt{\frac{\Sigma(x_i - \bar{x})^2}{n}}$$

where
$\bar{x}$ = the mean of the values for all individuals
x_i = the value for each individual
n = the number of individuals

The variance, σ^2, is also used as a measure of dispersion

$$\sigma^2 = \frac{\Sigma(x_i - \bar{x})^2}{n}$$

The formula for standard deviation in a sample is

$$s = \sqrt{\frac{\Sigma(x_i - \bar{x})^2}{n - 1}}$$

The variance in a sample is:

$$s^2 = \frac{\Sigma(x_i - \bar{x})^2}{n - 1}$$

Since most pest management calculations involve samples, the latter calculation is the primary one.

Testing Hypotheses and Levels of Significance. The purpose of experimentation is to test questions or hypotheses about the effectiveness of new techniques or products. For instance, you might ask whether a new pesticide (A) is as effective or more effective than the pesticide you are currently using. Because experiments normally involve a sample of only part of a pest population or crop, experiments do not allow us to say with absolute certainty that our hypothesis is correct or incorrect. However, statistical analysis allows us to determine the probability (*P*) that our hypothesis will be true for any given sample. Generally, agricultural research applies the null hypothesis, which assumes that the treatment will not have an effect. If observed differences between the treatment and some standard, such as no treatment or a standard treatment (the control), occur more often than would occur from chance, the null hypothesis is not true and the treatment must have had some effect.

If the treatment had a similar impact in 99 out of 100 samples, we can say that the probability is only 1% that the observed variation could be expected to occur by chance ($P = \leq 0.01$). If the probability is 5% or less that the observed variation among means could occur by chance we say that the differences are significantly different ($P = \geq 0.05$). Either of these levels of significance are considered adequate to assume that the experimental results are valid. Generally, if *P* is greater than 0.10, the conclusions of the experiment are considered to be less reliable. When reviewing research results, look for levels of significance to judge the reliability of the conclusions.

Confidence Limits. Confidence limits are another tool to help interpret the reliability of research results. Confidence limits give us a measure of how likely we are to find a sample that falls within a range of means. For instance, a 95% confidence level tells us that 95 times out of 100, the mean of our sample will occur within the interval between the upper and lower confidence limits of a mean, and that 5 times out of 100 it will not. Confidence limits are frequently used in interpreting *t* test statistics and regression analysis (see below).

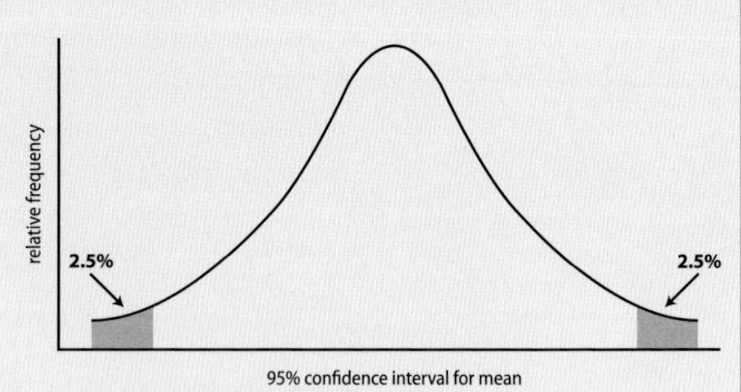

Standard Error. Standard deviations of various statistics are generally known as standard errors. The term *standard error* is used for various statistics, but when used alone, it conventionally implies the standard error (SE) of the mean or

$$SE = \frac{s}{\sqrt{\underline{n}}}$$

where
s = the standard deviation
n = the number of samples

Standard error (6SE) gives you a good sense of the variability in experimental results; it is a common statistic reported in experiments comparing treatments when replicates are involved. Standard errors are often drawn as little lines above and below means illustrated in graphs. In Figure 7-12, only the positive half (+SE) of the standard error is shown.

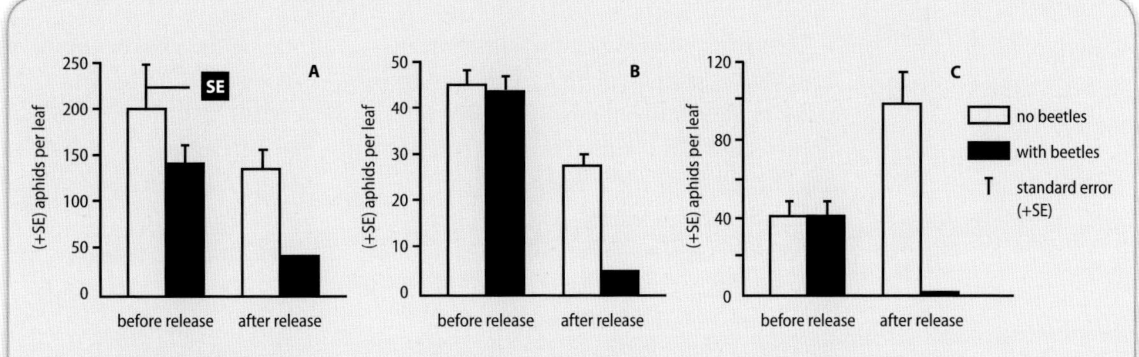

FIGURE 7-12

A graph showing standard error (SE). Three different trial dates are shown (A, B, C) for the release of adult lady beetles. The bars show the mean and SE of aphids per leaf before release and 3 days after release. Source: Dreistadt and Flint 1996.

Normal Distribution. Many statistical methods are based on the assumption that the population being sampled has a normal distribution and closely fits a mathematically defined curve called the normal frequency curve, typically a bell-shaped curve. The position of the curve on the horizontal axis is determined by the mean. The amount of spread on the bell curve (the Y-axis) is determined by the standard deviation. Figure 7-13 gives examples of differences in normal distributions due to differences in means (A) and standard deviations (B).

In reality, field collected data rarely fit the normal frequency curve as perfectly as shown here, but in most cases the fit is close when a very large number of samples is taken. For instance, if you were to sample a large number of potatoes for scab lesions and recorded the number of scabs per potato, you could then put that data in a frequency table that grouped potatoes according to a range of scab lesions (e.g., 0–3, 4–7, 8–11, 12–15, etc.) A histogram graph such as in C below could be drawn from this data and the normal curve fitted over it to see how the results vary from the expected normal distribution.

ANOVA, *t* tests, and *F* tests. In analyzing different treatments from an experiment, it is assumed that one of the treatments had an effect if differences are observed that otherwise would most likely not have occurred in a random sample. The next questions are whether the treatment effect is significant and whether the treatment is responsible for the difference.

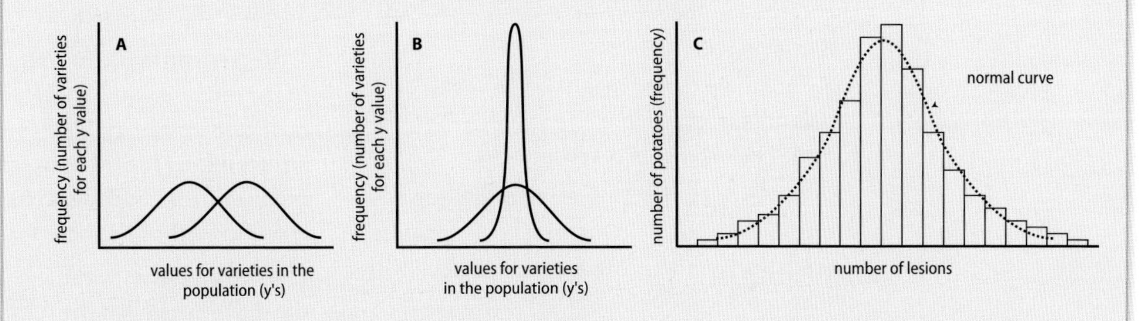

FIGURE 7-13

Normal frequency curves. Graph A shows an example where the means of the samples from two varieties are different, but the standard deviations are the same. Graph B shows a sample where the means of the two varieties are the same, but standard deviations are different. Graph C shows a histogram of field data for potato scab sampling fitted against a normal curve.

Tests of significance compare the means from several samples or analyze the variation between the treatment means and the population means. They also offer a systematic way of describing and determining the reliability of research results. A *t* test (or student's *t* test) expresses the difference between two means in terms of the standard error of the difference. They are typically used when only two treatment means are involved (e.g., when a treatment is compared with a control or expected outcome). To interpret the results of *t* tests, researchers use tables that provide the critical values of *t* and associate them with levels of significance. A significance level of 0.05 means that 95% of the time this result can be expected to be correct.

Analysis of variation (ANOVA, or ANOV) procedures, such as *F* tests, are used when means of more than two treatments are being compared. *F* tests are used to test the degree of variance and to determine whether the difference in means is due to the treatment or perhaps due to variability between samples. As with *t* tests, *F* tables that document the critical values used in analysis of variance are used by researchers.

Analysis of variance makes several assumptions (from Little and Hill 1978; see the "Resources" section):

1. The error terms are randomly, independently, and normally distributed.
2. The variances of different samples are homogeneous.
3. Variances and means of different samples are not correlated.
4. The main effects are additive.

If these assumptions are not met, the analysis may result in incorrect conclusions. Statistical methods are available to deal with these problems in many situations.

Mean Separation Tests. Once treatment differences are observed, the researcher might wish to determine whether the mean value of one of the treatments is significantly different from the mean value of another treatment, or to determine which means caused the difference observed in the ANOVA. One method tests for the *least significant difference* (LSD). LSD is the smallest difference between two group means that would yield a significant result if the two groups were compared using a *t* test.

Multiple range tests also test for mean separation and are very similar to LSD tests. Multiple range tests are used when the total number of treatments is large and several unrelated treatments are a part of the experiment. Mean separation tests are valid only if performed after an ANOVA test indicates that significant differences exist.

Tests for Homogeneity of Variance. The assumption that the variances between two samples are equal is an important precondition for several statistical tests, especially the ANOVA test described above. Tests for homogeneity of variance are simple tests used to verify the equality of several variances. This test is used whenever more than two variances are tested.

Regression Analysis. Regression analysis measures the cause and effect relationship between two variables. Simple regression is actually just a special case of ANOVA. In regression, the amount of change of one variable, X, (e.g., pesticide dose) brings about a change in the other variable, Y, (e.g., pest mortality). X is manipulated in an experiment to determine whether variation in Y is associated with changes in X.

Regression analysis also studies the degree of the relationship between variables, that is, how much one variable depends on or is influenced by the other variable (i.e., raising or lowering the independent variable X directly or indirectly influences Y).

The correlation coefficient (*r*) expresses the degree to which the relationship between X and Y is consistent on a scale of –1 to +1. Correlation coefficients above 0.90 show that variables are highly correlated. Once *r* drops below 0.80, the correlation becomes less reliable, and there are probably other factors that must be taken into account to explain changes in Y. Statistically, significant *r* values can be obtained at much lower values if the experiment is highly replicated. Negative values for the correlation efficient (e.g., $r = -0.90$) indicate that the factors correlate negatively (e.g., when *X* is high, *Y* is low).

Another important measurement in regression analysis is called the coefficient of determination; it is used to determine how closely the regression equation fits the observed data. The coefficient of determination is the square of the correlation coefficient (r^2). Values of r^2 over 0.70 usually indicate a significant correlation between two factors.

Linear correlation and regression are illustrated on a type of graph called a scatter diagram in which the data are plotted on the graph (see Figure 7-14). The line represents the closest fit of the straight line that will make the sum of squares of deviation as small as possible.

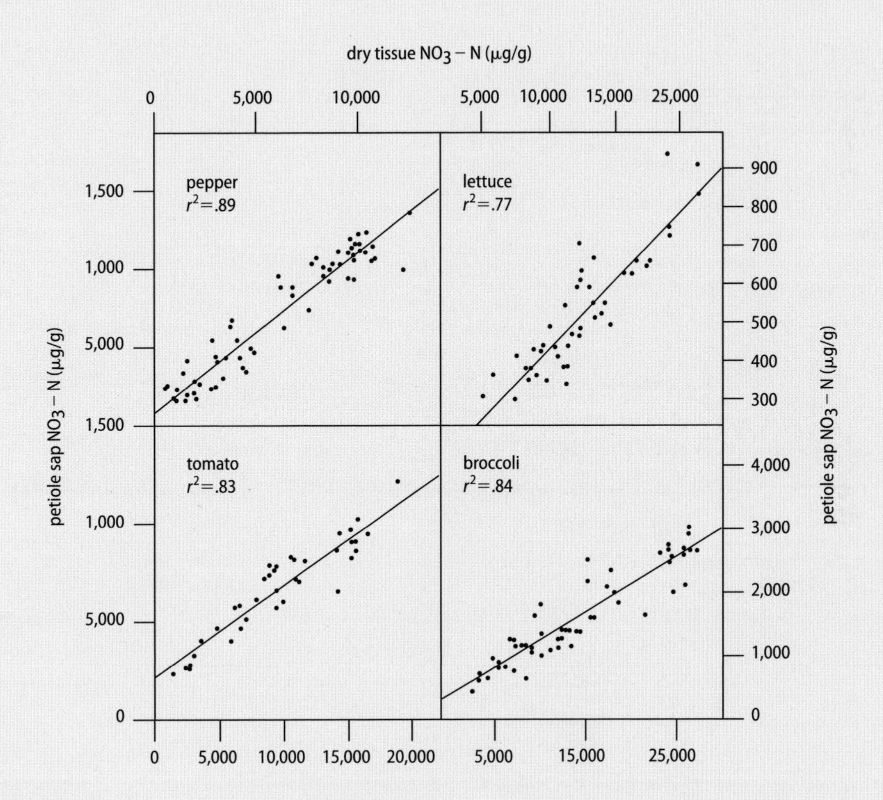

FIGURE 7-14

These regressions show the relationship between the nitrogen in plant sap and the nitrogen in dry tissue for various vegetable crops. All coefficients of determination (r^2) are above 0.77, showing significant correlation between these two measurements. Source: Hartz et al. 1994.

8 Health and Environmental Concerns Associated with Pesticide Use

Pesticides are important and necessary pest management tools for agriculture, urban landscapes, forestry, and public health, but they must be handled with appropriate caution to minimize exposure and avoid risks to human health and the environment. A pesticide is any substance or mixture of substances used to prevent, destroy, repel, or mitigate a pest. There are many factors to consider when choosing a pesticide; those that directly affect the selection of one pesticide over another, such as efficacy, formulation, and site considerations, have been discussed in Chapter 5. This chapter reviews the impacts of pesticides on the environment and living organisms, including people.

To use pesticides safely, you must know where in the environment pesticides may be found after an application; how they break down in the environment; and how they affect people, nontarget organisms, and beneficial organisms. This understanding aids in the selection of pesticides, improves application timing, and reduces the potential for environmental contamination and injury to nontarget organisms.

PESTICIDES IN THE ENVIRONMENT

The purpose of a pesticide application is to control a pest population. Ideally, pesticide applications should impact only the target organism and have little or no impact on other organisms in the environment. However, many pesticide applications have the potential to affect nontarget organisms and move beyond the application site.

Pesticide residues get into the environment as a result of application or by accident and may be found in air, water, and soil (Figure 8-1). The potential for a pesticide to contaminate the environment depends, in large part, on the nature of the pesticide, its ability to break down in a given substrate, type of formulation, application rate, frequency of application, and environmental conditions.

Air

Pesticides become airborne in many ways, including volatilization, drift, or through movement as dustborne particles.

Volatilization. The process by which a pesticide changes from a liquid form to a vapor is called volatilization. Volatilized pesticides leave the application site and move into the atmosphere, often travelling long distances. Compounds that vaporize or evaporate at low temperatures are said to be highly volatile. Volatility is a useful property that helps disperse a pesticide in the target area and increase exposure of pests to the pesticide, but it also can lead to exposure for nontarget organisms. The volatility of a chemical is measured by its *vapor pressure*, the pressure exerted by a material in its gaseous form. All pesticides are volatile to some degree, but they differ greatly in this property.

Although a pesticide's chemical volatility is a determining factor, the environment into which it is applied also affects the actual volatilization of a pesticide in the field. The tendency for the pesticide to be adsorbed, or to adhere to, the treated surface, whether leaf, stem, or soil, is important. Environmental conditions such as air movement, relative humidity, and temperature also influence volatilization. The ability of a pesticide to dissolve in water (solubility)

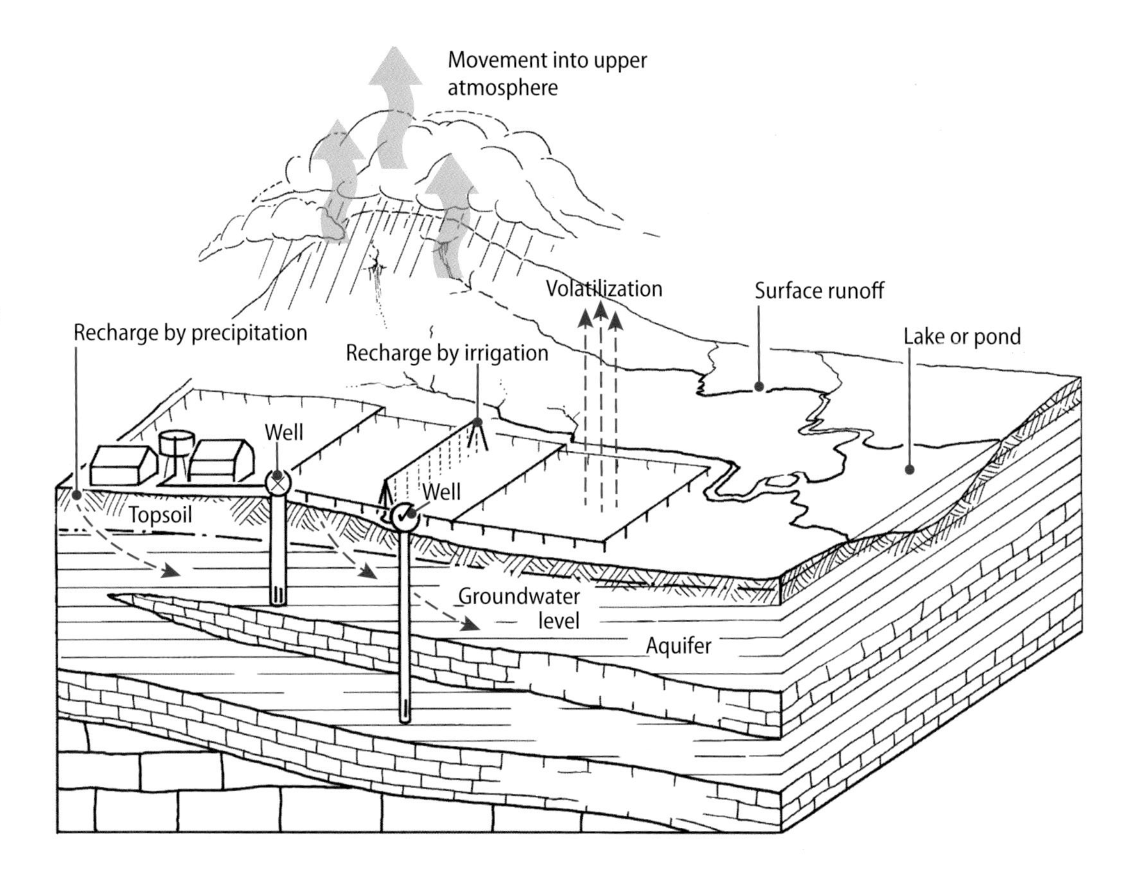

FIGURE 8-1

Pesticides can move through the environment in air, water, and soil to affect nontarget organisms many miles away.

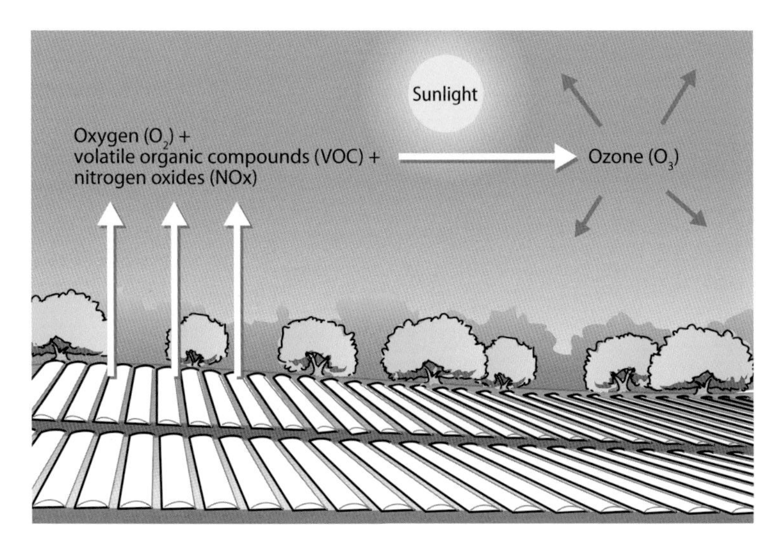

FIGURE 8-2

Ozone is formed when volatile organic compounds (VOCs), such as fumigants, move into the lower atmosphere in the presence of sunlight and react with nitrogen oxides.

or be rapidly taken up by plants also limits volatilization.

A number of pesticides, such as emulsifiable concentrate formulations and all of the fumigants, are classified as volatile organic compounds (VOCs) because they readily volatilize into the atmosphere. In the presence of sunlight, VOCs react with nitrogen oxides to produce ozone (Figure 8-2). Accumulations of ozone in the lower atmosphere contribute to smog, which can irritate the human respiratory system and injure plants. Regulatory restrictions have been placed on use of VOC pesticides to reduce damage to the ozone layer.

Practices that restrict movement of an applied pesticide, such as soil incorporation, tarping (Figure 8-3), or applying pesticides

FIGURE 8-3

Covering the soil with a polyethylene tarp immediately after injecting a soil fumigant helps reduce volatilization. Soil fumigation in California must be supervised by a qualified applicator who is certified in the area of field fumigation.

TABLE 8-1.

Factors influencing pesticide drift.

Pesticide

- ☐ Volatility of active ingredient
- ☐ Cormulation
- ☐ Carrier used to dilute pesticide in spray tank

Adjuvants

- ☐ Deposition aids, thickeners, and stickers (reduce drift by making droplets larger or less volatile)

Application equipment

- ☐ Sprayer type (e.g., conventional air blast vs. multi-fan, tower sprayer, or smart sprayers)
- ☐ Operating pressure of spraying system
- ☐ Nozzle type, orifice size
- ☐ Distance from nozzles to target surface
- ☐ Height from which spray is released
- ☐ Speed of travel of application equipment
- ☐ Application pattern and technique

Target surfaces

- ☐ Size of target area
- ☐ Location of target area
- ☐ Nature of target surfaces

Weather conditions

- ☐ Wind speed
- ☐ Wind direction
- ☐ Air temperature
- ☐ Humidity

at night, when there is less heat and wind and often higher humidity, reduce volatilization. Volatilization increases with practices such as disking that enhance soil movement after treatment. Information about the vapor pressure or volatility of a pesticide is available on the manufacturer's Material Safety Data Sheet *(MSDS)* under "Physical Properties" (see the section "Material Safety Data Sheets" later in this chapter).

Drift. Drift refers to the airborne movement of pesticides away from the treatment site during application. An important and costly problem, drift can damage susceptible plants away from the application site. Drift also reduces the effectiveness of a pesticide application and causes environmental contamination such as water pollution. It can also be responsible for unintended but illegal pesticide residues on nearby crops.

Many factors influence drift (Table 8-1). Drift is most serious when applications are made in windy conditions. The amount of pesticide lost from the target area and the distance it moves increase as wind velocity increases. Low relative humidity and high temperatures increase the potential for drift by causing spray droplets to evaporate faster.

Air temperature also contributes to pesticide drift by creating *inversion layers.* A temperature inversion is a layer of cool air near the soil surface trapped under a layer of warmer air. Inversion layers form when the air 20 to 100 feet (6 to 30 m) or more above the ground is warmer than the air below it (Figure 8-4). During a pesticide application, inversion layers cause fine spray droplets and pesticide vapors to become trapped, forming a concentrated cloud that can move from the treatment site.

Droplet size is another important factor in the movement of spray particles away from the application site. Small droplets fall through the air slowly and have a greater potential to drift; large droplets fall faster and are more likely to fall to the ground. The application method can also contribute to pesticide drift. Applications that release the pesticide as close to the target site as possible reduce drift. Spray pressure also affects drift by influencing the size of spray droplets;

FIGURE 8-4

A temperature inversion, as illustrated below, is a layer of warm air above cooler air close to the ground. This warm air prevents air near the ground from rising.

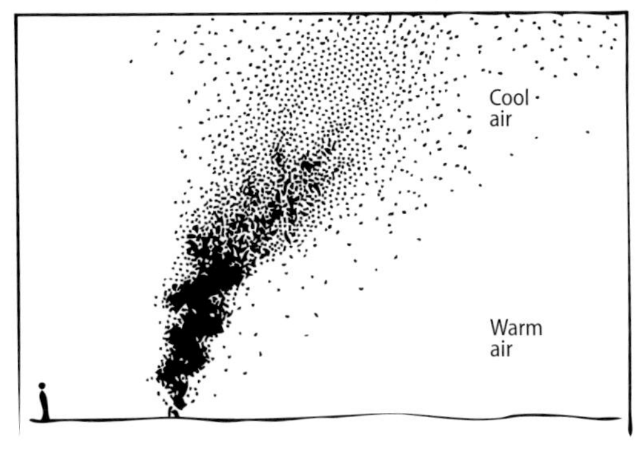

Normal condition—
Smoke rises and disperses.

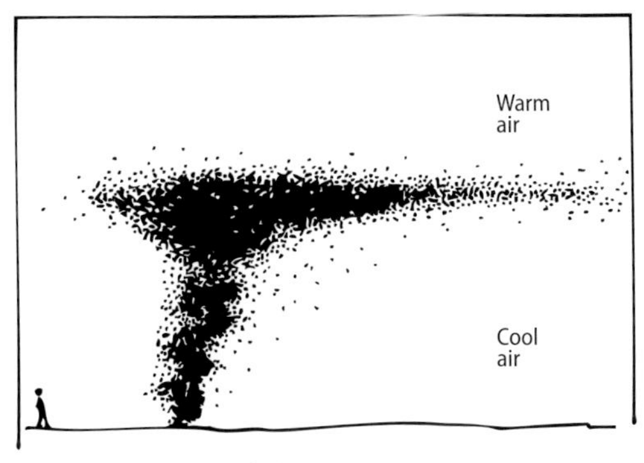

Inversion condition—
Smoke concentrates

FIGURE 8-5

Leaving an untreated buffer zone helps reduce the potential for drift from a pesticide application when the site adjoins locations where nontarget organisms, people, or structures may be exposed. Source: O'Connor-Marer 2000.

higher pressure decreases droplet size and increases drift.

Certain precautionary practices can reduce the potential for drift from pesticide applications. For example, pesticide applications should not take place when wind speed is high or when it is blowing toward sensitive or nontarget crops, dwellings, livestock, or water sources. Having some air movement (up to 5 mph) can be helpful for getting good coverage of treated surfaces, but if wind speeds rise above 7 to 10 miles per hour, drift may occur, depending on the product and situation. Using buffer zones (Figure 8-5) increases the distance that spray droplets have to move to reach adjacent crops, nontarget organisms, people, or structures. Spraying parallel to sensitive areas also reduces drift potential. When possible, use adjuvants to increase spray droplet size or reduce spray droplet evaporation.

Dustborne Particles. Pesticides move around in the air on dust particles in several ways. Fine particles of pesticide dust formulations may be moved in the air during the application process. Liquid pesticide droplets may also adhere to soil particles after application and later be carried into the atmosphere by wind.

The movement of dustborne pesticides can be minimized by avoiding the use of dust formulations whenever possible and by incorporating pesticides into the soil, either mechanically or by irrigation. Irrigating as soon as possible after the application increases the ability of some materials to adhere; check the label for recommendations on irrigation practices and incorporation. Avoid applying pesticides under windy conditions.

Water

Another potential fate of pesticide residues in the environment is movement into water. The potential for movement is greater for pesticides that have a long persistence rate. Other factors include tendency to *adsorb* to the soil and high water solubility.

Persistence is measured by the half-life of the pesticide. This is the period of time that it takes for half of the pesticide to break down in the environment. The longer the half-life, the greater the possibility for movement of the pesticide before it degrades. Adsorption refers to the tendency of pesticides to become attached to soil particles. Lower adsorption indicates a greater potential for pesticides to leach or move with water. However, some pesticides that adsorb to soil particles, such as pyrethroid insecticides, can be washed into surface water when soil and sediment erode.

The water solubility of a pesticide affects the ease with which it leaches into soil or moves with surface runoff water; it is measured in parts per million (ppm). Pesticides with low solubilities (less than 1 ppm) are likely to remain near the soil surface; soluble pesticides (greater than 100 ppm) are more likely to move dissolved in water.

Pesticides may contaminate surface water or groundwater. Surface water and groundwater contamination can be closely connected and water soluble pesticides can be a problem for both. In most situations, surface water recharges groundwater, but, in certain regions, groundwater flows toward surface water.

How Pesticides Get into Surface Water. Surface water contamination occurs through direct application (usually by accident) or through drift or runoff. Accidental spills can also contaminate water. The extent and duration of pesticide contamination from direct application to a body of water depend on environmental conditions and the nature of the pesticide. It is illegal to spray a body of water (unless for control of aquatic organisms) or to drain spray equipment into or near water sources. The same procedures used to reduce drift discussed above help prevent drift onto water.

Runoff is the movement of water and dissolved or suspended matter from the area of application into surface water or onto neighboring land. Less-soluble pesticides that adsorb on soil particles may also end up in streams, rivers, and other water bodies through runoff in soil sediment. Runoff is also influenced by the extent of pesticide use, soil properties, climate, topography, land use, and management practices.

Runoff is one of the most common ways that surface water can become contaminated. It is likely to occur when heavy rainfall or irrigation takes place after an application. Runoff of organophosphate or pyrethroid insecticides applied as dormant sprays in orchards during heavy winter rains is a major concern in California orchards. Crop irrigation can also increase runoff potential. For example, during flood irrigation, if excess water reaches the end of the field, water can flow into the drainage system,

FIGURE 8-6

In chemigation, small injection pumps are used to meter pesticides into irrigation water.

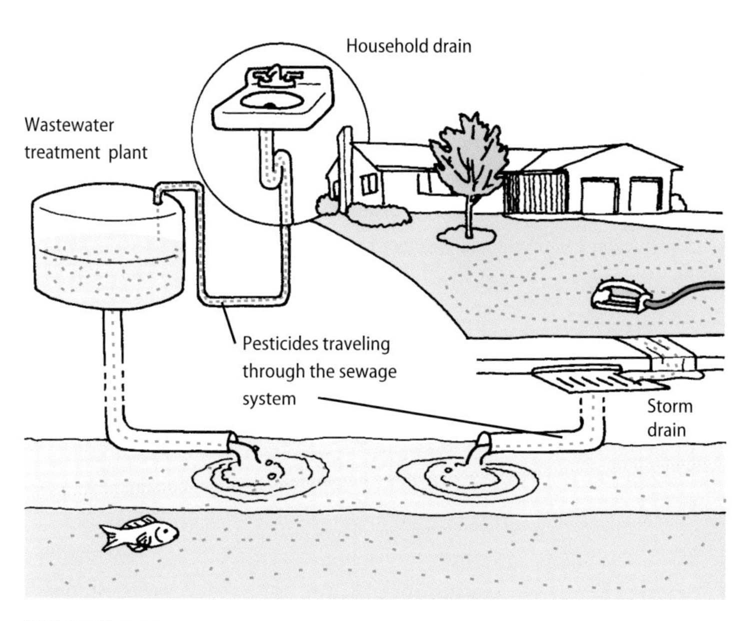

FIGURE 8-7

Irrigation or rain can wash pesticides applied on landscapes into storm drains that lead to creeks and rivers. Pesticides poured down indoor or outdoor drains also ultimately end up in surface water because water treatment plants cannot remove them.

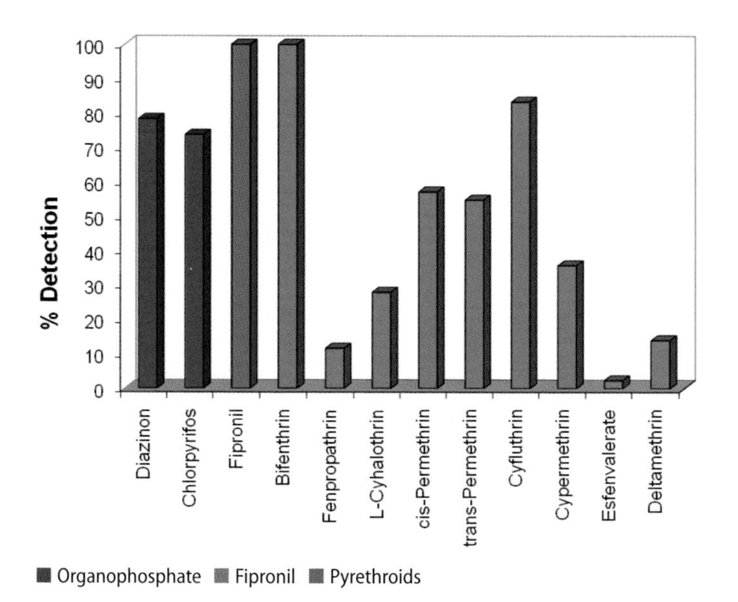

FIGURE 8-8

Frequency of detection of various insecticides in water running off residential landscapes in a neighborhood in Orange County, CA, in 2007-2008. Pesticides such as fipronil and bifenthrin were detected almost 100% of the time, often at levels toxic to aquatic invertebrates. Source: *Gan, Haver, and Oki 2010.*

where a pesticide may be transported in the tailwater. In rice production, pesticides are applied directly to flooded fields; runoff potential is high if the water is not held on the field for the amount of time specified on the label. Runoff can also occur when pesticides are applied through irrigation systems, a practice known as chemigation (Figure 8-6). In urban and suburban areas, pesticides applied to lawns, landscapes, and around the perimeter of houses may be washed off with rain or sprinklers into storm drains that lead to creeks, rivers, bays, and other water bodies (Figure 8-7). In many urban areas, pesticides regularly contaminate storm drains, often at levels that are toxic to aquatic invertebrates (Figure 8-8).

There are several ways to prevent the runoff of pesticides into surface water. Diligently monitoring and reducing water flow to prevent overflows into the drainage system can minimize tailwater runoff. Recirculating water in which a pesticide has directly been applied or increasing the water-holding time before draining the field can assist in pesticide breakdown and decrease the amount released into the water source. Also, the use of buffer strips, if feasible, between the crop and the water source can help trap pesticides and sediment. Choosing pesticides that are less likely to move or persist in water or nonchemical control methods may be the best solution in some circumstances.

To minimize contamination of the water source when using chemigation, water pumps must be equipped with a shutoff device to stop pesticides from being injected into the system when irrigation stops. Safety valves are also installed in the irrigation system to prevent possible contamination of the water source by backflow of irrigation water or pesticide.

Groundwater Contamination. Groundwater contamination occurs in various ways (see Figure 8-9). Pesticides contaminate groundwater through direct entry and by leaching through soil. Direct entry includes any opening in the soil that allows water (or contaminants in the water) to bypass the soil's natural filtration process. Such openings would include any cracks or holes in the ground from natural causes (plant roots, burrowing animals, etc.), geologic structures, functioning and abandoned wells, or cracked well casings. Spilling pesticides while mixing them near a well, pumping water into pesticide application equipment without using air gaps or backflow protection devices, and injecting pesticides into an irrigation system without using backflow protection devices are other ways in which pesticides can enter groundwater.

Leaching is the movement of a pesticide in water downward through soil. It occurs as rainwater or irrigation water percolates through the soil, carrying water-soluble pesticides with it. Applying pesticides in irrigation water (chemigation) can also contribute to leaching if uneven applications occur or too much material is applied.

A pesticide or other chemical can enter groundwater over time as a result of normal applications on a field, orchard, or other wide area. This is referred to as nonpoint source pollution. With nonpoint source pollution, only a small amount of the pesticide can enter the groundwater from any one location. A pesticide or other chemical can also enter groundwater as a result of mishandling during mixing, storage, and transportation, or from improperly constructed disposal sites or holding facilities. In these situations, larger quantities of contaminants can enter the groundwater at small, defined locations such as wells, drains, or spill locations. This is referred to as point source pollution (Figure 8-10)

Many factors, including pesticide properties, soil characteristics, environmental

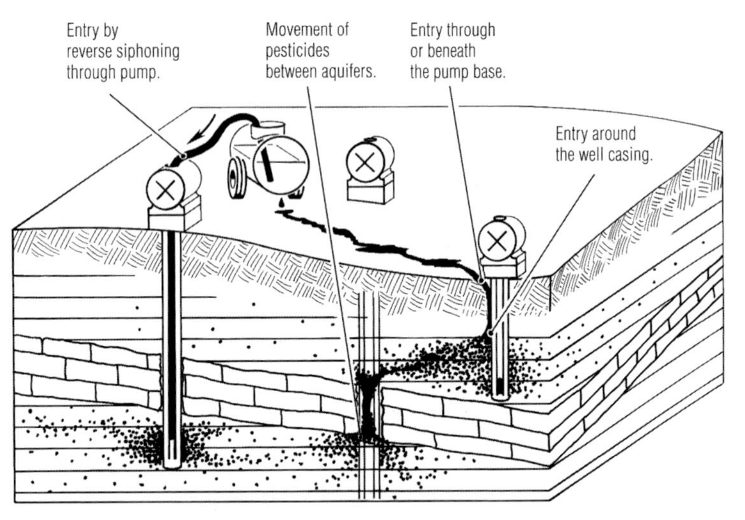

FIGURE 8-9

Pesticides and other contaminants can enter groundwater directly through water wells or direct channels into an aquifer, as shown by this illustration. Source: O'Connor-Marer 2000.

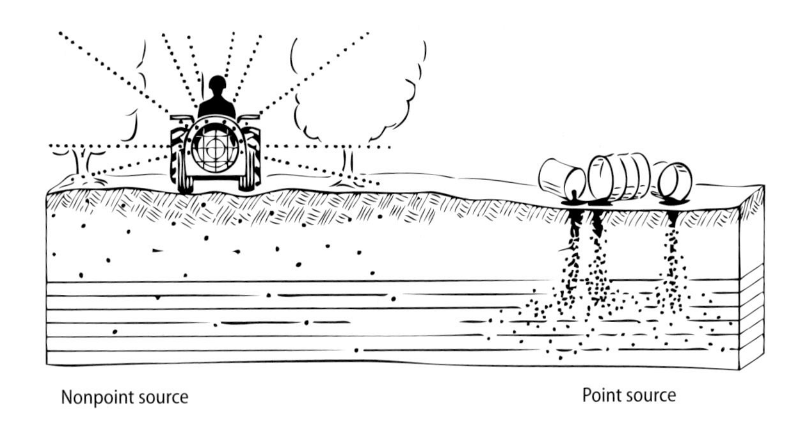

FIGURE 8-10

Point pollution sources are areas where large quantities of pesticide or other pollutants are discharged into the environment, such as from spills, waste discharge pipes, or dump sites. Nonpoint pollution sources arise from normal application where the pesticide or other material is applied over a large area. Source: O'Connor-Marer 2000.

TABLE 8-2.

Factors affecting pesticide leaching.

Agricultural practices

- ☐ Amount and type of pesticide used
- ☐ Method of pesticide application
- ☐ Irrigation practices at treatment site:
  - frequency of irrigation
  - timing of irrigation in relation to pesticide application

Geologic conditions of treatment area

- ☐ Slope
- ☐ Underlying formations
- ☐ Proximity of surface water channels such as ponds, lakes, rivers

Interaction of pesticides in soils

- ☐ Properties of pesticides:
  - water solubility
  - volatility
  - soil adsorption
  - persistence
- ☐ Soil influence on pesticides:
  - soil texture
  - soil organic matter content
  - soil water content

TABLE 8-3.

Pesticides regulated in California in 2009 to protect groundwater and summary of wells with verified detections 2008–2009. Of 3,691 wells sampled, 380 had detections.

Pesticide detected	Number of wells with detections	Range of concentrations (ppb)	Maximum contaminant level (ppb)
atrazine	3	0.055–0.46	1
bromacil	21	0.052–4.46	NE[1]
deethyl-atrazine[2]	14	0.053–1.33	NE
deethyl-simazine or desisopropyl atrazine[3]	60	0.055–0.537	NE
desmethyl-norflurazon[4]	32	0.05–0.803	NE
diamino-chlorotriazine[5]	58	0.05–4.98	NE
diuron	31	0.05–0.498	NE
norflurazon	14	0.055–0.537	NE
prometon	1	0.062–0.062	NE
simazine	42	0.053–0.32	4

1. NE = None established.
2. Degradate of atrazine.
3. Degradate of simazine or atrazine.
4. Degradate of norflurazon,
5. Degrate of atrazine or simazine.

Source: From DPR, July 2010.

TABLE 8-4.

Management practices that reduce the potential for groundwater contamination by pesticides.

- ☐ Whenever possible, select pesticides that have not been associated with groundwater contamination.
- ☐ Avoid excessive irrigation after an application and prevent irrigation runoff.
- ☐ In sandy soils and locations where groundwater contamination is known to occur, do not apply highly leachable materials.
- ☐ Triple-rinse containers, pour rinsate into the spray tank, and use the rinsate to treat a labeled site.
- ☐ Avoid spilling pesticides; if a spill occurs, clean it up immediately and remove contaminated soil. Dispose of contaminated soil as a hazardous waste.
- ☐ Never dump excess pesticides onto the soil or into water sources.
- ☐ Use proper methods of handling, storage, mixing, and loading.
- ☐ Store pesticides in enclosed areas on impermeable surfaces.

conditions, and depth to the water table, influence the amount of groundwater contamination that occurs from leaching (Table 8-2). The most important factor is the nature of the pesticide. Pesticides that are more mobile in the soil and are resistant to degradation are more likely to be found in groundwater. A large percentage of groundwater contamination cases in California over the last 10 years have been associated with a few chemicals (Table 8-3). Soil characteristics, such as the amount of organic matter, temperature, pH, moisture content, dissolved salts, and the quantity and type of soil organisms present, affect the stability of pesticides and their capacity to leach into groundwater.

Environmental and geological features also influence the potential for a pesticide application to contaminate groundwater. Shallow water tables beneath treated areas are more susceptible to contamination because pesticides pass through less soil and therefore do not degrade much. Layers of impermeable subsoil prevent leaching and the slope of an area can determine whether runoff or leaching will occur.

To reduce the potential for groundwater contamination from pesticide applications, promote management practices such as those listed in Table 8-4 to minimize pesticide movement into water.

Soil

Pesticides may be applied directly to the soil surface, incorporated into the top few inches of soil, or applied through chemigation. Pesticides can also enter soil unintentionally, through spills, as drift and atmospheric fallout, and in runoff. Once pesticides are present, the soil acts as a reservoir from which persistent pesticides can move into the bodies of invertebrates, be taken up by plants, pass into air or water, or break down.

The physical and chemical properties of a pesticide are major determinants of its persistence. After contact with the soil, pesticides are influenced by many factors, including adsorption rate, soil texture, the amount of organic matter in the soil, microorganisms, and the presence of water.

The type of formulation and the quantity of pesticide in the soil also influence pesticide persistence. Environmental conditions such as soil and air movement, light conditions, and cultural practices including cultivation and the presence of cover crops also determine how long a pesticide remains in the soil.

The soil type influences pesticide persistence and leaching. The tendency for pesticides to be adsorbed varies with the amount of clay and organic matter in the soil: the higher the percentage of clay and organic matter in soil, the greater the number of adsorption sites. Pesticides tend to stay longer in soils with high clay content and organic matter.

The amount of water in the soil also affects the persistence of pesticides. Water molecules are polar, and they compete with pesticides for adsorption sites on soil particles. Therefore, when more water is added, the water can release the pesticide from the soil particle and force it into solution.

HOW PESTICIDES BREAK DOWN IN THE ENVIRONMENT

Pesticides released into the environment have many different fates. A pesticide can move through the air and eventually end up in soil or water. Pesticides applied to soil can end up in nearby bodies of water or leach through the soil and into groundwater. Pesticide applications in soil or water can also find their way back to the air through volatilization or transport on wind-blown dust particles.

Partitioning in the Environment

The possibility of finding a pesticide or its breakdown products in the environment is determined, in large part, by the physical and chemical properties of the pesticide. Persistent pesticides are more likely to be found in soil, water, on plants, or on crops well after application. Volatile pesticides are most likely to be found in the air. Highly or moderately soluble pesticides are often easily leached from soils. Water-soluble pesticides are more likely to stay mixed in surface waters. Pesticides that strongly adsorb to soil particles are more likely to persist in the soil and are less likely to volatilize or otherwise be biologically active.

Within soils, pesticides distribute themselves between air, soil, water, and soil particles. Pesticides may break down in the soil substrate through various processes or leave the soil, either through volatilization into the air or through leaching or runoff into water. Figure 8-11 illustrates the behavior and fate of pesticides in the soil environment.

Transformation

After their release into the environment, pesticides undergo a series of reactions that transform the original compound into various degradation products. The transformation of a pesticide involves a change in its chemical structure, producing one or more new chemicals until the original pesticide disappears. The breakdown products of pesticides may be more toxic, less toxic, or equally as toxic as the parent compound.

When recommending pesticides, it is important to know the stability of the pesticide and what its breakdown products are. For example, in studies of the Sacramento River that monitored concentrations of pesticides and their transformation products after irrigation water was released from rice fields, one of the pesticides applied, methyl parathion, was not detected in the river. Methyl parathion appeared to have degraded while the water was held in the rice fields. However, further investigation showed that one of its breakdown products, paranitrophenol, was detected in all samples of river water; it appeared to be stable over a longer period of time than its parent compound. Although less toxic than methyl parathion, the breakdown product still had significant toxicity, which would not be detected in a standard test for methyl parathion. Likewise, the insecticide fipronil, which is widely used for managing ants, breaks down into three metabolites—fipronil sulfone, fipronil sulfide, and fipronil desulfinyl—that are as toxic or perhaps more toxic to aquatic invertebrates than fipronil itself, but they are also more stable in the aquatic environment.

Mechanisms of Pesticide Breakdown. Various mechanisms are at work in the transformation of pesticides. Chemically induced transformations of pesticides occur through hydrolysis, photodegradation, *microbial degradation*, and oxidation-reduction. These transformation processes help influence the persistence, toxicity, and other physiochemical properties of pesticides. Breakdown of most pesticides often involves more than one of these processes.

Hydrolysis is the chemical reaction of a pesticide with water, which results in the formation of smaller, more water-soluble segments and C-OH or C-H bonds. The transformation of many organophosphate and carbamate pesticides occurs through hydrolysis.

Photodegradation, or photolysis, is a chemical reaction induced by sunlight. Pesticide transformation occurs directly, as a result of absorption of sunlight energy by a pesticide, or indirectly, by reacting with another chemical in an excited state (or at a higher energy level). Transformation of pesticides through direct photodegradation occurs only for a few pesticides. This type of reaction depends on the chemical structure of the pesticide and certain environmental conditions.

Oxidation-reduction involves a transfer of electrons and is a chemically or biologically mediated reaction. This process requires the simultaneous reaction of two chemicals: one undergoes oxidation, losing one or more electrons, and the other undergoes reduction, gaining one or more electrons taken from the first chemical.

Chemical transformations result in structural changes in organic pesticides. These changes can also occur biologically. Microbial degradation is the breakdown of pesticides by living organisms using enzymes. Soils include a diverse population of microorganisms, including fungi, actinomycetes, and bacteria, that degrade organic chemicals such as pesticides, using them as a source of energy and carbon. All pesticides with carbon as a component of their structure can be degraded through microbial action.

Microbial degradation of pesticides is most likely to take place when soil conditions support high microbial activity. Warm, moist, well-aerated soils with high levels of organic matter enhance pesticide breakdown. Moisture and temperature also determine the species and viability of microorganisms present.

Soil depth and the aerobic (presence of oxygen) or anaerobic (absence of oxygen) condition of the soil also affect the action of microorganisms. Soil microbial populations increase near the soil surface, where light and oxygen are more available. These factors influence the rate of pesticide breakdown. Aerobic soil conditions are more likely to accelerate pesticide breakdown by microorgan-

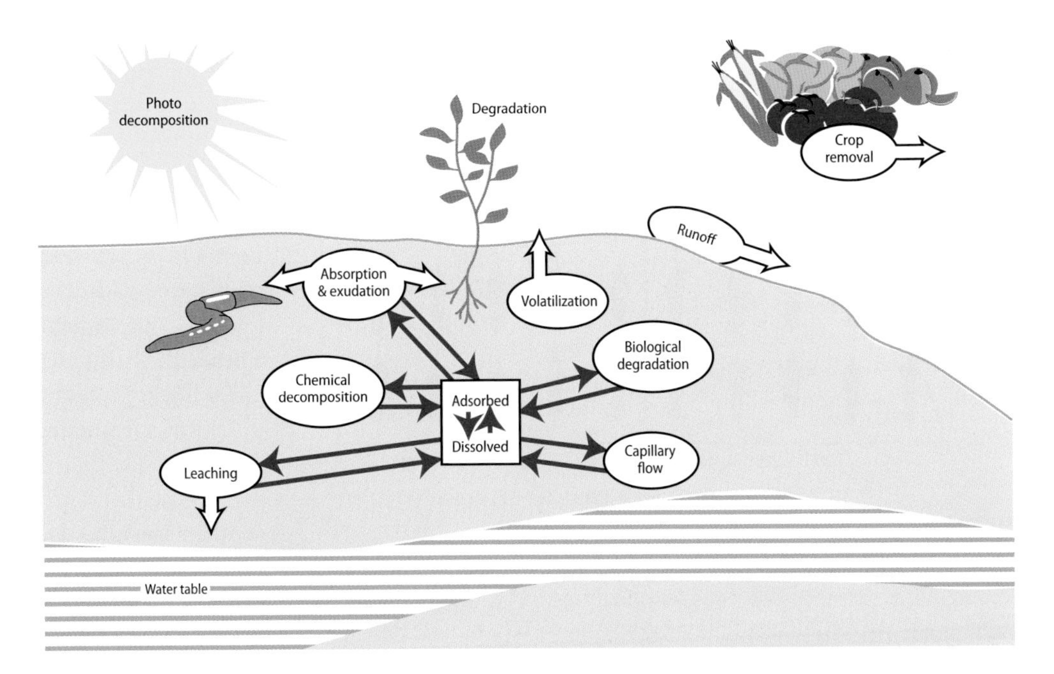

FIGURE 8-11

Processes that influence the behavior and fate of pesticides in the soil environment.

isms, especially when a high level of organic matter is present. Pesticide breakdown can also occur in anaerobic conditions, and some pesticides, such as the organochlorines, may require both aerobic and anaerobic conditions for complete degradation.

Effect of the Environment on Degradation

The rate at which a pesticide degrades is largely determined by a host of interrelated environmental factors. Soil type and texture, organic content, pH, soil moisture, and soil temperature influence microbial degradation. Soil acidity, as indicated by pH, is important in determining the rate and the magnitude of pesticide breakdown in hydrolysis and oxidation-reduction reactions. Temperature is also an important factor in most chemical and microbial reactions. A rise in temperature, for example, increases the rate of hydrolysis and oxidation-reduction. In photodegradation, although temperature is important in determining the rate of degradation, it is sunlight that is responsible for the chemical reaction.

GENERAL TOXICOLOGY

Pesticide applications can pose potential hazards for people and nontarget organisms. When selecting a pesticide, it is important to consider how people could be exposed to it and the various methods available to reduce that exposure. Likewise, direct and indirect impacts on wildlife and beneficial organisms and the means to reduce these impacts must be adequately considered.

Toxicology is the science that deals with poisons and their effect on living organisms. All pesticides (indeed, all chemicals) are toxic at some level. Some pesticides are more toxic than others and therefore present higher risks to users, nontarget organisms, and the environment. Toxicology is based on the premise that a relationship exists between the dose (the amount of a pesticide applied) and the response (the toxic reaction) and that there is almost always a dose below which no response occurs or is measurable. Once this relationship is understood, the risk of using a particular pesticide can be evaluated based on its capacity to cause harm and the potential for exposure to people and other nontarget organisms.

To better understand toxicology, several terms and concepts must be understood. These include toxicity, the difference between risk and hazard, acute versus chronic exposures, and problems associated with residues and persistence.

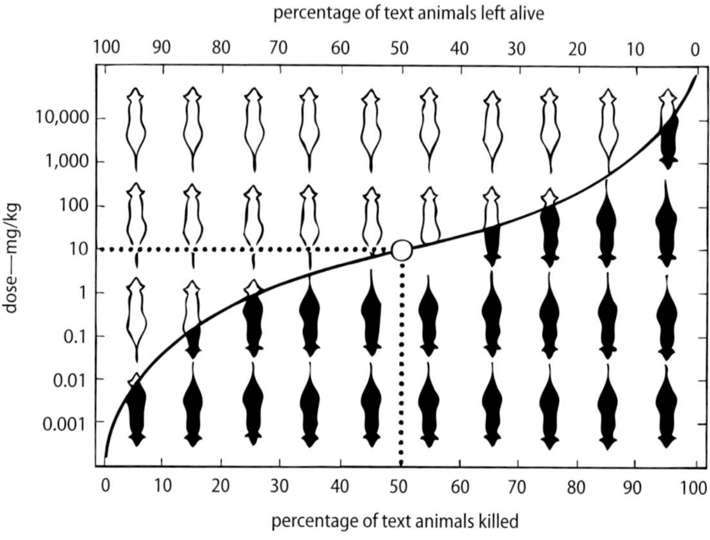

FIGURE 8-12

The amount of pesticide that will kill half a group of test animals is known as the LD_{50}. The smaller the LD_{50}, the more toxic the pesticide is considered to be. LD_{50} values are established for both oral and dermal exposure. In this illustration, the LD_{50} is 10 mg of pesticide per kg of body weight of test animal. Source: O'Connor-Marer 2000.

Toxicity

Toxicity is the general term used to describe the potential for a chemical to cause harm. The toxicity of a pesticide is an inherent quality of the chemical and does not change. Toxicity is commonly measured by the lethal dose (LD) that kills 50% of the animals tested (LD_{50}) (Figure 8-12). LD_{50} is expressed as milligrams (mg) of active ingredient (a.i.) of pesticide per kilogram (kg) of body weight of the test animal (mg/kg). A pesticide with a low LD_{50} (e.g., 5 mg/kg) is very toxic, whereas a pesticide with a high LD_{50} (e.g., 1,000 to 2,000 mg/kg) is less toxic to the tested animal.

Toxicity measurements, however, are only guidelines. They are used to compare the lethal potential of one dose of one pesticide with another. The LD_{50} does not provide information about long-term toxic effects. They measure one point on the dose response curve that reflects a pesticide's potential to cause death

TABLE 8-5.

Pesticide toxicity categories.

Exposure	Category I	Category II	Category III	Category IV
acute oral[1]	up to and including 50 mg/kg (< 1 teaspoon)	> 50 through 500 mg/kg (> 1teaspoon to 2 tablespoons)	> 500 through 5,000 mg/kg (> 2 tablespoons to 1 oz)	> 5,000 mg/kg (1 oz to almost nontoxic)
acute dermal[1]	up to and including 200 mg/kg	> 200 through 2,000 mg/kg	> 2,000 through 5,000 mg/kg	> 5,000 mg/kg
acute inhalation[1,2]	up to and including 0.05 mg/liter	> 0.05 through 0.5 mg/liter	> 0.5 thru 2 mg/liter	> 2 mg/liter
primary eye irritation	corrosive (irreversible destruction of ocular tissue) or corneal involvement or irritation persisting for more than 21 days	corneal involvement or other eye irritation clearing in 8 to 21 days	corneal involvement or other eye irritation clearing in 7 days or less	minimal effects clearing in less than 24 hours
primary skin irritation	corrosive (tissue destruction into the dermis and/or scarring)	severe irritation at 72 hours (severe erythema or edema)	moderate irritation at 72 hours (moderate erythema)	mild or slight irritation at 72 hours (no irritation or slight erythema)
signal word	Danger	Warning	Caution	Caution

[1] Probable lethal oral dose for a 170-pound human.
[2] 4-hour exposure.
Source: U.S. EPA Label Review Manual Chapter 7: Precautionary Statements, 2007.

in one species of laboratory animal. While a pesticide with a low LD_{50} may pose a higher risk than one with a large LD_{50}, it can often be used safely provided precautions are taken to protect nontarget organisms. To evaluate the potential risks associated with a particular pesticide, it is necessary to know about nonlethal effects as well.

Another measurement of toxicity is the lethal concentration of a pesticide in the air or water that will kill 50% of a test animal population (LC_{50}). It is expressed in micrograms (1 gram = 1,000,000 micrograms) per liter (l) of air or water (µg/l). LC_{50} is commonly used to measure the toxicity of fumigants to people and the toxicity of pesticides to fish and other aquatic life. Besides LD_{50} and LC_{50}, toxicity is also indicated by the potential of the pesticide to affect the eyes and skin. Table 8-5 shows the toxicity and potential for causing injury to people for the three toxicity categories (Danger, Warning, and Caution).

The half-life, or time it takes for a given quantity of pesticide to break down to half of its original quantity, is a measurement of the persistence of a pesticide. Half-life is affected by the chemical nature of the pesticide, its formulation, soil microbes, ultraviolet light, quality of water used in mixing, and impurities combined with the pesticide. Combining different pesticides or other chemicals can alter the half-life or change toxicity.

The threshold limit value (TLV) gives an indication of the airborne concentration of the chemical that produces no adverse effects over a period of time. It is measured in parts per million (ppm). TLVs are commonly established for workers exposed for 8 hours per day for 5 consecutive days to low-level concentrations of a toxic chemical. The TLV is set at a level to prevent minor adverse reactions such as skin or eye irritation.

The no observable effect level (NOEL) is the largest or highest dose or exposure level tested that produces no noticeable toxic effect on test animals. NOEL is used as a guide to establish maximum exposure levels for people and residue tolerance levels on pesticide-treated produce. Maximum exposure levels are commonly set from 100 to 1,000 times less than the NOEL to provide a wide margin of safety.

Relationship of Dose to Toxicity. The most important variable in determining the toxicity of a pesticide is the dose. The dose of a pesticide determines the degree of effect that it produces. The more pesticide an organism is exposed to, the greater the toxic response. There is a dose level at which no effect is observed (the NOEL), and there is a dose level that produces a response (Figure 8-13).

Risk versus Hazard

Hazard, the inherent ability of a pesticide to produce an adverse affect, is a constant for a particular pesticide product. Risk, in toxicology, is the probability that a chemical will cause harm. In the case of pesticides, risk is a function of hazard and dose. The degree of risk can depend on the concentration of the pesticide, the way it is handled, and the duration of exposure.

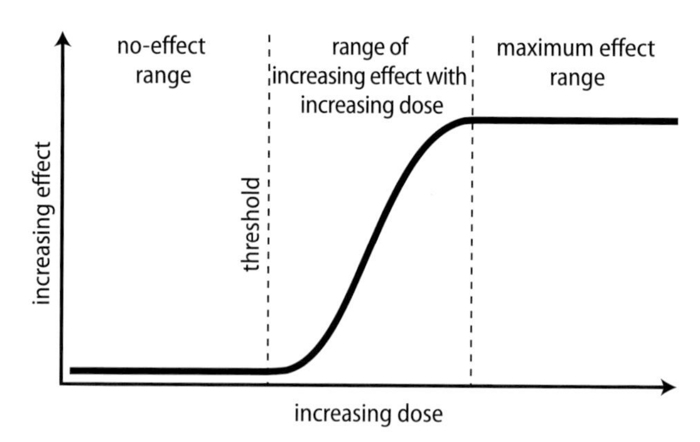

FIGURE 8-13

The relationship between dose and response. As the dose increases, the response increases; once a maximum response is reached, further increases in the dose will not result in an increased effect.

FIGURE 8-14

Closed mixing systems can reduce exposure to pesticides when mixing pesticides with the signal word Danger. *Closed mixing systems enable accurate measuring of pesticides and rinsing of empty containers.*

Many factors impact risks associated with the use of a pesticide product. They include

- **Formulation.** Different pesticide formulations can substantially influence the potential for hazard of a given pesticide. Formulation may affect its percentage of active ingredient, propensity to drift, phytotoxicity, or persistence in the environment. It may also affect a pesticide's potential for dermal uptake, inhalation, or ability to pass through the skin (oil versus water).
- **Environmental Conditions.** Environmental conditions can increase the potential for injuring living organisms. Heat, light, and precipitation can speed up or slow down the breakdown of toxic residues. Weather conditions, for example, that are favorable to drift or inversion layers can cause pesticides to move beyond the target site, increasing the risk of exposure or environmental contamination. The type of soil at the application site can increase risk of groundwater or surface water contamination by easing movement of the pesticide through the soil.
- **Proximity of Sensitive Organisms.** The closer a person or organism gets to an application or mixing activity, the greater the chance that they will receive an exposure (dose) that will cause harm. The risk of exposure to people and other sensitive organisms depends on their proximity to the application site during and after the application. For example, the risk of exposure to the applicator is highest during mixing, loading, and application of the material. The risk of exposure to the fieldworker from residues on the crop increases if fieldworkers are in the field during or following an application. Exposure

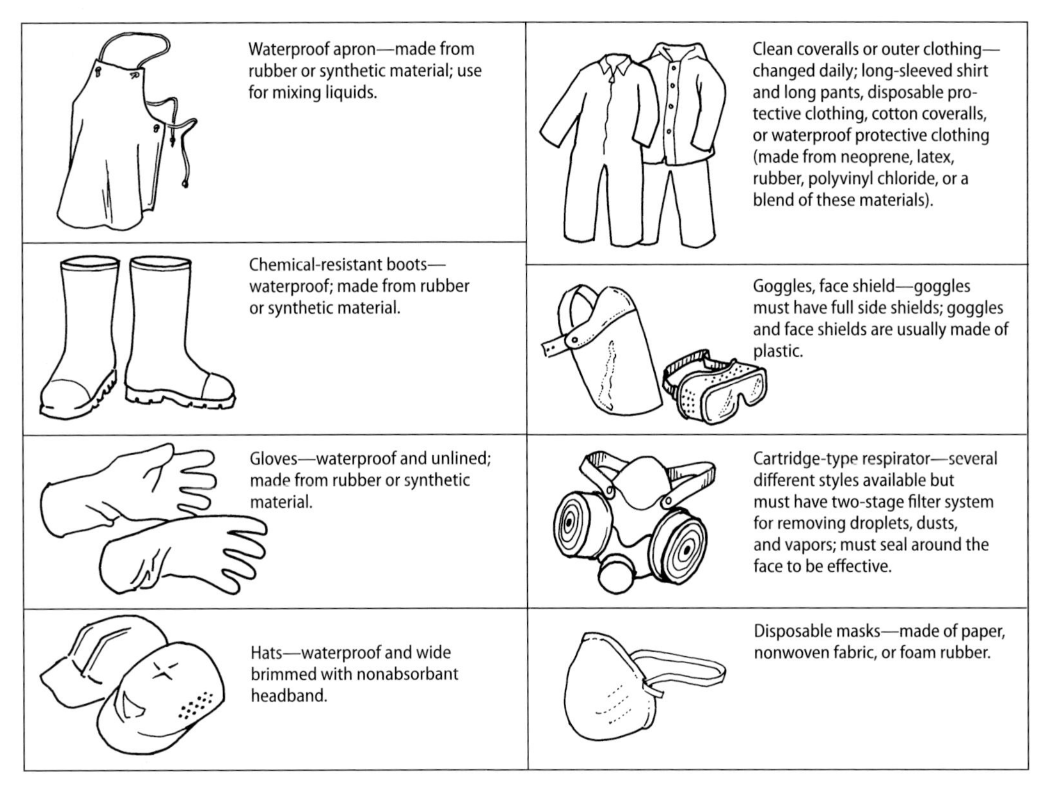

FIGURE 8-15

Personal protective equipment.

to people or nontarget organisms can be increased if they are in or near the application site and weather conditions prevail that move the pesticide off-site. Applications that take place close to sensitive areas such as buildings, waterways, or roadways may also increase risk of exposure to people and nontarget organisms.

Ways to Reduce Risk. Many steps can be taken to reduce the potential risk of pesticides to people. For example, closed mixing systems (Figure 8-14) allow for the safe handling of toxic liquid pesticides because applicators do not have to handle opened pesticide containers, thereby eliminating accidental contact. The use of water-soluble packaging also reduces the risks of exposure during mixing and loading because direct contact is eliminated.

Wearing protective clothing and equipment reduces the risk of exposure when handling pesticides or entering an application site before the expiration of the restricted-entry interval (REI). Personal protective equipment (PPE) includes all or some of the following: waterproof apron, chemical-resistant boots, gloves, head protection, clean coveralls or outer clothing, goggles, face shield, and dust- or mist-filtering respirator or an organic/vapor-removing respirator (Figure 8-15). Personal protective equipment prevents pesticides from contacting the body and eyes and can also protect against inhaling pesticide dusts, droplets, and vapors. Personal protective equipment requirements are listed on the pesticide label.

Entering or working in a treated area soon after an application increases the risk of exposure to harmful levels of pesticide residues. Observing the established restricted-entry interval is an important requirement to reduce this type of exposure (Figure 8-16). The pesticide label provides guidelines for restricted-entry intervals. These intervals must be noted on the pesticide recommendation.

The choice of material and formulation can also reduce risks. When possible, select

FIGURE 8-16

Establishment of restricted-entry intervals following the application of highly toxic or hazardous pesticides has helped to reduce farmworker injury. Treated fields, like the one shown, are posted to warn workers not to enter during the restricted-entry period.

the least-toxic material to prevent injury to people, nontarget organisms, and the environment. Certain formulations are also safer to handle than others. For example, water-dispersible granules are generally safer than wettable powders because they are less likely to become airborne. They are safer for applicators to mix and apply and less likely to drift from the application site. These formulations are also not as phytotoxic to sensitive plants.

Single versus Multiple Exposures. In some situations, a single exposure to a sufficient dose of certain pesticides will produce illness. In other cases, one exposure will not cause illness, but the effects of repeated, low-level exposures may accumulate in the body so that a person may not develop an illness or other symptoms. A toxic dose of pesticide may produce immediate symptoms of illness or injury, or the symptoms may not appear for several days or weeks after the exposure (delayed onset). Sometimes an exposure may produce a combination of immediate and delayed symptoms.

Acute versus Chronic Injury. A pesticide-related illness or injury may have a sudden onset and last for a short duration before the person recovers. This is acute illness. Usually, the exposed person recovers as a result of medical treatment, but with a severe exposure, the person may die. Exposure to some pesticides produces chronic health effects. These are long-lasting and often permanent disorders such as cancer, permanent nerve damage, sterility, and blindness.

One good source of information to assess the potential health hazard from a pesticide is the Material Safety Data Sheet (MSDS). Some pesticide labels also have important information on health hazards. All labels have one of the three signal words that indicate the level of acute toxicity (based on animal studies) or other hazard posed by the pesticide product. The signal word indicates the degree of risk to the user and is usually measured in terms of oral or dermal toxicity.

SIGNAL WORD	TOXICITY
DANGER	HIGHLY TOXIC
WARNING	MODERATELY TOXIC
CAUTION	SLIGHTLY TOXIC OR RELATIVELY NONTOXIC

Along with the signal words, labels include

precautionary statements about the most common route of entry (usually dermal or oral) and specific actions that must be taken to avoid exposure.

Residues and Persistence

Even after a pesticide is applied, the risk of exposure may remain because residues may stay on the treated surface for a period of time. Residues are important and necessary in some types of pest control because they provide continuous exposure to the pest, improving the chances of control. They are undesirable, however, when they expose people and other nontarget organisms to unsafe levels of pesticides. Pesticides can remain as residues on or in the plant material and in soil, water, or on surfaces in nontarget areas.

To protect the public from harmful residues on harvested produce, federal law establishes pesticide residue tolerances for all materials registered for use on agricultural products. These tolerances are based on laboratory and animal testing with a margin of safety to allow for differences in people and to protect children. These tolerances, also known as maximum residue limits (MLRs), are established for each commodity the pesticide is used on. Preharvest intervals prohibit application of some pesticides during a specified time before harvesting a commodity. These are established to allow residues to break down before reaching consumers.

HOW PEOPLE ARE EXPOSED AND METHODS TO REDUCE HUMAN EXPOSURE

Pesticides can enter the body through four major routes (Figure 8-17): dermal (absorption by the skin), ocular (through the eyes), oral (ingestion through the mouth), and inhalation (breathing into the lungs).

Dermal exposure is the most frequent type of pesticide exposure by those using pesticides. Certain pesticides injure the skin, while others may pass through the skin and affect internal organs. Absorption into the body starts as soon as the pesticide touches the skin and continues as long as there is contact. The ability of a pesticide to be absorbed through the skin depends on the chemical characteristics of the pesticide and its formulation. Pesticides that are more soluble in oil or petroleum solvents penetrate the skin more easily than those that are soluble in water. Wear personal protective

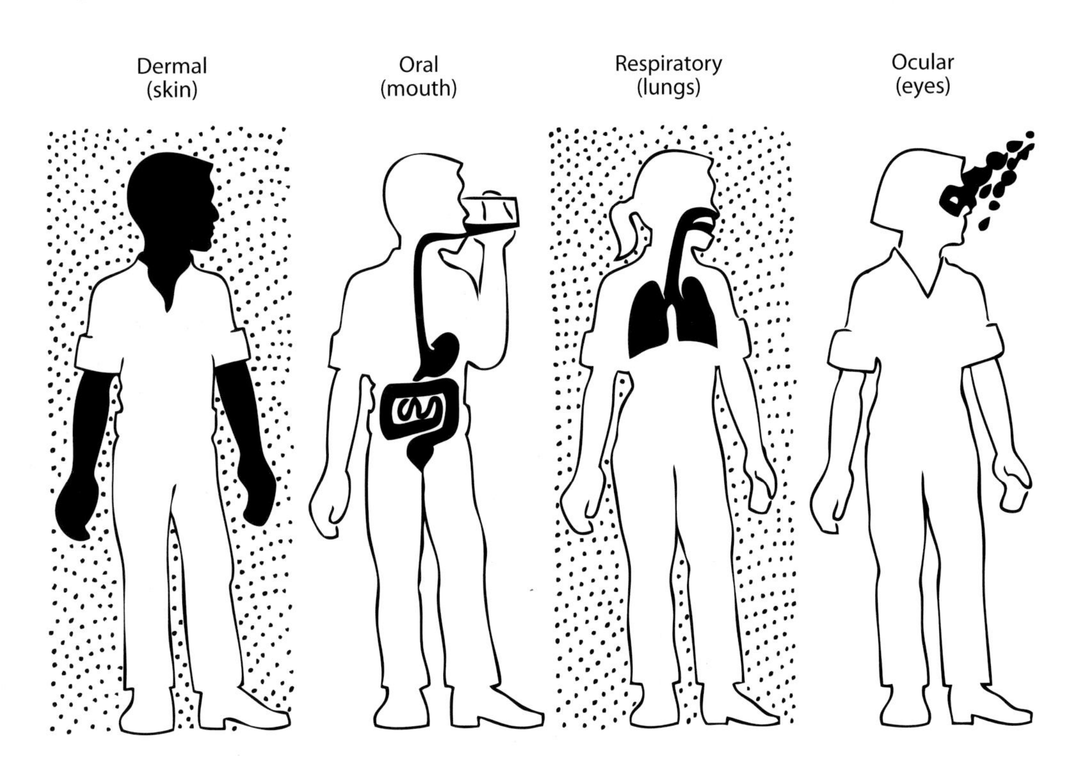

FIGURE 8-17

The most common routes for pesticide exposure are through the skin (dermal), through the mouth (oral), through the lungs (respiratory), and through the eyes (ocular).

equipment when working with pesticides to prevent dermal exposure; avoid contact with treated surfaces. Wash thoroughly after an application or if a pesticide accidentally contacts the skin.

Ocular exposure is the second most frequent type of pesticide injury. Some pesticides can cause serious damage if they splash into the eyes. Direct injury to the eye itself can be a serious consequence, but the eyes also provide another route of entry for pesticides to get into the bloodstream. Wearing protective eyewear can protect the eyes.

Oral exposure usually occurs accidentally, for example, by drinking out of a container that has had a pesticide stored in it. It also occurs through the splashing of spray materials or pesticide dust into the mouth during mixing or application, by eating or drinking contaminated foods or liquids, or by smoking while handling pesticides. Wearing protective equipment such as a respirator or face shield can minimize the risk of getting pesticides into the mouth. Wash hands thoroughly and change contaminated clothing before eating, drinking, or smoking. Keep foods and liquids away from areas where pesticides are being applied, mixed, or stored, and never put pesticides in food or beverage containers.

Inhalation exposure occurs if dusts, mists, or vapors are inhaled during mixing or application. Inhalation or respiratory exposure is particularly hazardous because pesticides are quickly absorbed by the lungs and transported in the bloodstream to other areas of the body. If inhaled in sufficient amounts, pesticides can cause serious damage to nose, throat, and lung tissue. Breathing vapors and very small particles pose the most serious risks. To reduce respiratory exposure, wear the label-required respirators during mixing, application, and other handling activities.

How People Can Be Exposed

People can become exposed to pesticides in several ways, but the most common is during mixing and application and by entering treated areas during an application or before the expiration of the restricted-entry interval. Applicators are most at risk from direct exposure to the pesticides they handle and are also the most susceptible to exposure from accidents occurring during mixing and application. However, anyone entering treated fields before the expiration of the restricted-entry interval without the proper protective equipment and early-entry worker training also risks exposure. Drift of spray materials onto adjacent sites also can expose workers and others who happen to be near the application area.

People are also at risk from exposure to excessive pesticide residues that may be on food and in water when pesticides have been improperly used. Soil residues of some pesticides, particularly persistent ones, can be taken up by plants and can accumulate in edible portions. Contamination of drinking water is another way in which people can be exposed to pesticides: improper use or disposal or leaching of pesticides through the soil can lead to the contamination of water supplies.

Methods to Reduce Human Exposure

Following the instructions on the pesticide label, understanding and following the regulations that govern pesticide use, and taking a few simple precautions can substantially reduce the risk of accidental exposure to pesticides. PCAs are not necessarily involved in the application process, but by recommending the least-toxic material that will effectively control the pest and by designating the safest application time, they can reduce risk of exposure to applicators, fieldworkers, and other individuals in and around the application site. In situations where there is a high risk of human exposure, these risks should be noted on the written recommendation.

The pesticide formulation that is recommended, such as a liquid concentrate instead of a dust, can limit exposure. The use of closed-system mixing equipment and water-soluble packaging, when applicable, reduces direct contact of highly toxic pesticides for handlers. Using procedures to minimize drift can also reduce the risk of exposure to pesticides. Following plantback restrictions and preharvest intervals protects consumers from exposure to excessive residues on agricultural commodities.

Whenever possible, know who is making or supervising the pesticide application. A certified applicator often has a good understanding of the hazards of pesticides and how to avoid injury on a particular property. Communicate with the applicator about the application process and any potential problems resulting from the application. Also, ask for the applicator's observations regarding site conditions and application effectiveness.

The Worker Protection Standard. Many state and federal regulations are in place to protect employees working in urban landscapes, farms, forests, nurseries, and greenhouses from occupational exposure to pesticides. The federal regulation that governs all aspects of agricultural pesticide use is the Federal Insecticide, Fungicide, and Rodenticide Act (FIFRA). The Worker Protection Standard (WPS) is an amendment to FIFRA designed to set standards for basic workplace protections. The WPS is intended to reduce the risk of illness and injury from occupational exposures for agricultural workers and pesticide handlers. It sets requirements for employers of workers in production agriculture, commercial greenhouses and nurseries, and forests to protect their workers. Its requirements include

- **Protection during applications.** Prohibits the application of pesticides in a way that would cause exposure; requires that workers be restricted from sites while pesticides are being applied.
- **Restricted-entry intervals (REIs).** Establishes time-limited restrictions for all agricultural pesticides depending on toxicity; during this time, workers are excluded from entering pesticide-treated sites, with only narrow exceptions.
- **Personal protective equipment (PPE).** Requires that all handlers wear the proper protective clothing and equipment during mixing and application, cleaning and repairing contaminated equipment, flagging, and handling open containers; also applies to individuals entering the treated site before the REI has expired.
- **Notification of workers.** All employees that may inadvertently enter the application site or are working within ¼ mile of the treatment area must be notified so that they can avoid treated areas.
- **Decontamination supplies.** An ample supply of soap, water, and towels must be available for routine washing and emergency decontamination.
- **Emergency assistance.** If a worker becomes poisoned or injured, transportation must be provided to a medical care facility; information about the pesticide must be provided to the medical personnel.
- **Pesticide safety training.** Prior to handling pesticides, all handlers must receive safety training for all pesticides that they handle, and workers must receive training on how to avoid exposure and what to do in case they are exposed; a poster on pesticide safety may be displayed.
- **Access to labeling and site-specific information.** Pesticide label requirements and Material Safety Data Sheets (MSDSs) must be available to all workers and handlers; central posting of recent pesticide applications is required.

In California, pesticide regulations are part of the California Code of Regulations (CCR); FIFRA provisions are incorporated into the CCR. The CCR, therefore, becomes the operative regulation for PCAs; follow the CCR to be in compliance with California and federal laws. Be aware that California requirements may be more restrictive than federal requirements. For instance, restricted-entry intervals for some pesticides in California are substantially longer than required by federal law. The DPR website, www.cdpr.ca.gov, has up-to-date information about legal requirements for using pesticides in California. In other states, check the website of the pesticide regulatory agency.

There may also be local regulations on pesticide use and restrictions that apply to pesticide use in sensitive areas. Be aware of conditions that could increase the risk of exposure to people in or near the area. Check

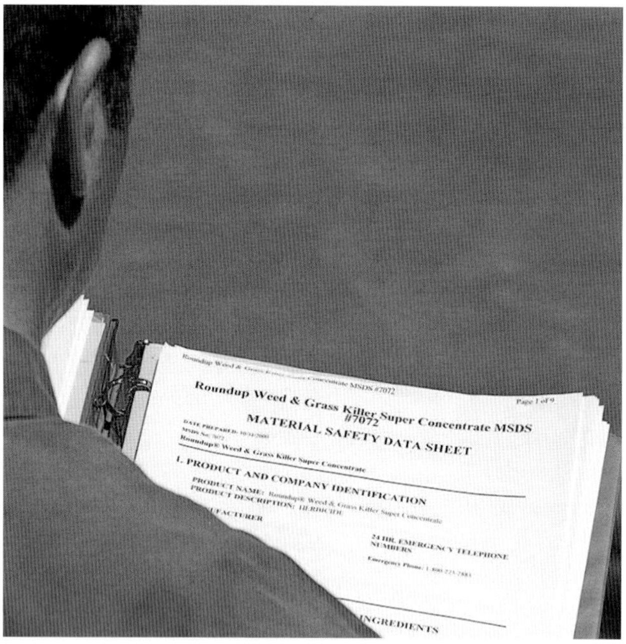

FIGURE 8-18

The Material Safety Data Sheet (MSDS) provides valuable information about pesticide hazards.

FIGURE 8-19

Honey bees may be poisoned if certain pesticides are applied while bees are foraging for nectar or pollen. Avoid the use of materials toxic to bees when crops or weeds are in bloom. Recommend that applications be made early in the morning, late in the afternoon, or at night to reduce chances of killing foraging bees. Also, recommend the use of pesticides that have a low toxicity to bees.

with the local agricultural commissioner's office for information regarding pesticide use specific to the application site.

Material Safety Data Sheets. Material Safety Data Sheets (MSDSs) are another tool that provides information about the potential hazards from using a particular pesticide. An MSDS (Figure 8-18), prepared by the manufacturer, is available for every labeled pesticide. It describes the chemical characteristics of active and other hazardous ingredients and lists fire and explosion hazards, health hazards, reactivity and incompatibility characteristics, along with types of protective equipment required for handling. Storage information and emergency spill or leak cleanup procedures are described. LD_{50} and LC_{50} ratings are given for various test animals. Emergency telephone numbers of the manufacturer are also listed on the MSDS.

Workers who handle pesticides or work in fields where pesticides are applied must be told about the potential hazards in the workplace and must be provided with information on how to avoid these hazards. They must also be trained on what to do in case they are exposed to pesticide sprays or residues. The MSDS for each material that they may be exposed to must be available and accessible to workers at their request.

IMPACTS ON NONTARGET ORGANISMS

Nontarget organisms include all plants, vertebrates, invertebrates, and microorganisms in or near a treated area that are not the intended target of a pesticide application. Pesticides can injure nontarget organisms directly or indirectly. Direct poisoning harms nontarget organisms that encounter pesticides during application or directly afterward by coming into contact with harmful residues. For example, pesticides applied while bees are foraging in the treated area (Figure 8-19) can harm them and other pollinators near the application site; Table 8.6 lists examples of insecticides that are highly toxic to bees and those with low or no toxicity. Migrating birds and other vertebrates can be harmed by improper applications of granular pesticides. Drift from an application site can harm nearby wildlife, livestock, pets, and sensitive plants. Residues remaining on crop plants may kill natural enemies, pollinators, birds, or other wildlife immigrating into the site after the application.

It may be several years after a problem arises before a cause and solution can be identified. For example, dormant sprays of the organophosphate parathion were commonly applied in California orchards in the 1980s. Early reports from the Department of Fish and Game suggested these sprays might have adverse effects on raptors. After several years of research, these suspicions were confirmed, and mortality of hawks was directly related to dormant applications of parathion in 1989. Hawks absorbed the pesticide through their feet when landing on treated trees. These findings led to regulatory changes and changes in pest management practices. Use of reduced rates and alternate materials in the 1990s eliminated the problem.

Runoff from pesticide applications into

TABLE 8-6.

Impacts of insecticides on honey bees: some examples.

Has at least one formulation highly toxic	Generally very low or no toxicity
abamectin/avermectin	amitraz
azinphosmethyl	*Bacillus thuringiensis*
bendiocarb	*Beauveria bassiana*
bifenthrin	buprofezin
carbaryl	clofentezine
carbofuran	cryolite
chlorpyrifos	Cydia pomonella granulosis virus
chlothianidin	dicofol
cyfluthrin	diflubenzuron
cypermethrin	fenbutatin oxide
diazinon	hexythiazox
dimethoate	kaolin clay
esfenvalerate	methoxyfenozide
fenoxycarb	potassium salts of fatty acids (soaps)
fenpropathrin	pyriproxyfen
fenvalerate	granular formulations of many insecticides
imidacloprid	
lambda-cyhalothrin	
malathion	
methamidophos	
methidathion	
methyl parathion	
naled	
permethrin	
phosmet	
propoxur	
spirodiclofen	
thiamethoxam	
zeta-cypermethrin	

Source: Riedel et al. 2006.

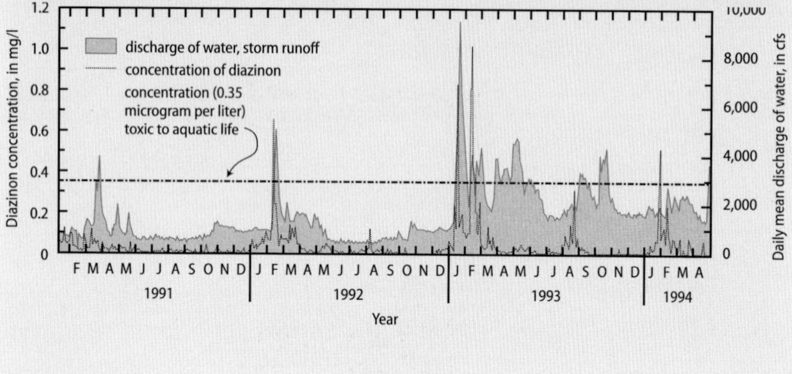

FIGURE 8-20

Applications of pesticides during the dormant spray season may run off the orchard floor into creeks and rivers, resulting in toxicity to aquatic invertebrates. This graph shows total discharge of water into the San Joaquin River and concentrations of diazinon measured near Vernalis in the years 1991–94. Measurements of the diazinon concentration in the water show that levels during winter runoff in January (J) and February (F) frequently exceeded concentrations toxic to aquatic life (0.35 µg per liter). Although diazinon use as a dormant application has been reduced, dormant season runoff of other pesticides, including organophosphates and pyrethroids, is still a concern. Source: Dubrovsky et al. 1998.

nearby ponds, streams, and lakes can harm aquatic animals and plants. For example, dormant spray applications of certain organophosphate or pyrethroid insecticides are typically made to stone fruit orchards during January and February, often the wettest time of the year. When it rains soon after an application, elevated concentrations of organophosphate insecticides have been detected in storm drains, rivers, and other waterways. The concentration of these insecticides in waterways is of particular concern because of potential toxicity to some aquatic organisms at relatively low concentrations. From 1992 to 1995, the U.S. Geological Survey found that concentrations of diazinon in the Merced, Tuolumne, and San Joaquin rivers frequently exceeded levels that are toxic to aquatic life after winter storms (Figure 8-20). Pesticide runoff in orchards was reduced in the 2000s by treating later in the year and/or switching to sprays of reduced-risk materials (*Bacillus thuringiensis* or spinosad) that do not run off into water or pose threats to aquatic invertebrates. Pesticide runoff continues to be a threat to aquatic wildlife, however, especially with the use of pesticides applied to control ants and other pests in urban areas.

Pesticide applications can destroy habitat and food sources on which nontarget organisms depend. For example, the use of herbicides to keep soil vegetation-free, in some cases has resulted in soil eroding into creeks and rivers. Herbicides can also reduce plant diversity and remove food and shelter necessary for the survival of some wildlife. Preserving wildlife habitat should be a consideration whenever pesticides are used.

Applications of persistent pesticides can also lead to secondary poisoning of nontarget organisms. This phenomenon, called *bioaccumulation* (Figure 8-21), occurs when certain pesticides gradually build up within the tissues of living organisms after feeding on other organisms (pest or nontarget) containing smaller amounts of these pesticides. Animals higher up on the food chain accumulate greater amounts of these pesticides in their tissues as time passes. Although this phenomenon is mostly associated with long-banned organochlorine insecticides, recent studies show that DDT, toxaphene, and chlordane continue to accumulate and build up in tissues of clams, fish, and other aquatic organisms relative to values in the 1970s and 1980s. These sediment-bound organochlorine insecticides continue to arrive in rivers through soil erosion. Secondary kill may also occur when carnivores feed on dead or dying rodents that have consumed rodenticide baits. Other indirect problems associated with pesticide use include the loss of soil fertility due to mortality of soil microorganisms, pesticide resistance, and the buildup of secondary pests that occurs when their biological control is disrupted.

Reducing Pesticide Impacts

The impact of pesticides on nontarget organisms can be reduced by choosing pesticides that are less toxic to nontarget organisms, applying at times when nontarget organisms are least likely to be harmed, using lower rates (when possible), and using spot treatments or other methods to selectively place pesticides (Table 8-7). For example, pesticide applications made in the early morning, late afternoon, or at night are less harmful to bees because they are not foraging at these times. There are regulatory restrictions for applying certain pesticides while bees are foraging; check with the county agricultural commissioner for information on application cutoff

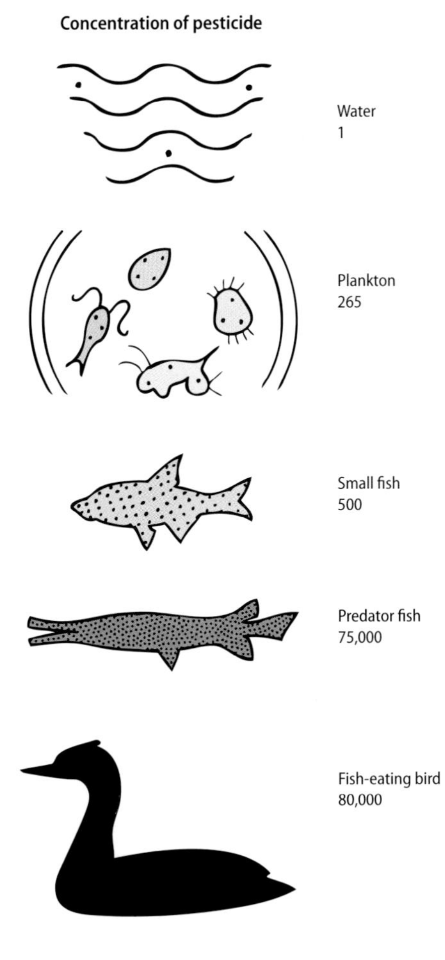

FIGURE 8-21

Some pesticides can be accumulated through the biological food chain. Microorganisms and algae containing pesticides are eaten by small invertebrates and hatching fish, and larger fish and birds eat these organisms in turn. Each passes greater amounts of pesticide to the larger animal.

TABLE 8-7.

Examples of pesticide application methods that conserve natural enemies.

Pests	Natural enemies	Pesticide application method
citrus thrips	*Euseius tularensis* (predatory mite) and others	spray only the outer tree canopy
cutworms infesting field crops	many parasites and predators	apply bait in plant rows
elm leaf beetle	*Erynniopsis antennata* (tachinid parasite)	bark banding
Homoptera-tending ants	many parasites and predators	enclosed bait, soil spray, or basal trunk spray
mealybugs in grapes	parasites and predators of mealybugs	leave at least 1 out of every 10 acres untreated to provide a refuge for natural enemies
spider mites on apples	predaceous mites and general predators	close bottom nozzles on sprayers and direct apple-thinning sprays away from tree centers to provide refuge for natural enemies
spider mites on cut roses	*Phytoseiulus persimilis* (predatory mite)	spray only upper canopy or marketed parts
various insects and pest mites in orchards	spider mite destroyer, predatory mites, and other natural enemies	spray only every other row in an orchard, so only half of trees are treated on each date

Federally Listed Species in California

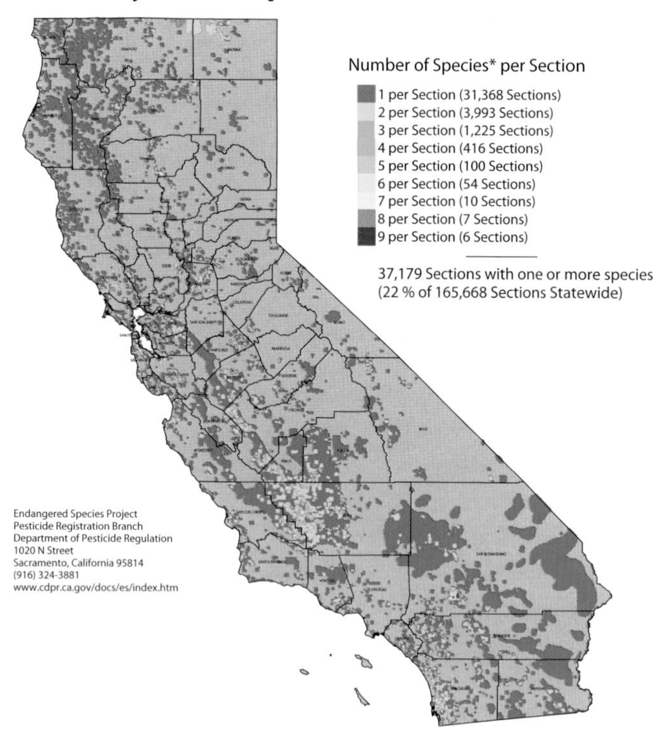

*Federally Listed Endangered, Threatened, Proposed Endangered, Proposed Threatened and Category 1 Candidate Species per California Department of Fish and Game Natural Diversity Database (August, 1997)

FIGURE 8-22

Federally listed endangered, threatened, proposed endangered, proposed threatened, and Category 1 candidate species per the California Department of Fish and Game Natural Diversity Database (August 1997).

dates and which pesticides are most toxic to bees.

The pesticide label also warns when the material is potentially harmful to nontarget organisms. Warning statements on labels provide information about problems that could occur if the material drifts onto blooming crops or if bees are visiting the treatment site. Likewise, statements such as "This product is toxic to fish" or "Keep out of lakes, streams, rivers" helps you evaluate the potential impact of a material on nontarget organisms. Information regarding environmental impact is also available on the product's MSDS.

Protection of Endangered Species. Federal and state laws are in place to protect endangered species from extinction. An endangered species is any rare or vulnerable animal or plant species that is in danger of becoming extinct. To ensure that endangered species are protected, the use of certain pesticides in areas where endangered species are known to exist may be highly restricted.

Before recommending the use of any pesticide, read the pesticide label carefully for precautions to protect endangered species and for the location of restricted areas. In California, about 22% of land sections contain one or more endangered species that must be protected (figure 8-22). An excellent source of information about pesticides and endangered species in California is available on the California DPR website, www.cdpr.ca.gov. An endangered species database identifies listed species in specific locations and provides a list of pesticides that have limitations for those species.

9 Setting Up an IPM Program

The heart of an integrated pest management program is the way in which pest management decisions are made and communicated. Biological and management information necessary to make informed decisions has been discussed in earlier chapters. There are, however, other important factors to consider when developing IPM programs. These include professionalism, client expectations, the concept of risk, the use of control action thresholds, the prevention of pest damage, and effective ways of communicating with growers or clients.

PROFESSIONALISM

As a PCA, you assume personal and professional responsibility when you make a pest control recommendation. By virtue of being licensed and providing a service on which others rely, PCAs are considered experts in their field. In the eyes of the law, PCAs are as responsible for giving sound pest management advice as doctors are responsible for giving sound medical treatment. A PCA can be held legally responsible for recommendations that are unlawful or unsound.

In California, PCAs can make recommendations only in categories in which they are licensed. For example, a PCA licensed only in Weed Control cannot make a recommendation to control aphids. To make an aphid management recommendation, the PCA would also have to be licensed in Insect, Mites, and Other Invertebrates.

When making pest control recommendations, PCAs can be held accountable if the pest management method is ineffective or harmful to the crop or surrounding areas. For example, if crop loss due to pest damage continues after a pest control recommendation has been followed or if the plants have a phytotoxic reaction to the application, the PCA who made the recommendation can be held liable for the damage. Personal liability can also be incurred if the wrong pesticide is recommended, if the application timing is incorrect, or if the recommended pesticide kills beneficial insects such as honey bees. If bees are pollinating a crop, liability can also be incurred for the loss of the crop. To prevent crop damage liability claims, PCAs should anticipate what could go wrong and take measures to prevent problems from occurring.

When a pesticide is recommended, it is the PCA's responsibility to be certain that the pesticide is labeled for the crop for which it is being recommended. Illegal residues can make a crop unmarketable, and the PCA can be held responsible for any loss to the grower, including fines and penalties. Likewise, the PCA is responsible if residues over MRL tolerance levels are detected on a crop as a result of an incorrect recommended rate. It is also the PCA's responsibility to note worker restricted-entry intervals and posting requirements when providing a recommendation to a grower or other client.

How to Reduce Potential Liability

To reduce potential liability, write complete and accurate recommendations. When writing pesticide recommendations, be very familiar with the label. Know the track record of the product. Where possible, suggest products that have been shown to be effective against a specific pest and safe in university research. Do not rely solely on manufacturer's claims. In most situations, the manufacturer's representative can be contacted when more product information is needed. Most are willing to meet at the site, evaluate the specifics of the situation, and participate in the decision to use a product. County Cooperative Extension advisors or agents are also a good resource.

Know the performance expectations of the product; test it on limited sites. Many variables, such as weather, application timing and method, soil conditions, and adjacent crops can affect the performance of a material. Resist the enthusiasm of a client who would have a material applied on the entire crop without pretesting. Failure of a product to perform can result in liability.

Take and keep field notes; they can provide backup documentation to support the pest management recommendation. Likewise, always check the performance of a material as soon as possible after the application. It is usually readily apparent if crop damage occurred or the product failed to perform correctly. If a problem is evident, take pictures of the symptoms and the application equipment. Talk with the client and pesticide applicator about the timing and conditions of the application; take detailed notes of observations and conversations. It may be necessary to take soil and tissue samples and aerial photographs. Contact the product manufacturer for help in determining whether the product or the application failed. Obtain a copy of the pesticide use report and have copies of the written recommendation. Accurate field records and complete written recommendations are the best defense should a problem arise from a pest management recommendation.

Be aware of local regulations and restrictions and state and federal regulations before recommending the use of any pesticide. Also be aware of any additional restrictions placed on a product by the purchaser or processor of the crop. It is the PCA's responsibility to know which pesticides are not allowed on a crop and to be certain that local statutes are not being violated and that written recommendations are in compliance with the law. Restrictions on the use of certain pesticides may be in place due to endangered species limitations, pesticide management zones, or county permit conditions. There are, for example, many federal, state, and local restrictions that impact pesticide use for vertebrate pest control; know whether any of these conditions apply before making a recommendation.

Many PCAs subscribe to pesticide data management software services for up-to-date label information and to assist in writing recommendations. These services automatically check to ensure that the recommendation is in accordance with the label and complies with state and federal label regulations. They provide information on crop rotation and other restrictions of the individual site or field and check for conflicts with previously applied material. Although these services guarantee that the information they provide is up to date and accurate, always double-check any information that may be doubtful. The bottom line is that if the recommendation is written incorrectly, the PCA is liable.

CLIENT EXPECTATIONS

PCAs work with clients in cropping systems, forests, rights-of-way, parks, and urban landscapes. In all these situations, they must know and fulfill the expectations of clients, educate them about realistic expectations, and inform them of their legal and ethical options. Expectations vary widely among different people, depending on their understanding of IPM and their direct practical experience with IPM programs. Expectations may range from a desire to completely eliminate pesticide use to an interest in using pesticides and other control options in an effective manner to control the pest. For many growers, expectations primarily involve improving the quality and economic return of the crop. In landscape situations, pest management goals may be based on the public's perception of the aesthetics of the landscape, or it may be a public health issue or a question of plant health. In most urban landscapes, expectations vary among the different types of landscape (city hall versus greenbelts) or the different types of turf plantings (golf courses versus turf strips along a roadside).

Before setting up a program, discuss with the client exactly what the expectations are for the pest management program. In one cropping system, the grower may have low tolerance for pest damage and demand control by any means possible at the first sign of pest infestation. Another grower may wish to consider several management strategies, recognizing that some pest levels can be tolerated, and allow for the use of pesticides as needed. Other growers may have limitations imposed upon them by their food processor. In an urban landscape, the attitude of the general public may include the expectation that an IPM program will reduce the amount of pesticides used or completely eliminate the use of highly toxic pesticides. In any case, know what the client expects. Over time, through trust, education, and by example, even the most conservative clients may be convinced to raise their tolerance for pests and try out innovative management practices.

THE CONCEPT OF RISK

From the client's perspective, one of the most important roles of the PCA may be to add to their peace of mind. An integrated approach to pest management reduces risk by increasing the reliability of the pest management information provided. Monitoring, *scouting*, and record-keeping activities combined with biological knowledge of the pest organism and specifics of the site reduce the potential for error in pest management decisions.

For some growers, implementing IPM will represent a shift away from practices to which they have grown accustomed; any change in pest management strategy can be perceived as a risk. Minimize the perception of risk by developing an IPM program that is flexible and builds on the grower's production practices. IPM programs that are suited to local conditions and are adapted to the grower's equipment, irrigation capabilities, management practices, local climatic conditions, marketing strategies, and labor constraints are often perceived as less risky. Increase the client's confidence in new methods by providing publications documenting their reliability or giving background information on pests being managed. Cooperative Extension leaflets and university Internet sites are good sources of IPM information.

CONTROL ACTION THRESHOLDS

Control action thresholds are a basic IPM strategy for providing information about when to treat to avoid losses from pests. When based on biological and economic information, thresholds can help improve profits, reduce unwanted environmental impacts, and reduce the risk of pest damage. The use of control action thresholds acknowledges that not all pests require management and that some levels of pests are tolerable.

Control action thresholds based on university research, however, exist for only a few pests, primarily insects in agricultural crops, and have been used only on a limited basis in disease and weed management. Where thresholds do exist, they offer a practical assessment of pest status, taking into account host, site, and pest specifics as well as the effects of other management actions on the pest population. When using thresholds, be flexible in their application. Agree with the grower or client in advance on acceptable thresholds and be aware of any management practices, such as irrigation timing or marketing standards, that may impact threshold levels.

Control action thresholds help assess when a pest management action is necessary to prevent economic damage. Where control action thresholds based on university research do not exist, work with clients to develop provisionary action thresholds to help determine when a pest management action is needed. By doing the background research necessary to develop these thresholds, the client's confidence increases and the risk of unacceptable losses is reduced.

Start by following the procedures described in Chapter 7 for setting up a monitoring program. Research biologies and life cycles of pests and know their potential damage. Discuss with the client special concerns about pest damage, including the market potential for the crop. Investigate the relative costs of different pest management options. Keep records to help in discussions and management recommendations. Establish monitoring and sampling programs to coordinate with provisionary thresholds. Keep written records of monitoring results, success of treatments, pest damage, yields, and economic losses over the season. Use this information to refine provisionary thresholds in subsequent years.

HOW TO COMMUNICATE THE DECISION TO THE CLIENT

IPM programs are much easier to implement when the client is made aware of the benefits of an integrated approach to pest management. Regular meetings to discuss field observations and pest development reassure the client that approaches are being considered with the specifics of the site in mind. To communicate pest management decisions, it is ideal to meet face-to-face (Figure 9-1). The more direct the communication, the more at ease the client will be with the IPM strategy, especially when recommending a control option that the client has not used before. If necessary, present supporting documentation (i.e., research, references, articles, other crop examples) to demonstrate the validity of the approach being recommended. In some situations, it may be necessary to demonstrate a strategy on a small section of the crop or landscape with a field trial (see the section "Field Trials" in Chapter 7).

Written communication is another effective tool for informing the grower of pest management decisions. The most common method of communication is the written recommendation, which must be completed whenever a pesticide or nonpesticide control practice is recommended. In addition, growers and clients benefit from weekly or monthly updates on monitoring, surveying, and field observations whether treatment is necessary or not. Many successful PCAs provide seasonal summaries with graphs and maps of pest activities in clients' fields, orchards, or landscapes.

FIGURE 9-1

A good PCA must be able to communicate well with clients.

The Written Recommendation

There are many reasons to put a pest control recommendation in writing. According to the California Food and Agricultural Code (§ 12003), all recommendations concerning any agricultural use must be in writing; this includes all recommendations for any type of pest management, chemical and nonchemical. (For instance, a written recommendation would be required for release of beneficial insects to control a pest.) Additionally, the Food and Agricultural Code specifies that the operator of the property receive a copy of the written recommendation prior to the pest control action being taken. If a pesticide use is recommended, a copy of the written recommendation must be provided to the pesticide dealer and applicator before the application can take place. In addition, PCAs must keep a copy of their pest control recommendations for 1 year. Written recommendations must contain specified information (see Sidebar 9-1), but they may be submitted in various formats. You may design

Pest Control Recommendation

1. Operator of the Property

2. Recommendation Expiration Date

Address City County

3. Location to be Treated

4. Commodity to be Treated

5. Acres or Units to be Treated

6. Method of Application:
☐ Air ☐ Ground ☐ Fumigation ☐ Other ______

7. Pest(s) to be Controlled

8. Name of Pesticide(s)	Rate Per Acre or Unit	Dilution Rate	Volume Per Acre or Unit

9. Hazards and/or Restrictions:
- ☐ 1. Highly toxic to bees
- ☐ 2. Toxic to birds, fish and wildlife
- ☐ 3. Do not apply when irrigation or run-off is likely to occur
- ☐ 4. Do not apply near desirable plants
- ☐ 5. Do not allow to drift onto humans, animals or desirable plants
- ☐ 6. Keep out of lakes, streams and ponds
- ☐ 7. Birds feeding on treated area may be killed
- ☐ 8. Do not apply when foliage is wet (dew, rain, etc.)
- ☐ 9. May cause allergic reaction to some people
- ☐ 10. This product is corrosive and reacts with certain materials (see label)
- ☐ 11. Closed system required
- ☐ 12. Restricted use pesticide (California and/or EPA)
- ☐ 13. Hazardous area involved (see map and warnings)
- ☐ 14. Other (see attachment)

10. Schedule, Time or Conditions

11. Surrounding Crop Hazards

12. Proximity of Occupied Dwellings, People, Pets or Livestock

13. Non-Pesticide Pest Control, Warnings and Other Remarks

14. Criteria Used for Determining Need for Pest Control Treatment:
☐ Sweep Net Counts ☐ Leaf or Fruit Counts ☐ Preventative
☐ Field Observation ☐ Pheromone or Other Trap ☐ Soil Sampling
☐ History ☐ Other

15. Crop and Site Restrictions:
- ☐ 1. Worker reentry interval ____ days
- ☐ 2. Do not use within ____ days of harvest/slaughter
- ☐ 3. Posting required ☐ Yes ☐ No
- ☐ 4. Do not irrigate for at least ____ days after application
- ☐ 5. Do not apply more than ____ application(s) per season
- ☐ 6. Do not feed treated foliage or straw to livestock
- ☐ 7. Plantback restrictions (see label)
- ☐ 8. Other (see attachment)

16. I certify that alternatives and mitigation measures that would substantially lessen any significant adverse impact on the environment have been considered and, if feasible, adopted.

Adviser Signature Date

Adviser License Number

Employer

Employers Address

N

W E

S

FIGURE 9-2

A sample pest control recommendation form. Source: California Department of Pesticide Regulation.

SIDEBAR 9-1

The Written Pest Control Recommendation

Follow these guidelines to fill out a complete and accurate pest control recommendation similar to the one in Figure 9-2. Most of this information is required by the Food and Agricultural Code § 12003 or the California Code of Regulations Division 6, § 6556. Asterisked (*) items are not required but are recommended because they are required to get a restricted agricultural use permit. Number 2, "Recommended Expiration Date," is not required, but it is advisable.

1. **Operator of the Property.** This is the person responsible for seeing that the recommendation is carried out. It could be the owner of the property or a representative, a firm, or an agency. Misunderstandings can be avoided by finding out who is responsible for carrying out the recommendation and always delivering the written recommendation that person.

2. **Expiration Date.** Recommendations are made for specific pests under specific circumstances. Therefore, the recommendation is valid only for a specified period of time. Some PCAs incorporate a "window of opportunity" into their recommendation, giving their client some time to take action. For some pests, a control action has to be enacted immediately. Application timing is important if a pest control action is to be successful: be specific and precise as to when the optimal control will be achieved.

3. **Location to Be Treated.** This is the site where the actual application is to take place. It can, and often does, differ from the address of the property owner. Be precise. If spot treatments are part of the recommendation—treating the outer rows of the NW perimeter for mite control, for instance—put it in writing here.

4. **Commodity to Be Treated.** List exactly what is being treated (apples, oranges, turf, etc.). It is especially important to note the commodity when more than one crop is grown on the site. In landscape situations, note whether the application is for turf or a specific ornamental plant.

5. **Acres or Units to Be Treated.** In addition to identifying the site, include the total acreage that needs to be treated. For many pest species, the total acreage will not need to be treated. In landscape situations, individual plants or small areas may be treated instead of acreages; if so, list the number of plants or square footage to be treated.

6. **Method of Application.*** Indicate your recommendation for the best application method (air, ground, fumigation, other) for the treatment. Consider the application equipment available, the client's resources, the severity of the pest, the location, and environmental conditions that might affect the success of the application.

7. **Pest(s) to Be Controlled.** Identify the pest(s) to be controlled by the recognized common name. This is an important component of a pest control recommendation; if the pest identification is incorrect, the recommended control may not work.

8. **Name of Pesticide(s).** If a pesticide treatment is being recommended, list the name of the pesticide(s) by common or trade name. If recommending the addition of an adjuvant, it also should be listed. In addition, recommend the rate per acre or unit (how much of the material, for example, 2 oz/acre; on a tree in a landscape, specify how much material should be applied per tree); the dilution rate (how much material per gallon of water); and the volume per acre or unit (the total amount of diluted material to be applied).

9. **Hazards and/or Restrictions.** Check off the environmental hazards and restrictions that apply to the use of the recommended material. It is the responsibility of the PCA to note all adverse or potentially adverse situations that can be encountered from the use of the material being recommended. Much of this information is available on the label or MSDS, such as toxicity to bees and adverse effects on wildlife. If the material is a restricted-use pesticide or a closed system is required, note it here.

   Some of the information listed should always or nearly always be checked, such as "Do not allow drift onto humans, animals or desirable plants" (because drift is not supposed to occur); "may cause allergic reactions to some people" (because chemical sensitivity is so specific to the individual, it is impossible to know whether the use of a material will cause an allergic reaction, so it is best to be on the alert for unusual possibilities here); "keep out of lakes, streams and ponds" (unless the application is being made to a body of water, applications should always avoid contact with water).

   Specific conditions will apply to specific materials and should be noted here, such as not applying a specific herbicide near desirable plants (as opposed to some herbicides, for instance, that can be applied directly over desirable plants); avoiding irrigation or moist conditions; taking into account the corrosiveness or reactivity of a material; and preventing birds from feeding on treated materials.

   In addition to warnings noted on the label, hazardous conditions that are observed from monitoring and field surveys should also be noted here and, if feasible, detailed on the field map.

10. **Schedule, Time, or Conditions.** Specify the scheduling, timing, and/or conditions that relate to the application. Note temperature limitations (such as not to apply if temperature exceeds 90°F), optimal time of day for the application (some materials are best applied at night due to less wind and low temperatures), or whether the material needs to be incorporated, either mechanically or with water. Any label restrictions on the use or disposition of the crop or crop by-product should also be noted here.

11. **Surrounding Crop Hazards.*** A material applied to one crop can cause severe problems if it drifts onto a neighboring crop. For instance, rice and cotton have severe restrictions on the use of certain materials when these two crops are grown in close proximity. Always be aware of the neighboring crops when making a pest control recommendation. Also note potential impacts to wildlife or water resources adjacent to the crop site. If the label does not cite

any limitations with neighboring crops or other environmental impacts, check with the county agricultural commissioner to be sure that a conflict does not exist.

12. **Proximity of Occupied Dwellings, People, Pets, or Livestock.*** If there are houses on or close to the application site, or if people, pets, or livestock are likely to be in or near the application site, noting it here will make the client or applicator aware of potential risks. It is always a good idea to include buildings on the map.

13. **Nonpesticide Pest Control, Warnings, and Other Remarks.** Make non-pesticide recommendations here. Explain in full detail what the property owner or manager is to do. Don't be brief—include additional paperwork if necessary. If a pesticide application is recommended, a warning as to the possibility of damages that could result from the application and that should reasonably be known be the PCA should be included here. Any other warning or remarks that pertain to the recommendation should also be written here.

14. **Criteria Used for Determining Need for Pest Control Treatment.** It is important to leave a record showing how the decision for the recommendation was made. Note whether the information came from monitoring (sweep net counts, leaf counts, pheromone counts), field observation, field history, or if the action being recommended is preventive. This will aid in building a field history and also provides documentation should something go wrong with the recommendation.

15. **Crop and Site Restrictions.** Give information to the owner or operator of the property to provide for safety and reduce exposure hazards. Much of this information is available on the label, such as the worker restricted-entry interval (note the number of days), harvest interval, number of applications recommended per season, plantback restrictions, and information regarding feeding treated foliage to livestock. Highlighting this information reduces potential problems.

16. **Signature.** This is where you sign and date the recommendation and provide your PCA license number. Below the signature is a statement that says by signing, you certify that you have considered all alternatives and mitigation measures that would substantially lessen any significant impact on the environment and, if feasible, that you have adopted their use. It is important to understand the significance of this statement when signing a recommendation. Signing the recommendation states that the most environmentally sound solution to the problem is being recommended. This means that the recommendation includes *all* pest management recommendations, not just pesticide recommendations. A signature also signifies that you have considered all hazards, warnings, and alternatives, and that no significant environmental impact should result from this recommendation. In effect, this statement means that your recommendation is a scaled-down environmental impact report.

your own, or if you use an agrichemical software service, they may provide templates for written recommendations as well. A sample written recommendation form is shown in Figure 9-2.

PCAs should communicate with their clients regarding each potential pest control situation, whether a control action is needed or not. The written recommendation can aid communication by serving notice on how a pest management decision was reached and provides an avenue to highlight special concerns and mitigation possibilities for the site. Should a problem arise, such as an ineffective control method or damage to a neighboring crop, the written recommendation is the PCA's best protection from liability.

Where to Get Information

The pesticide label is the primary source of information for the correct use and application of a pesticide. It is a legal document. The label identifies the classification of the pesticide and indicates whether it is a general or restricted-use material. The label also identifies the site (crop, plant type, animal, or location) where the material can be used; it is illegal to apply a pesticide to a site not identified on the label. The format for pesticide labels is established by federal regulations. Sidebar 9-2 lists the information that must be on a label.

When making a pesticide recommendation, you may need information that may

SIDEBAR 9-2

What Pesticide Labels Contain

Brand or Product Name. The specific name given to the product by the manufacturer. This is the name used for advertising and promotion.

Chemical Name. Describes the chemical structure of the pesticide.

Common Name. If an approved common name for the active ingredient exists, it may be listed.

Formulation. The formulation type, such as emulsifiable concentrate, wettable powder, or soluble powder. Often listed as a suffix in the brand name. For example, in Dimilin 25W, the W indicates a wettable powder formulation.

Ingredients. The active and inert ingredients listed in percentages by weight. The names of inert ingredients sometimes are not stated, but the label must indicate their percentage of the total contents.

Contents. The net contents contained in the package by weight or liquid volume.

Manufacturer. The name and address of the manufacturer of the product.

Registration Number. Number assigned by EPA and DPR as proof that the product has been approved for sale in the marketplace.

Establishment Number. The site of manufacture or repackaging of a pesticide.

Signal Word. The signal words *Caution, Warning, and Danger* (in order of increasing toxicity) indicate the relative toxicity of the active ingredients to humans. The word *Poison* and the skull and crossbones symbol are associated with the signal word *Danger*.

Precautionary Statements. The hazards and precautions associated with a chemical. Includes hazards to people and domestic animals, listing the adverse effects that may occur if people become exposed; also includes requirements specifying the type of protective equipment that must be worn. Describes environmental hazards with indicators of toxicity to nontarget organisms and describes physical and chemical hazards, including risk of explosion or hazards from fumes.

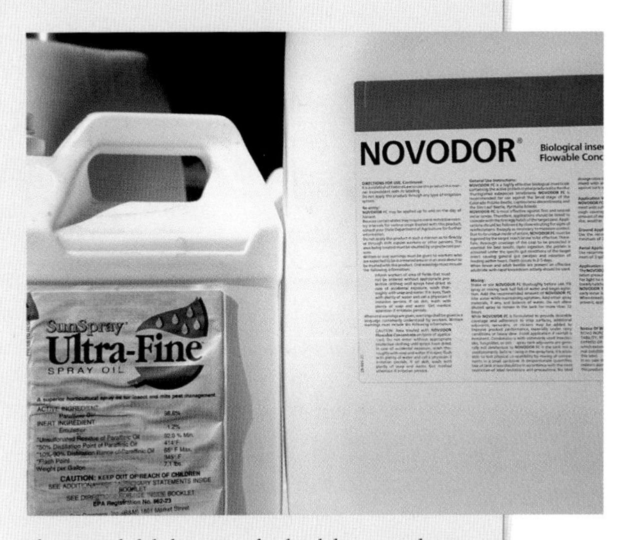

The pesticide label is a complex legal document that must be read and understood before a pesticide is applied. U.S. EPA regulations set the format for pesticide labels and prescribe the information that they must contain. In some cases, pesticide packages are too small to have all the necessary information printed on them. In these cases, supplemental labels are attached.

Statement of Practical Treatment. Describes emergency first-aid measures to take when the pesticide contacts skin, splashes into eyes, is swallowed, or is inhaled; tells you when to seek medical attention. Also contains a "Note to Physicians" describing the appropriate medical procedure for poisoning cases and may also indicate an antidote.

Classification Statement. *General use or restricted use,* depending on the hazards, intended use, and effect on the environment. Classification is based on the potential of the pesticide to cause harm to people, animals, or the environment. Pesticides classified as California restricted materials will not be so indicated on federal labels.

Directions for Use. Instructions for applying the pesticide. Lists the target pests and the permitted application site (crop animal, location, etc.); provides mixing instructions, the rate of application, and when and where to apply. May include preharvest intervals and special restrictions such as plantback restrictions.

Agricultural Use Requirements. Refers to the Worker Protection Standard; provides information on personal protective equipment (PPE) required for early-entry workers.

Restricted-Entry Statement. Indicates how much time must pass before a person can enter a treated area without personal protective equipment and special training. For some pesticides, California REIs are more restrictive than federal REIs.

Misuse Statement. Reminds users to apply according to label directions.

Storage and Disposal Statement. Directions for the proper storage and disposal of pesticide and empty pesticide containers; may include special requirements. Local regulations may be more restrictive.

Warranty. Manufacturer warranty and disclaimer; gives information on the rights of the purchaser and limits the liability of the manufacturer.

not be on the label. California restricted-use materials, established by the director of DPR, are not noted on the label; a list of these materials can be obtained from the county agricultural commissioner's office. Specific local restrictions, county permit conditions, information on endangered species, or groundwater protection areas may not be listed on the label. Drift considerations to protect neighboring crops or local regulations to protect honey bees may not be listed. Always check with the local county agricultural commissioner's office for conditions specific to the area. The local county agricultural commissioner's office will also be informed of recent label changes or changes in the law that could affect a recommendation.

COLLABORATING WITH OTHER PCAS AND GROWER GROUPS

When making pest management decisions, PCAs must consider pest activity and development in neighboring locations as well as those on-site. Pest organisms are not restricted to plots of land; pests often migrate from outside regions. Weed seeds and disease spores can be blown in from neglected fields, and highly mobile insects can colonize from distant areas.

To make effective, long-lasting pest management decisions, address regional concerns by sharing information on pest development, movement, and the effectiveness of control options with other PCAs. Working together and sharing up-to-date information on pest status, the occurrence of pesticide resistance, and alternative methods provide more timely information that can be translated into more effective decision making. Collaboration with other PCAs and grower groups is the most effective means for gathering the regional information necessary for developing a timely field strategy. Join professional organizations and participate in meetings. Also keep in touch with local Cooperative Extension agents or farm advisors. They may sponsor field meetings or have newsletters with up-to-date information on pest status in the area. For some crops and areas, weekly status reports are available from commercial services that can help you focus your activities.

There are many examples in which collaboration among PCAs and growers enhances IPM. One of the earliest examples involved the citrus growers and pest managers who formed the Fillmore Protection District in southern California in 1922 to combat California red scale, *Aonidiella aurantii*. Beneficial insects were reared in their central *insectary* and released to control California red scale and other pests in the region's citrus orchards for over 60 years. In another successful effort in the 1990s, growers and PCAs formed the Randall Island Project, a research and demonstration project in the northern San Joaquin Valley that collaborated with the public sector to discover and evaluate alternative control strategies for codling moth in pears. Working as a partnership, the members of the Randall Island Project were successful in developing, refining, and implementing codling moth mating disruption technology in their fields.

PCAs must be aware of other types of collaboration that are regulatory in nature but are beneficial to growers and advantageous to IPM programs. An important preventive management tool for the control of some diseases is the use of host-free periods. Host-free periods are mandated by the Food and Agriculture Code and are designed to be used as a tool to interrupt the life cycle of a pest that has not been effectively controlled by conventional means. For example, in celery production, a host-free period of at least 2 to 3 months (in which celery is not being grown in the area) greatly reduces the incidence of celery mosaic. In 2010 and 2011, grape growers in the Napa Valley banded together with the Farm Bureau, UC Cooperative Extension, the county agricultural commissioner's office and others to join forces in an effort to eliminate an invading vineyard pest, the European grapevine moth. Such area-wide collaborations are often essential for battling invasive exotic pests.

FIGURE 9-3

Your local Cooperative Extension office will have publications such as these IPM manuals from the University of California that can help you identify key pests on your crop.

STEPS FOR SETTING UP AN IPM PROGRAM

The success of an IPM program depends on following reasonable and logical procedures for each pest management situation and communicating effectively with clients. The previous chapters in this book have discussed the individual components of IPM. Steps for developing an IPM program are outlined below.

Step 1. Identify the key pests, understand their biology, and recognize the damage symptoms associated with each pest. Every crop has one or a few key pests around which pest management practices must be designed. Recognize the key pests and be able to identify them to the species level. Management practices depend on the species present. For instance, an undesirable plant is not just a weed; it is a particular species with unique biological and ecological requirements and susceptibility to management tools. Know the type of damage associated with each pest and when the crop is most severely affected. Identify natural enemies and management practices or environmental conditions that may impact the pest. Consult university and other resources (Figure 9-3) and consult with other experts working in the crop. For more information on pest identification, refer to Chapter 4.

Step 2. Establish monitoring guidelines for each pest. Effective monitoring is essential for a successful IPM program. Monitoring tools and techniques vary from pest to pest, but all involve regular checking for pests, searching for evidence of pests' presence such as damage symptoms or other clues, and field surveys. Also monitor for the presence of natural enemies. For more information on monitoring guidelines, refer to Chapters 6 and 7.

Step 3. Establish injury levels and action thresholds for each pest species. Determine the infestation levels that will be intolerable or that will cause unacceptable damage at various times of the year or during specific plant growth stages. Devise a monitoring plan to detect these pest levels and to aid in determining when a management action is necessary. For more information on control action guidelines, refer to Chapter 3. For more information on how to develop injury levels and action thresholds, refer to the sections "Relating Monitoring Results to Treatment Thresholds" in Chapter 6 and "Step 3. Establish Injury Levels and Action Thresholds for Each Pest Species" in Chapter 7.

Step 4. Develop a list of acceptable management strategies for each pest species. Preferred management strategies in an IPM program prevent pest problems. These strategies may include modifying the habitat to be less conducive to pest survival, using pest-resistant or pest-tolerant cultivars, and using cultural, mechanical, and physical controls to discourage pests. Encouraging naturally occurring biological controls as well as using traps or barriers can be important tools. For more information on pest management strategies, refer to Chapter 5. For each pest species, develop a list of pesticides that are registered for the site and pest and that are effective but the least disruptive to the environment. Also list biological control agents and nontarget organisms and know when each pesticide must be applied to be most effective. Encourage the use of low rates, spot treatments, and other selective application techniques as appropriate. For more information on considerations when using pesticides, refer to the section "Using Pesticides in an IPM Program" in Chapter 5; also refer to Chapter 8.

Step 5. Develop specific criteria for selection of pest management methods. For each pest species, consider methods and materials that suppress or permanently restrict the pest. Preventive methods are often the most effective over the long term. Consider integrating various pest management options, such as the introduction and establishment of beneficial organisms, the removal of breeding sites and refuges, sanitation, and crop rotation, to reduce the pest's ability to survive and reproduce. For more information on integrating management techniques, refer to Chapter 3.

Step 6. Develop guidelines to aid in selecting the most effective pesticide for use in an IPM program. In choosing a pesticide for use in an IPM program, consider materials that are least disruptive of natural controls, least hazardous to human health, least toxic to nontarget organisms, least damaging to the environment, and the most cost effective. In addition, check the label to ensure that the pesticide being considered is appropriate for the intended use. Be sure that the recommendation addresses all mitigating factors. For information on using pesticides in an IPM program, refer to Chapter 5; for safety considerations, refer to Chapter 8.

Step 7. Develop a procedure for creating a written recommendation. When filled out properly, the written pest control recommendation (see Figure 9-2) can be one of the best lines of communication with the operator of the property. Be thorough; accurately identify the location and commodity to be treated and the pest or pests to be controlled. If a pesticide application is being recommended, identify (by brand name) the material, the rate, volume per acre or unit, and the dilution rate. Be aware of and note on the recommendation all potential hazards, such as nearby crops, farm ponds, waterways, bee hives, endangered species, wildlife, or livestock that could be damaged by the chemical being applied. Note special precautions that should be taken due to the proximity of residential areas, schools, parks, and other public areas in the treatment area. Also note precautions to reduce the risk of exposure to people.

Step 8. Establish a record-keeping system. In addition to the written recommendation, ensure that comprehensive field notes, maps, and monitoring records are part of the record-keeping system. Electronic databases provide a convenient way to store and retrieve records. Good records are essential for evaluating and improving an IPM program. They also serve as backup documentation to support recommendations. Refer to Sidebar 9-1 for more information on how to complete a written recommendation. For more information on record keeping, refer to Chapters 6 and 8.

Step 9. Develop a list of resources. Know where to go when you need help with pest identification and pesticide recommendations and where to find information on pest management, pesticides, and pesticide-handling emergencies. Refer to the "Resources" section for a list of references. For a regularly updated list of resources, check the DPR and the UC IPM websites. See the "Resources" section for website addresses.

Step 10. Build flexibility into the program. Pest management situations change from area to area, field to field, and year to year. A pest management program that worked well one year may not be as successful the next. A well-designed IPM program recognizes the variability in pest organisms and the agroecosystem and can easily adjust to accommodate such changes. For more information on the fluctuations that can occur within an agroecosystem, refer to Chapter 2.

Step 11. Develop procedures to regularly evaluate the effectiveness of management actions. It is essential to constantly evaluate the effectiveness of the pest management program. Follow-up monitoring is essential to assess the effectiveness of management practices. Formal evaluation programs help in making better decisions about adopting new materials and practices, and they also help in detecting pesticide resistance. Every farm and landscape is unique. Practices that are effective on one plot of land may need modification to be effective in other locations. For information on post-treatment monitoring and setting up field trials, refer to Chapter 7.

Step 12. Establish communication procedures that work for clients. The job of a pest management professional is to keep clients informed about the status of pests on their property. Ultimately, the client makes the final decision on what pest management practices will be implemented. It is to your benefit to increase clients' understanding of pest biology and management options so they will make good management decisions and follow through on your recommendations in a timely manner. Different methods of communication work better with different clients. Supplement written recommendations with monthly or seasonal summaries, establish regular meetings, or create a newsletter; find out what works best for the client. For more information, refer to the section "How to Communicate the Decision to the Client" earlier in this chapter.

Setting Up an IPM Program for a Public Agency

When working as a PCA for a public agency or school, there are other important points to consider when setting up a successful IPM program. Pest management programs in public agencies rely on the coordinated activities of many individuals. Many public agencies don't have PCAs on staff but do have people who are involved in pest management decisions. Often several departments or supervisors may be involved in activities that affect pests and their management. All must be enlisted in a program that shares common goals and approaches.

In addition, public agencies must be accountable and responsive to the citizens of their community. The public often wants justification for the use of certain types of pesticides or demands to know why the agency isn't doing a better job controlling organisms that they consider pests. A written IPM policy and program enhance an agency's ability to carry out successful pest management and their responsiveness to public concerns. Recommendations for setting up an IPM policy for a public agency are outlined below.

Adopt written policies and procedures. Adopting written policies and procedures provides an agency with a standard that details how to carry out and improve the agency's internal decision-making process, resulting in more efficient, effective, and safer resolution of pest problems. Written policies and procedures give staff direction when responding to questions from the public. They also provide a means of involving agency staff and the public in the development of pest management policies and educating them on the potential benefits and hazards of pest management practices. For each pest that can be a problem, write out all the information outlined in the steps for setting up an IPM program, above. Put this information in a binder or computer database so it can be readily accessed if the pest becomes a problem or if a query from the public is received.

Determine the goals of the pest management program. To establish a more effective IPM program, determine the policy goals that will give the agency a solid foundation on which to base pest management decisions. Policy goals provide a set of priorities to work from and include political, educational, and public relations goals as well as operational goals. Pest management policy goals will differ with the function of the agency, but all goals should establish procedures for keeping the public and employees informed regarding the pest management tactics used. Goals should establish procedures for educating the public and employees about pest management and solutions. Operational goals include establishing plant and pest surveys, monitoring programs, pesticide use reporting, and record keeping for all pest management activities.

Define the responsibilities of all staff involved in pest management. Specific job duties for staff vary among agencies. A public agency should designate someone to be in charge of the program and function as an IPM coordinator to facilitate communication and program oversight and assume responsibility for pest management decision making (Figure 9-4). Define the role of the site manager, the groundskeeper, and all maintenance personnel. Clearly designate who is responsible for receiving complaints about pests, who makes pest management decisions for different sites, who carries them out, and who evaluates their effectiveness. Recognize that many people often not normally thought of as involved in pest control—janitors, kitchen staff, mowing or pruning crews, and their supervisors—have important contributions to make in an effective IPM program. Often, these are the people who are the first to observe a pest problem and, in many cases, are the ones who have been initiating control. Implement a method to constructively involve them and enlist their support of the program.

Establish procedures for selecting a pest management contractor. Some public agencies have no staff or a limited staff to devote to pest management activities. For this reason and others, agencies sometimes hire an outside contractor for pest management services.

To assist in hiring a pest management contractor proficient in the principles of IPM, prepare a request for qualifications (RFQ). An RFQ allows for the prescreening of prospective pest control contractors and ensures that only qualified contractors submit proposals for the bid process. Next, prepare a request for proposals (RFP) that details the terms of the agency's IPM policy. Evaluate the responses to the RFP according to the contractor's ability to adhere to the terms of the IPM policy. After a contractor has been selected, develop a procedure for continuing communication. Program evaluation should continue through informal conversations and a process for handling complaints.

As part of the pest management contract, develop a quality assurance form (QAF). A QAF should be filled out by the contractor as each service is provided and should detail information on pest sightings, sanitation and structural concerns, pesticides or traps used, monitoring stations installed, and any additional pest management concerns. Additionally, the pest manager or site manager can use the QAF to comment on the quality of service and communication between the contractor and staff.

Mandate staff training in IPM. For IPM to be successful, the cooperation of all personnel impacted by the program is required. Educational training for all staff before implementation of the program helps in understanding IPM and defines everyone's role in managing pests in the landscape and work environment. Additionally, pest management personnel, site managers, and maintenance workers (building or landscape) should be provided with continuing education to assist them in obtaining the information, skills, and equipment they must carry out the policy.

Form an IPM advisory committee. An advisory committee composed of interested members of the public, employee representatives, persons with toxicological and pest management expertise, members of environmental organizations, and worker health advocates can assist with the initial review of procedures and changes in agency policies. They can also evaluate the effectiveness of the IPM program if the public has expressed a great interest in being involved in the program. They can help review goals, define worker responsibilities, aid in the selection of outside contractors, and provide staff training.

The advisory committee should review the IPM program regularly to make sure it is meeting the needs of the community, is effective, and is up to date. An IPM program must be flexible and able to adjust to changing needs and technologies. Maintaining an advisory committee requires a substantial commitment by all parties, but it may be critical in keeping the program current with changes in technology and assuring the public that their input is valued.

Once an IPM program and policy have been adopted by a public agency, staff must help carry them out. Change never comes easily, and substantial leadership may be required to help staff or contractors overcome barriers to successful IPM implementation. These barriers may be technological, psychological, or institutional, and the IPM coordinator or pest management program supervisor must be prepared to meet each challenge constructively.

Problems may occur because of the normal factors associated with any new program—resistance to change, anticipated difficulty in learning the new technology, or fear that the program may mean that pesticides can never be used and that pest control will be less effective or more expensive. Often there is a lack of in-house expertise on IPM. Most concerns can be overcome with education or facilitated discussion. Often these concerns have a legitimate base; staff are able to point out problems in the program that hadn't been identified previously. Once their input is included and they see that they have an important role in the program, they are likely to become more enthusiastic supporters.

For additional information on establishing an IPM program in a public agency, refer to *Establishing Integrated Pest Management Policies and Programs: A Guide for Public Agencies*, UC ANR Publication 21513.

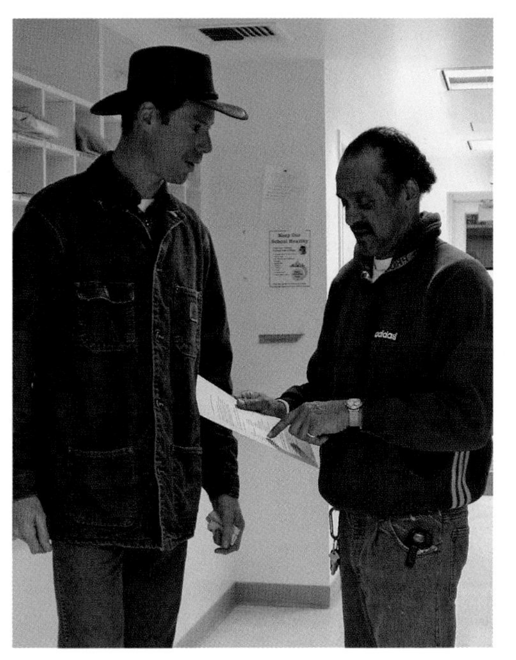

FIGURE 9-4

Schools and public agencies should designate an IPM coordinator to facilitate communication and program oversight. Photo by Cheryl Reynolds.

HOW TO EVALUATE ALL THE FACTORS

Pest management is but one consideration in a complex ecosystem. Whether the ecosystem is an agricultural crop, urban landscape, school building, right-of-way, or forest, pest management activities must be coordinated with other management practices to achieve economical protection from pest injury while minimizing hazards to the plant host, human health, and the environment.

Successful IPM programs require professional attention and dedication. Pest management consultants must stay informed about the latest methods, techniques, and protection programs available. Up-to-date information is essential—build a network with other crop consultants, member organizations, Cooperative Extension advisors, and university experts. Access the Internet for current information on weather, diagnostic tools, pest population forecasting models, pesticides, and other pest management information. Use computer databases to track information and keep records. Read journals, published reports, and newsletters for current information. Actively participate in field days and seminars to keep up to-date on the most effective pest management strategies.

Pest management systems are dynamic and constantly changing. The threat to the host from a pest can change from year to year or area to area. New pests invade. Economics or changes in marketing strategy can change the profitability of managing a pest. Putting it all together for effective pest management involves intelligence, skill, and information gathered from monitoring, field visits, and historical records for each site, along with knowledge of the host, the pest, and the ecosystem. Effective pest management is a hands-on, one-on-one, field-by-field service provided by trained professionals using an integrated approach.

Resources

SUGGESTED READING AND WEBSITES OF INTEREST

This suggested reading list includes outstanding publications that will broaden your understanding of the various aspects of integrated pest management. The list includes books that offer detailed information on specific types of pests; pest management tools; or IPM programs for various agricultural crops, landscape trees and shrubs, and home gardens. See the "References" for additional publications, including those that were cited in the text.

Also included in this list are relevant websites. The Web allows electronic access to a vast spectrum of university, government, industry, and private resources on pest management. Although any printed list of online sites is immediately out of date or incomplete, several relevant sites and their online addresses (URLs) are included.

Biological Control

Biological Control. 1996. R. G. Van Driesche and T. S. Bellows. Chapman and Hall, New York.

Biological Control of Weeds: A Handbook for Practitioners and Students. 1992. K. L. S. Harley and I. W. Forno. Inkata Press, Melbourne, Australia.

Cornell University Biological Control Guide. www.biocontrol.entomology.cornell.edu.

Natural Enemies: An Introduction to Biological Control. 2004. A. E. Hajek. Cambridge University Press, Cambridge, UK.

Natural Enemies Handbook. 1998. M. L. Flint and S. H. Dreistadt. Univ. Calif. Div. Agric. Nat. Res. Publ. 3386, Oakland.

UC IPM Natural Enemies Gallery. www.ipm.ucdavis.edu/PMG/NE.

Chemical Control and Pesticide Safety

California-EPA/Department of Pesticide Regulation Product/Label Database Queries. www.cdpr.ca.gov/docs/label/labelque.htm.

FRAC. Fungicide Resistance Action Committee. www.frac.info/frac.

HRAC. Herbicide Resistance Action Committee. www.hracglobal.com.

IRAC. Insecticide Resistance Action Committee. www.irac-online.org.

National Pesticide Information Center (NPIC). http://npic.orst.edu.

The Safe and Effective Use of Pesticides. 2nd ed. 2000. P. J. O'Connor-Marer. Univ. Calif. Div. Agric. Nat. Res. Publ. 3324, Oakland.

U.S. EPA Pesticide Information. www.epa.gov/pesticides.

Arthropods

An Introduction to the Study of Insects. 2004. N. Johnson and C. A. Triplehorn. Brooks/Cole, Belmont, CA.

California Insects. 1979. J. A. Powell and C. L. Hogue. University of California Press, Berkeley.

Common Names of Insects and Related Organisms. Continuously updated. Committee on Common Names of Insects, Entomological Society of America. www.entsoc.org/common-names.

Entomology and Pest Management. 6th ed. 2008. L. P. Pedigo. Prentice-Hall, Upper Saddle River, NJ.

How to Know the Insects. 3rd ed. 1978. R. Bland and H. Jaques. Brown, Dubuque, IA.

The Insects: An Outline of Entomology. 4th ed. 2010. P Gullan and P. Cranston. Wiley-Blackwell, Chichester, West Sussex, UK.

Plant Pathogens

American Phytopathological Society. www.apsnet.org.

Plant Pathology. 5th ed. 2005. G. N. Agrios. Academic Press, San Diego.

The Disease Compendium Series. Ongoing. The American Phytopathological Society. APS Press, St. Paul.

Vertebrate Pests

A Field Guide to Western Birds. 3rd ed. 1990. R. T. Person. Houghton Mifflin, Boston.

Prevention and Control of Wildlife Damage. Vol. 1 and 2. S. E. Hygnstrom, R. M. Timm, and G. E. Larson, eds. 1994. University of Nebraska, Lincoln. http://icwdm.org/handbook/index.asp.

Wildlife Pest Control Around Gardens and Homes. 2006. T. P. Salmon, D. A. Whisson, and R. E. Marsh. Univ. Calif. Div. Agric. Nat. Res. Publ. 21385, Oakland.

Weeds

The Jepson Manual: Vascular Plants of California. 2d ed. 2012. B. G. Baldwin, et al., eds. University of California Press, Berkeley.

Principles of Weed Control in California. 3rd ed. 2002. California Weed Conference. Fresno: Thomson Publications.

University of California Weed Research and Information Center. http://wric.ucdavis.edu.

Weed Ecology Implications for Management. 2nd ed. 1997. S. Radosevich, J. Holt, and C. Ghersa.Wiley, New York.

Weeds of California and Other Western States. 2007. J. M. Ditomaso and E. A. Healy. Univ. Calif. Agric. Nat. Res, Publ. 3350, Oakland.

Weeds of the West. 6th ed. 2002. T. D. Whitson, L. C. Burrill, S. A. Dewey, D. W. Cudney, B. E. Nelson, R. D. Lee, and R. Parker, eds. Western Society of Weed Science, Newark, CA.

IPM, General

Concepts in Integrated Pest Management. 2003. R. F. Norris, E. P. Caswell-Chen, and M. Kogan. Prentice Hall, Upper Saddle River, N.J.

Ecologically Based Pest Management: New Solutions for a New Century. 1995. National Academy Press, Washington, D.C.

Integrated Pest Management: Concepts, Tactics, Strategies and Case Studies. 2009. E. Radcliffe, W. D. Hutchison, and R. E. Cancelado, eds. Cambridge University Press, Cambridge, UK.

National Integrated Pest Management Network, North Carolina Component. http://cipm.ncsu.edu.

The integrated control concept. 1959. V. M. Stern, R. F. Smith, R. van denBosch, and K. S. Hagen. *Hilgardia* 29(2): 81-101.

University of California Statewide IPM Project. www.ipm.ucdavis.edu.

Western Integrated Pest Management Center. www.wripmc.org.

IPM for Specific Crops and Situations

Grape Pest Management. 2nd ed. 1992. D. L. Flaherty, L. P. Christensen, W. T. Lanini, J. J. Marois, P. A. Phillips, and L. T. Wilson, eds. Univ. Calif. Div. Agric. Nat. Res. Publ. 3343, Oakland.

Integrated Pest Management for Alfalfa Hay. 1985. M. L. Flint. Univ. Calif. Div. Agric. Nat. Res. Publ. 3312, Oakland.

Integrated Pest Management for Almonds. 2002. L.L. Strand and B. L. P. Ohlendorf. Univ. Calif. Div. Agric. Nat. Res. Publ. 3308, Oakland.

Integrated Pest Management for Apples and Pears. 2nd ed. 1999. B. L. P. Ohlendorf. Univ. Calif. Div. Agric. Nat. Res. Publ. 3340, Oakland.

Integrated Pest Management for Avocados. 2008. S.H. Dreistadt. Univ. Calif, Div. Agric. Nat, Res. Publ. 3503. Oakland.

Integrated Pest Management for Citrus. 3rd ed. 2012. S. H. Dreistadt. Univ. Calif. Div. Agric. Nat. Res. Publ. 3303, Oakland.

Integrated Pest Management for Cole Crops and Lettuce. 1991. M. L. Flint. Univ. Calif. Div. Agric. Nat. Res. Publ. 3307, Oakland.

Integrated Pest Management for Cotton. 2nd ed. 1996. B. P. Ohlendorf, P. A. Rude, and M. L. Flint. Univ. Calif. Div. Agric. Nat. Res. Publ. 3305, Oakland.

Integrated Pest Management for Floriculture and Nurseries. 2001. S. H. Dreistadt. Univ. Calif. Div. Agric. Nat. Res. Publ. 3462, Oakland.

Integrated Pest Management for Onions. 1996. M. P. Hoffmann, C. H. Petzoldt, and A. C. Frodsham. Cornell University Cooperative Extension, Ithaca, NY.

Integrated Pest Management for Potatoes. 1986. L. L. Strand. Univ. Calif. Div. Agric. Nat. Res. Publ. 3316, Oakland.

Integrated Pest Management for Rice. 3rd ed. 2012. L. L. Strand. Univ. Calif. Div. Agric. Nat. Res. Publ. 3280, Oakland.

Integrated Pest Management for Small Grains. 1990. L. L. Strand. Univ. Calif. Div. Agric. Nat. Res. Publ. 3333, Oakland.

Integrated Pest Management for Stone Fruits. 1999. L. L. Strand and M. L. Flint. Univ. Calif. Div. Agric. Nat. Res. Publ. 3389, Oakland.

Integrated Pest Management for Strawberries. 2009. L. L. Strand. Univ. Calif. Div. Agric. Nat. Res. Publ. 3351, Oakland.

Integrated Pest Management for Tomatoes. 4th ed. 1998. P. A. Rude and L. L. Strand. Univ. Calif. Div. Agric. Nat. Res. Publ. 3274, Oakland.

Integrated Pest Management for Walnuts. 3rd. 2003. L.L. Strand. Univ. Calif. Div. Agric. Nat. Res. Publ. 3270. Oakland.

Pests of the Garden and Small Farm: A Grower's Guide to Using Less Pesticide. 2nd ed. 1999. M. L. Flint. Univ. Calif. Div. Agric. Nat. Res. Publ. 3332, Oakland.

Pests of Landscape Trees and Shrubs: An Integrated Pest Management Guide. 2001. S. H. Dreistadt. Univ. Calif. Div. Agric. Nat. Res. Publ. 3359, Oakland.

UC IPM Pest Management Guidelines. Revised continuously. www.ipm.ucdavis.edu/PMG/crops-agriculture.html.

UC IPM Pest Notes for Home, Garden and Landscape. Revised continuously. www.ipm.ucdavis.edu/homegarden.

Urban Pest Management of Ants in California. 2010. J. Klotz, L. Hansen, H. Field, M. Rust, D. Oi, and K. Kufer. Univ. Calif. Div. Agric. Nat. Res Publ. 3524, Oakland.

Statistics and Sampling

Agricultural Experimentation. T. M. Little and F. J. Hills. 1978. Wiley, New York.

Ecological Methods. T. R. E. Southwood and P. A. Henderson. 2000. Blackwell Science, London.

Introduction to Biostatistics. R. R. Sokal and F. J. Rohlf. 1973. W. H. Freeman, San Francisco.

Sampling: Design and Analysis. S. L. Lohr. 2010. Duxbery Press, Pacific Grove, CA.

Sampling Techniques. W. G. Cochran. 1977. Wiley, New York.

Statistical Procedures for Agricultural Research. 2nd ed. K. A. Gomez and A. A. Gomez. Wiley, New York.

Laws and Regulations

California Environmental Protection Agency (Cal-EPA). www.calepa.ca.gov.

Cal-EPA/Department of Pesticide Regulation (DPR). www.cdpr.ca.gov.

Cal-EPA/DPR Endangered Species Project. www.CDPR.CA.Gov/docs/es/index.htm.

California Food and Agricultural Code. www.cdpr.ca.gov/docs/inhouse/calcode/3ccrovr.htm.

Laws and Regulations Study Guide. 2011. California Department of Pesticide Regulation, Sacramento.

U.S. Environmental Protection Agency. www.epa.gov.

COOPERATIVE EXTENSION SERVICE

Land grant universities in each state are partners with the USDA Cooperative State Research, Education, and Extension Service. Local Cooperative Extension offices provide information about pest management, and they also have specialists who can answer your questions or refer you to someone who can. Additionally, most Cooperative Extension offices produce newsletters and sponsor workshops, seminars, and other educational events.

Contact your local county Extension office (in the Local Government section of a telephone directory or via an online search), a land grant university, or the Cooperative State Research, Education, and Extension Service, U.S. Department of Agriculture, Washington, D.C. 20250-0900, http://npic.orst.edu/pest/countyext.htm.

Listings for Cooperative Extension offices in California can also be found on the University of California Agriculture and Natural Resources (ANR) website, http://ucanr.org/.

COUNTY DEPARTMENTS OF AGRICULTURE

Most counties have a local department of agriculture that can give you information about pesticide regulations and help with the identification of pests. In California, county agricultural commissioners are also involved in local biological control programs. California county agricultural commissioners' addresses can be found at the CDFA website, www.cdfa.ca.gov/exec/county/county_contacts.html.

PROFESSIONAL ORGANIZATIONS

Many professional organizations regularly publish journals or newsletters that provide current information on pest management. Some important organizations are listed below.

American Phytopathological Society, www.apsnet.org, www.scisoc.org.

Association of Applied Insect Ecologists, www.aaie.com.

California Agricultural Production Consultants Association, www.capca.com.

California Weed Science Society, www.cwss.org.

Certified Crop Adviser, American Society of Agronomy, www.agronomy.org.

Entomological Society of America, www.entsoc.org.

National Alliance of Independent Crop Consultants, www.naicc.org.

Society of Nematologists, http://ianrwww.unl.edu/ianr/plntpath/nematode/son/sonhome.htm.

Weed Science Society of America, www.wssa.net.

References

Agrios, G. N. 2005. *Plant Pathology.* 5th ed. San Diego: Academic Press.

Ahrens, W. H., ed. 1994. *Herbicide Handbook.* 7th ed. Champaign, IL: Weed Science Society of America.

Ali, A. D., and C. L. Elmore. 1989. *Turfgrass Pests.* Oakland: Univ. Calif. Div. Agric. Nat. Res. Publ. 4053.

Anderson, W. P. 1996. *Weed Science: Principles and Applications.* 3rd ed. Minneapolis-St. Paul: West.

Ashton, F. M., and W. A. Harvey. 1987. *Selective Chemical Weed Control.* Oakland: Univ. Calif. Div. Agric. Nat. Res. Bulletin 1919.

Bartkowiak, D., M. Pepple, J. Troiano, D. Weaver, and SRWRCB staff. 1997. *Sampling for Pesticide Residues in California Well Water.* Sacramento: California Environmental Protection Agency, Environmental Hazards Assessment Program, Publication EH98-04.

Bedding, R., R. Akhurst, and H. K. Kaya, eds. 1993. *Nematodes and the Biological Control of Insect Pests.* East Melbourne: CSIRO Australia.

Beers, E. H., J. F. Brunner, M. J. Willett, and G. M. Warner, eds. 1993. *Orchard Pest Management: A Resource Book for the Pacific Northwest.* Yakima, WA: Good Fruit Grower.

Benbrook, C. M., E. Groth III, J. M. Halloran, M. K. Hansen, and S. Marquardt. 1996. *Pest Management at the Crossroads.* Yonkers, NY: Consumer Union.

Bohmont, B. L. 1983. *The New Pesticide User's Guide.* Fort Collins, CO: Reston.

Borner, H., ed. 1994. *Pesticides in Ground and Surface Water.* New York: Springer-Verlag.

Borror, D. J., and R. E. White. 1970. *A Field Guide to Insects of America North of Mexico.* Boston: Houghton Mifflin.

Borror, D. J., C. A. Triplehorn, and N. F. Johnson. 1989. *An Introduction to the Study of Insects.* 6th ed. Philadelphia: Saunders College.

Brandenburg, R. L., and M. G. Villani, eds. 1995. *Handbook of Turfgrass Insect Pests.* Lanham, MD: Entomological Society of America.

Briggs, S. A. 1992. *Basic Guide to Pesticides: Their Characteristics and Hazards.* Washington: Taylor and Francis.

Burn, A. J., T. H. Coaker, and P. C. Jepson, eds. 1987. *Integrated Pest Management.* San Diego: Academic Press.

California Department of Pesticide Regulation. 2010. Sampling for Pesticide Residues in California Well Water. 2009 Update of the Well Inventory Database. Sacramento: Calif. Dep. of Pesticide Regulation. www.cdpr.ca.gov/docs/emon/grndwtr/wellinv/wirmain.htm.

———. 2011. Endangered Species Project. www.cdpr.ca.gov/docs/endspec.

———. 2011. *Laws and Regulations Study Guide.* 2nd ed. Sacramento: Calif. Dep. of Pesticide Regulation. www.cdpr.ca.gov/docs/license/liccert.htm.

California Fertilizer Association, Soil Improvement Committee. *Western Fertilizer Handbook: Horticulture Edition.* 1990. Danville, IL: Interstate Publishers.

California Weed Conference. 1985. *Principles of Weed Control in California.* 2nd ed. Fresno: Thomson Publications.

Chittenden, F. H. 1921. *The Beet Leaf-Beetle and Its Control.* Washington, D.C.: U.S. Dept. Agric. Farmer's Bull. 1193.

Clark. J. P. 1996. *Vertebrate Pest Control Handbook.* Sacramento: California Dept. of Food and Agric., Div. of Plant Industry.

Cohen, S., M. L. Flint, and N. Hines. 2010. *Lawn and Residential Landscape Pest Control: A Guide for Maintenance Gardeners.* Pesticide Application Compendium 8. Univ. Calif. Statewide IPM Program. Oakland: ANR Publication 3510.

Cohen, S., T. A. Martin, and M. L. Flint. 2009. *Field Fumigation.* Pesticide Application Compendium 9. Univ. Calif. Statewide IPM Program. Oakland: ANR Publication 9005.

Committee on Common Names of Insects. 1997. *Common Names of Insects and Related Organisms.* College Park, MD: Entomological Society of America. www.entsoc.org/common-names.

Cooksey, D., J. Jacobsen, D. Baldridge, A. Baquet, J. Baringer, J. Bauder, H. Bowman, D. Inglis, G. Jensen, M. Juhnke, J. Nelson, and J. Riesselman. 1988. *Integrated Crop and Pest Management Record Keeping System.* Bozeman: ICPM Montana State University EB 12, 13.

Craigmill, A. L. 1982. *Toxicology, The Science of Poisons.* Oakland: Univ. Calif. Div. Agric. Nat. Res. Publ. 21221.

Davidson, N. A., J. E. Dibble, M. L. Flint, P. J. Marer, and A. Guye. 1992. *Managing Insects and Mites with Spray Oils.* Oakland: Univ. Calif. Div. Agric. Nat. Res. Publ. 3347.

Davidson, R. H., and W. F. Lyon. 1979. *Insect Pests of Farm, Garden, and Orchard.* 7th ed. New York: Wiley.

Davis, D. W., S. C. Hoyt, J. A. McMurtry, and M. T. AliNiazee. 1979. *Biological Control and Insect Pest Management.* Oakland: Univ. Calif. Div. Agric. Nat. Res. Publ. 1911.

Debach, P., and D. Rosen. 1991. *Biological Control by Natural Enemies.* 2nd ed. New York: Cambridge University Press.

Dent, D. 1991. *Insect Pest Management.* Wallingford, Oxford, U.K.: C.A.B. International.

Ditomaso, J., and E. A. Healy. 2007. *Weeds of California and Other Western States.* Oakland: Univ. Calif. Agric. Nat. Res, Publ. 3350.

Dreistadt, S. H. 2001. *Pests of Landscape Trees and Shrubs: An Integrated Pest Management Guide.* Oakland: Univ. Calif. Div. Agric. Nat. Res. Publ. 3359.

Dreistadt, S. H., and M. L. Flint. 1996. Melon aphid control by inundative convergent lady beetle release on chrysanthemum. *Environmental Entomology* 25:688–697.

Dreistadt, S. H., J. P. Newman, and K. L. Robb. 1998. *Sticky Trap Monitoring of Insect Pests.* Oakland: Univ. Calif. Div. Agric. Nat. Res. Publ. 21572.

Dubrovsky, N. M., C. R. Kratzer, L. R. Brown, J. M. Gronberg, and K. R. Burow. 1998. *Water Quality in the San Joaquin–Tulare Basins, California, 1992–95.* U.S. Geological Survey Circular 11595.

Duncan, J. M., and L. Torrance, eds. 1992. *Techniques for the Rapid Detection of Plant Pathogens.* Boston: Blackwell.

Ebeling, W. 1975. *Urban Entomology.* Berkeley: University of California Press.

Edwards, C. A. 1973. *Persistent Pesticides in the Environment.* 2nd ed. Boca Raton, FL: CRC Press.

Elmore, C. L., J. J. Stapleton, C. E. Bell, and J. E. DeVay. 1997. *Soil Solarization: A Nonpesticidal Method for Controlling Diseases, Nematodes, and Weeds.* Oakland: Univ. Calif. Div. Agric. Nat. Res. Publ. 21377.

Feenstra, G., C. Ingels, and D. Campbell. 1997. *What is Sustainable Agriculture?* University of California Sustainable Agriculture Research and Education Program. www.sarep.ucdavis.edu.

Flint, M. L. 1985. *Integrated Pest Management for Alfalfa Hay.* Oakland: Univ. Calif. Div. Agric. Nat. Res. Publ. 3312.

———. 1987. *Integrated Pest Management for Cole Crops and Lettuce.* Oakland: Univ. Calif. Div. Agric. Nat. Res. Publ. 3307.

———. 1990. *Pests of the Garden and Small Farm.* Oakland: Univ. Calif. Div. Agric. Nat. Res. Publ. 3332.

Flint, M. L., and S. H. Dreistadt. 1998. *Natural Enemies Handbook: The Illustrated Guide to Biological Pest Control.* Oakland: Univ. Calif. Div. Agric. Nat. Res. Publ. 3386; Berkeley: University of California Press.

Flint, M. L., and B. Kobbe. 1993. *Integrated Pest Management for Walnuts.* Oakland: Univ. Calif. Div. Agric. Nat. Res. Publ. 3270.

Flint, M. L., and P. A. Roberts. 1988. Using crop diversity to manage pest problems: Some California examples. *American Journal of Alternative Agriculture* 3(4): 163–167.

Flint, M. L., and R. van den Bosch. 1981. *Introduction to Integrated Pest Management.* New York: Plenum Press.

Flint, M. L., S. Daar, and R. Molinar. 2003. *Establishing Integrated Pest Management Policies and Programs: A Guide for Public Agencies.* Oakland: Univ. Calif. Div. Agric. Nat. Res. Publ. 12.

Flint, M. L., P. B. Goodell, and L. Godfrey. 1995. *Whiteflies in California: A Resource for Cooperative Extension.* Oakland: Univ. Calif. Div. Agric. Nat. Res. Publ. 19.

Flint, M. L., J. M. Lyons, J. P. Madden, M. N. Schroth, A. R. Weinhold, J. L. White, and F. G. Zalom. 1992. *Beyond Pesticides: Biological Approaches to Pest Management in California: Executive Summary.* Oakland: Univ. Calif. Div. Agric. Nat. Res. Publ. 21512.

Forster, L. D., R. F. Luck, and E. E. Grafton-Cardwell. 1995. *Life Stages of California Red Scale and Its Parasitoids.* Oakland: Univ. Calif. Div. Agric. Nat. Res. Publ. 21529.

Frank, W. 1943. The entomological control of St. Johns wort (*Hypericum perforatum* L.) with particular reference to the insect enemies of the weed in southern France. *Australian Council of Science and Industrial Research Bulletin* 169:1–88.

Fryer, T. D., and S. A. Evans, eds. 1968. *Weed Control Handbook.* Vol. 1. *Principles.* Oxford, England: Blackwell.

Gan, J., D. Haver, and L. Oki. 2010. Water quality—Contaminants in runoff from urban landscapes. *UCIPM Green Bulletin* 1(1):3. www.ipm.ucdavis.edu/greenbulletin.

Georghiou, G. P., and T. Saito. 1983. *Pest Resistance to Pesticides*. New York: Plenum Press.

Georghiou, G. P., and C. E. Taylor. 1976. Pesticide resistance as an evolutionary phenomenon. In *Proc. 15th Int. Congr. Entomol*. College Park, MD.: Entomological Society of America. 159–785.

———. 1977. Operational influences in the evolution of insecticide resistance. *Journal of Economic Entomology* 70(5): 653–658.

Gomez, K. A., and A. A. Gomez. 1984. *Statistical Procedures for Agricultural Research*. 2nd ed. New York: Wiley.

Goodell, P. B., R. E. Plant, T. A. Kerby, J. F. Strand, L. T. Wilson, L. Zelinkski, J. A. Young, A. Corbett, R. D. Horrocks, and R. N. Vargas. 1990. CALEX/Cotton: An integrated expert system for cotton production and management. *California Agriculture* 44(5): 18–20.

Grafton-Caldwell, B., L. Godfrey, and P. Goodell. 1996. Maximizing natural enemies in San Joaquin Valley cotton. *California Cotton Review* 40:7–9.

Grafton-Cardwell, E. E., T. J. Dennehy, J. Granett, and S. M. Normington. 1987. *Managing Dicofol and Propargite Resistance in Spider Mites Infesting San Joaquin Valley Cotton*. Oakland: Univ. Calif. Div. Agric. Nat. Res. Publ. 5.

Hake, S. J., T. A. Kerby, and K. D. Hake. 1996. *Cotton Production Manual*. Oakland: Univ. Calif. Div. Agric. Nat. Res. Publ. 3352.

Haney, P. B., J. G. Morse, R. F. Luck, H. Griffiths, E. E. Grafton-Cardwell, and N. V. O'Connell. 1992. *Reducing Insecticide Use and Energy Costs in Citrus Pest Management*. Oakland: Univ. Calif. Div. Agric. Nat. Res. Publ. 15.

Harley, K. L. S., and I. W. Forno. 1992. *Biological Control of Weeds: A Handbook for Practitioners and Students*. Melbourne, Australia: Inkata Press.

Harrington, H. D., and L. W. Durrell. 1957. *How to Identify Plants*. Denver: Sage Press.

Hartz, T., R. Smith, K. F. Schulbach, and M. LeStrange. 1994. On-farm nitrogen tests improve fertilizer efficiency, protect ground water. *California Agriculture* 48(4): 29–32.

Hickman, J. C., ed. 1993. *The Jepson Manual: Higher Plants of California*. 1993. Berkeley: University of California Press.

Higley, L. G., and L. P. Pedigo, eds. 1996. *Economic Thresholds for Integrated Pest Management*. Lincoln: University of Nebraska Press.

Hoffmann, M. P., and A. C. Frodsham. 1993. *Natural Enemies of Vegetable Insect Pests*. Ithaca: Cornell University Press.

Hokkanen, H. M. T., and J. M. Lynch. 1995. *Biological Control, Benefits and Risks*. Cambridge: Cambridge University Press.

Holloway, J. K. 1957. Weed control by insect. *Scientific American* 197(1): 56–62.

Holt, J. S., and S. R. Radosevich. 1989. Plants. In *Principles of Weed Control in California*. 2nd ed. Fresno: Thompson Publications.

Horn, D. J. 1988. *Ecological Approach to Pest Management*. New York: Guilford Press.

HRAC. Herbicide Resistance Action Committee website. www.hracglobal.com/.

Huffaker, C. B., and P. S. Messenger. 1976. *Theory and Practice of Biological Control*. San Diego: Academic Press.

Hygnstrom, S. E., R. M. Timm, and G. E. Larson, eds. 1994. *Prevention and Control of Wildlife Damage*. Vols. 1 and 2. Lincoln: University of Nebraska Cooperative Extension.

Ingels, C. A., R. L. Bugg, G. T. McGourty, and L. P. Christensen, eds. 1998. *Cover Cropping in Vineyards*. Oakland: Univ. Calif. Div. Agric. Nat. Res. Publ. 3338.

IRAC. Insecticide Resistance Action Committee website. www.irac-online.org/.

Jarvis, W. R. 1992. *Managing Diseases in Greenhouse Crops*. St. Paul, MN: APS Press.

Johnson, W. T., and H. H. Lyon. 1976. *Insects That Feed on Trees and Shrubs: An Illustrated Practical Guide*. New York: Comstock.

Johnston, A., and C. Booth, eds. 1983. *Plant Pathologist's Pocketbook*. 2nd ed. Slough, Eng.: Commonwealth Mycological Institute.

Kabashima, J. N., J. D. MacDonald, S. H. Dreistadt, and D. E. Ullman. 1997. *Easy On-Site Tests for Fungi and Viruses in Nurseries and Greenhouses*. Oakland: Univ. Calif. Div. Agric. Nat. Res. Publ. 8002. http://anrcatalog.ucdavis.edu.

Kamrin, M. A., ed. 1997. *Pesticide Profiles, Toxicity, Environmental Impact, and Fate*. Boca Raton, FL: CRC/Lewis Publishers.

Kaya, H. K. 1993. *Contemporary Issues in Biological Control with Entomopathogenic Nematodes*. Taipei City, Taiwan: Food and Fertilizer Technology Center Extension Bulletin 375.

Kempen, H. M. 1989. *Growers Weed Management Guide*. Fresno: Thomson Publications.

Klotz, J., L. Hanse, H. Field, M. Rust, D. Oi, and K. Kufer. 2010. *Urban Pest Management of Ants in California*. Oakland: Univ. Calif. Agric. Nat. Res. Publ. 3524.

Kobbe, B., and S. H. Dreistadt. 1991. *Integrated Pest Management for Citrus*. Oakland: Univ. Calif. Div. Agric. Nat. Res. Publ. 3303.

Kogan, M. 1998. Integrated pest management: Historical perspectives and contemporary developments. *Annual Review of Entomology* 43:243–70.

Kogan, M., ed. 1986. *Ecological Theory and Integrated Pest Management*. New York: Wiley.

Kogan, M., and D. C. Herzog, eds. 1980. *Sampling Methods in Soybean Entomology*. New York: Springer-Verlag.

Kono, T., and C. S. Papp. 1977. *Handbook of Agricultural Pests: Aphids, Thrips, Mites, Snails and Slugs*. Sacramento: California Department of Food and Agriculture.

LaRue, J. H., and R. S. Johnson. 1989. *Peaches, Plums, and Nectarines: Growing and Handling for Fresh Market*. Oakland: Univ. Calif. Div. Agric. Nat. Res. Publ. 3331.

Leslie, A. R., ed. 1994. *Handbook of Integrated Pest Management for Turf and Ornamentals*. Boca Raton, FL: Lewis Publishers.

Leslie, A. R., and G. W. Cuperus, eds. 1993. *Successful Implementation of Integrated Pest Management for Agricultural Crops*. Boca Raton, FL: Lewis Publishers.

Little, T. M., and F. J. Hills. 1978. *Agricultural Experimentation Design and Analysis*. New York: Wiley.

Long, R., J. Gan, and M. Nett. 2005. *Pesticide Choice: Best Management Practice (BMP) for Protecting Surface Water Quality in Agriculture*. Oakland: Univ. Calif. Agric. Nat. Res. Public 8161.

Lowrance, R., B. R. Stinner, and G. J. House. 1984. *Agricultural Ecosystems, Unifying Concepts*. New York: Wiley.

Lumsden, R. D., and J. L. Vaughn, eds. 1993. *Pest Management: Biologically Based Technologies*. Conference Proceedings. Beltsville, MD: American Chemical Society.

Maggenti, A. R. 1981. *General Nematology*. New York: Springer-Verlag.

Mahr, D. L., and N. M. Ridgway. 1993. *Biological Control of Insects and Mites: An Introduction to Beneficial Natural Enemies and Their Use in Pest Management*. Madison, WI: North Central Regional Coop. Ext. Publ. 481.

Marer, P. J., M. Grimes, and R. Cromwell. 1995. *Forest and Right of Way Pest Control*. Oakland: Univ. Calif. Div. Agric. Nat. Res. Publ. 3336.

McEwen, F. L., and G. R. Stephenson. 1979. *The Use and Significance of Pesticides in the Environment*. New York: Wiley.

McKenry, M. V., and P. A. Roberts. 1985. *Phytonematology Study Guide*. Oakland: Univ. Calif. Div. Agric. Nat. Res. Publ. 4045.

Metcalf, R. L., and W. H. Luckman, eds. 1994. *Introduction to Insect Pest Management*. 3rd ed. New York: Wiley.

Metcalf, R. L., and R. A. Metcalf. 1993. *Destructive and Useful Insects*. 5th ed. New York: McGraw-Hill.

Miller, T. L. 1993. *Movement of Pesticides in the Environment*. EXTOXNET Oregon State University. http://ace.orst.edu/info/extoxnet.

Mills, W. D, and A. A. La Plante. 1954. *Diseases and Insects in the Orchard*. Ithaca, NY: Cornell Ext. Bull. 711.

National Research Council. 1986. *Pesticide Resistance Strategies and Tactics for Management*. National Research Council, Board on Agriculture, Committee on Strategies for the Management of Pesticide Resistant Pest Populations. Washington, D.C.: National Academy Press.

———. 1989. *Alternative Agriculture*. National Research Council, Board on Agriculture, Committee on the Role of Alternative Farming Methods in Modern Production Agriculture. Washington, D.C.: National Academy Press.

Nechols, J. R., L. A. Andres, J. W. Beardsley, R. D. Goeden, and C. G. Jackson, eds. 1995. *Biological Control in the Western United States: Accomplishments and Benefits of Regional Project W-84, 1964-1989*. Oakland: Univ. Calif. Div. Agric. Nat. Res. Publ. 3361.

Niederholzer, F. 2009. *Spray Adjuvants: What's in a Name? UC Cooperative Extension Topics in Subtropics* 7(2). cekern.ucdavis.edu/newsletterfiles/Topics_in_Subtropics17746.pdf.

Norris, R. F., E. P. Caswell-Chen, and M. Kogan. 2003. *Concepts in Integrated Pest Management*. Upper Saddle River, NJ: Prentice-Hall.

O'Connor-Marer, P. J. 1999. *The Illustrated Guide to Pesticide Safety, Instructor's Edition*. Oakland: Univ. Calif. Div. Agric. Nat. Res. Publ. 3347.

———. 2000. *The Safe and Effective Use of Pesticides*. 6th ed. Oakland: Univ. Calif. Div. Agric. Nat. Res. Publ. 3324.

Odum, E. P. 1971. *Fundamentals of Ecology*. 3d ed. Philadelphia: Saunders.

Ogawa, J. M., and H. English. 1991. *Diseases of Temperate Zone Tree Fruit and Nut Crops*. Oakland: Univ. Calif. Div. Agric. Nat. Res. Publ. 3345.

Ohlendorf, B. P. 1985. *Integrated Pest Management for Almonds*. Oakland: Univ. Calif. Div. Agric. Nat. Res. Publ. 3308.

———. 1999. *Integrated Pest Management for Apples and Pears*. 2nd ed. Oakland: Univ. Calif. Div. Agric. Nat. Res. Publ. 3340.

Ohlendorf, B. P., P. A. Rude, and M. L. Flint. 1996. *Integrated Pest Management for Cotton in the Western Region of the United States*. Oakland: Univ. Calif. Div. Agric. Nat. Res. Publ. 3305.

Olkowski, W., S. Daar, and H. Olkowski. 1991. *Common Sense Pest Control*. Newtown: Taunton Press.

Orloff, S. B., and H. L. Carlson. 1997. *Intermountain Alfalfa Management*. Oakland: Univ. Calif. Div. Agric. Nat. Res. Publ. 3366.

Palti, J., and R. Ausher, eds. 1986. *Advisory Work in Crop Pest and Disease Management*. New York: Springer-Verlag.

Pedigo, L. P. 2010. *Entomology and Pest Management*. 6th ed. Upper Saddle River, NJ: Prentice Hall.

Pedigo, L. P., and G. D. Buntin, eds. 1994. *Handbook of Sampling Methods for Arthropods in Agriculture*. Boca Raton, FL: CRC Press.

Person, R. T. 1990. *A Field Guide to Western Birds*. 3rd ed. Boston: Houghton Mifflin.

Pimentel, D. 1991. *CRC Handbook of Pest Management in Agriculture*. Vol. 2. Boca Raton, FL: CRC Press.

Powell, J. A., and C. L. Hogue. 1979. *California Insects*. Berkeley: University of California Press.

Prather, T. S. 1996. *Herbicide Resistance: The Problem and Management Strategies*. Paper presented at Resistance Management Workshop, UC Extension, Davis, November, 1996.

Radosevich, S., J. Holt, and C. Ghersa. 1997. *Weed Ecology Implications for Management*. 2nd ed. New York: Wiley.

Ramos, D. E., ed. 1981. *Prune Orchard Management*. Oakland: Univ. Calif. Div. Agric. Nat. Res. Publ. 3269.

Ratcliffe, E. B., W. D. Hutchison, and R. E. Cancelado. 2009. *Integrated Pest Management: Concepts, Tactics, Strategies and Case Studies*. UK: Cambridge University Press.

Raupp, M. J., R. G. Van Driesche, and J. A. Davidson. 1993. *Biological Control of Insect and Mite Pests of Woody Landscape Plants*. College Park: University of Maryland Press.

Reuveni, R. ed. 1995. *Novel Approaches to Integrated Pest Management*. Boca Raton, FL: Lewis Publishers.

Riedl, H., E. Johansen, L. Brewer, and J. Barbou. 2006. *How to Reduce Bee Poisoning from Pesticides*. Pacific Northwest Extension Publication 591.

Ross, M. A., and C. A. Lembi. 1999. *Applied Weed Science*. 2nd ed. Upper Saddle River, NJ: Prentice Hall.

Rude, P. A., and L. L. Strand. 1998. *Integrated Pest Management for Tomatoes*. Oakland: Univ. Calif. Div. Agric. Nat. Res. Publ. 3274.

Salmon, T. P., D. A. Whisson, and R. E. Marsh. 2006. *Wildlife Pest Control Around Gardens and Homes*. 2nd ed. Oakland: Univ. Calif. Div. Agric. Nat. Res. Publ. 21385.

Schulze, L., R. Grisso, and R. Stougaard. February, 1996. *Spray Drift of Pesticides*. Lincoln: Univ. Nebr. Coop. Ext., Inst. of Agric. and Nat. Res., NebGuide G90-1001-A. http://ianrwww.unl.edu/pubs/pesticides/g1001.htm.

Sokal, R. R., and F. J. Rohlf. 1987. *Introduction to Biostatistics*. 2nd ed. New York: Freeman.

Southwood, T. R. E. 1989. *Ecological Methods with Particular Reference to the Study of Insect Populations*. 2d ed. London: Chapman and Hall; New York: Wiley.

Stern, V. M., R. F. Smith, R. van den Bosch, and K. S. Hagen. 1959. The integrated control concept. *Hilgardia* 29(2): 81–101.

Strand, L. L. 1990. *Integrated Pest Management for Small Grains*. Oakland: Univ. Calif. Div. Agric. Nat. Res. Publ. 3333.

———. 1999. *Integrated Pest Management for Stone Fruits*. Oakland: Univ. Calif. Div. Agric. Nat. Res. Publ. 3389.

———. 2002. *Integrated Pest Management for Almonds*. Oakland: Univ. Calif. Div. Agric. Publ. 3308.

———. 2006. *Integrated Pest Management for Potatoes*. Oakland: Univ. Calif. Div. Agric. Nat. Res. Publ. 3316.

———. 2008. *Integrated Pest Management for Strawberries*. Oakland: Univ. Calif. Div. Agric. Nat. Res. Publ. 3351.

———. 2012. *Integrated Pest Management for Rice*. 3rd ed. Oakland: Univ. Calif. Div. Agric. Nat. Res. Publ. 3280.

Summers, C. G. 1976. Population fluctuations of selected arthropods in alfalfa: Influence of two harvesting practices. *Environmental Entomology* 5:103–110.

Tardiff, R. G. ed. 1992. *Methods to Assess Adverse Effects of Pesticides on Non-target Organisms*. New York: Wiley.

Tattar, T. A. 1978. *Diseases of Shade Trees*. New York: Academic Press.

Taylor, L. R. 1961. Aggregation, Variance and the Mean. *Nature* 189:732–735.

Thomsen, C. D., W. A. Williams, M. P. Vayssieres, C. E. Turner, and W. T. Lanini. 1996. *Yellow Starthistle: Biology and Control*. Oakland: Univ. Calif. Div. Agric. Nat. Res. Publ. 21541.

Tivy, J. 1990. *Agricultural Ecology*. New York: Wiley.

Turner, C. E., L. W. Anderson, P. Foley, R. D. Goeden, W. T. Lanini, S. E. Lindow, and C. O. Qualset. 1992. Biological approaches to weed management. In *Beyond Pesticides: Biological Approaches to Pest Management in California*. Oakland: Univ. Calif. Div. Agric. Nat. Res. Publ. 3354. 36–67.

University of California Agricultural Extension Service. 1998. *The Grower's Weed Identification Handbook*. Oakland: Univ. Calif. Div. Agric. Nat. Res. Publ. 4030.

U.S. Congress. 1990. *Beneath the Bottom Line: Agricultural Approaches to Reduce Agrichemical Contamination of Groundwater*. Office of Technology Assessment, OTA-F-418. Washington, D.C.: U.S. Government Printing Office.

———. 1995. *Biologically Based Technologies for Pest Control*. Washington, D.C.: Office of Technology Assessment.

van den Bosch, R., P. S. Messenger, and A. P. Gutierrez. 1982. *An Introduction to Biological Control*. New York: Plenum.

Vanderplank, J. E. 1984. *Disease Resistance in Plants*. San Diego: Academic Press.

Van Driesche, R. G., and T. S. Bellows. 1996. *Biological Control*. New York: Chapman and Hall.

Ware, G. W. 1983. *Pesticides: Theory and Application*. San Francisco: W. H. Freeman.

Weed Science Society of America, Herbicide Handbook Committee. 2007. *Herbicide Handbook*. 9th ed. Champaign, Ill.: Weed Science Society of America.

Welch, S. M., and B. A. Croft. 1979. *The Design of Biological Monitoring Systems for Pest Management*. New York: Wiley.

Whitson, T. D., L. C. Burrill. S. A. Dewey, D. W. Cudney, B. E. Nelson, R. D. Lee, and R. Parker, eds. 1992. *Weeds of the West*. 5th ed. Newark: The Western Society of Weed Science.

Wright, R. J., J. F. Witkowski, and L. D. Schulze. 1996. *Best Management Practices for Agricultural Pesticides to Protect Water Resources*. Lincoln: Univ. Nebr. Coop. Ext., Inst. of Agric. and Nat. Res, NebGuide G95-1260A. www.ianr.unl.edu/PUBS/water/g1182.htm.

Zalom, F. G., and W. E. Fry, eds. 1992. *Food, Crop Pests and the Environment*. St. Paul, Minn.: APS Press.

Zalom, F. G., P. B. Goodell, L. T. Wilson, W. W. Barnett, and W. J. Bentley. 1983. *Degree-Days: The Calculation and Use of Heat Units in Pest Management*. Oakland: Univ. Calif. Div. Agric. Nat. Res. Leaflet 21373.

Glossary

Note: Glossary terms are italicized at first mention in the text.

abiotic. Nonliving factors such as wind, water, temperature, and minerals.

abiotic disorders. Noninfectious diseases induced by adverse environmental conditions, often as a result of human activity.

absolute samples. Samples that count every individual in a population in a given area.

acaricides. Pesticides used to control mites.

action threshold. A population level at which some control action must be taken in order to avoid economic or aesthetic injury. Synonymous with *treatment threshold*.

adjuvants. Materials added to a pesticide formulation to enhance its performance, customize the application to site-specific needs, or to compensate for local conditions.

adsorb. To take up and hold on a surface, such as pesticides that become attached to soil particles.

aesthetic injury level. The level of pest damage or pest populations the general public will tolerate on an ornamental plant.

aestivation. Dormancy during summer or periods of high temperatures or drought.

age distribution. The proportion of individuals in a population within each age group.

agricultural pesticides. Any substance or mixture intended for preventing, destroying, repelling, killing, or mitigating problems caused by insects, rodents, weeds, nematodes, fungi, or other pests; and any other substance or mixture intended for use as a plant growth regulator, defoliant, or desiccant.

agroecosystem. An ecosystem managed for agricultural purposes.

allelopathy. The release of substances by one plant that inhibit the growth of another plant species.

alternate hosts. Plants that support the survival of a pest when its main host, usually a crop plant, is not available.

annuals. Plants that normally complete their life cycle of seed germination, vegetative growth, reproduction, and death in a single year.

antagonists. Organisms that kill or inhibit the activity or growth of other organisms (especially pests).

antibiosis. The secretion by one organism of substances that inhibit vital activities of other organisms.

antibiotic. A substance produced by a microorganism, such as fungi or bacteria, that is effective in the suppression or destruction of other microorganisms. Used as pesticide as well as medicine.

anticoagulant. A type of rodenticide that causes death by preventing normal blood clotting.

apparent resistance. Also referred to as **disease escape**; involves plants that fail to become infected with a pathogen because all the requirements for infection or disease development (i.e., environmental conditions, host stage and quality, or inoculum level and distribution) are not present.

arthropod. An animal having jointed appendages and an external skeleton, such as an insect, spider, mite, crab, or centipede.

auger. A tool used for boring holes.

augmentation. In biological control, supplementing the numbers of naturally occurring biological control agents with releases of laboratory-reared or field-collected natural enemies.

autotroph. The base of the food chain—producers (green plants).

bactericide. A pesticide that kills bacteria.

bacterium. A microscopic one-celled organism that does not produce chlorophyll and that belongs to the kingdom Procaryotae (plural, *bacteria*).

beneficial organisms. Organisms that provide a benefit to crop production or human existence, applied especially to natural enemies of pests and to pollinators such as bees.

biennial. A plant that completes its life cycle in 2 years and usually does not flower and reproduce until the second season.

bioaccumulation. The gradual buildup of certain pesticides or other toxic substances in the tissues of living organisms through the food chain as larger organisms feed on other organisms containing smaller amounts of these toxins.

bioassay. A laboratory test that compares the effect of a pesticide or other substance on a field-collected organism to a standard laboratory culture with a known reaction. Used for assessing pesticide resistance in field populations of pests.

biodiversity. The number of different species of plants, animals, and microorganisms within an ecosystem.

biofix. An identifiable point in the life cycle of a pest when degree-day accumulation or management action begins.

biogeochemical cycle. The circular movement of inorganic elements and compounds through the ecosystem from the nonliving to the living and back to the nonliving.

biological control. Any activity of one species that reduces the adverse effect of other species.

biosphere. The part of the earth and its atmosphere in which living things exist.

biotic disease. Disease caused by a living pathogen, such as a bacterium, fungus, mycoplasma, or virus.

biotypes. Physiological races of a species that differ in their reaction to certain environmental or biological factors. For instance, some biotypes may be able to survive applications of pesticides that kill other individuals of their species, or they may be able to survive on different hosts.

block. An area that includes one plot or replication of each experimental treatment.

botanicals. Pesticides that are derived from plants or plant parts.

broadleaves. Plants with flat rather than needlelike leaves and two leaves within the seed; dicotyledons.

broad-spectrum insecticide. A pesticide that kills a large number of unrelated species.

calibration. To standardize or correct the measuring devices on instruments; to adjust nozzles on a spray rig properly.

carnivores. Animals that eat other living animals.

cephalothorax. The head and the thorax combined, as in a spider.

certified seed. Seeds, tubers, or young plants certified by a recognized authority to be free of or to contain less than a minimum number of specified pests or pathogens.

characteristic abundance. The population level typical for a species within a community.

chlorophyll. Any of several green substances found in plants in which photosynthesis takes place.

chlorosis. Yellowing or bleaching of normally green plant tissue, usually caused by the loss of chlorophyll.

classical biological control. Also referred to as importation; involves the deliberate introduction and establishment of natural enemies into areas where they did not previously exist.

climate. Meteorological conditions such as temperature, humidity, precipitation, and other atmospheric conditions over a long period of time. Can refer to local, regional, or global conditions.

clumped distribution. A population distribution in which individuals of a species tend to appear close together in colonies in a few or many areas throughout a field rather than being found evenly distributed throughout the field.

cohort. A group of individuals of the same species that are the same age.

coleoptile. The first leaf to emerge in germinating grasses.

common name. The vernacular (English, Spanish, etc.) name used to identify a species. Common names for the same species often differ from region to region or country to country, as contrasted with the universally recognized scientific name, or Latin binomial, specifying each unique organism. For instance, housefly is the common name used in the United States for *Musca domestica*.

community. All the populations of plants, animals, and microorganisms that share the same habitat and environment and interact directly or indirectly with one another.

competition. Competition between organisms for limited supplies of essential resources, such as food in the case of animals, and water, nutrients, and light in the case of plants.

conservation tillage. Any conservation practice that retains at least 30% plant residue cover from the previous crop on the soil surface; includes no-till, ridge-till, strip-till, mulch-till, and other tillage systems that meet this requirement.

contact poison. A pesticide that kills target pests when they come in physical contact with it; does not need to be eaten.

control action guidelines. Guidelines developed to help decide whether and when management actions, including pesticide applications, are needed to avoid eventual loss from pest damage.

cotyledons. The first leaves to emerge in broadleaf plants.

cover crops. Noncrop plant species either grown concurrently with the host crop (usually perennial plants) or planted in rotation with annual crops; they are generally not harvested.

crop rotation. The intentional planting of specific crop sequences to improve crop health.

cultivar. A plant variety or strain developed and grown under cultivation.

cultural controls. The modification of normal crop or landscape management practices to decrease pest establishment, reproduction, dispersal, survival, or damage.

data logger. The storage unit of a computerized weather station consisting of a microprocessor, storage memory, internal battery backup, transceiver, and power supply control unit.

decomposers. Organisms that break down plants and animals into organic substances.

degree-day. The amount of heat that accumulates over a 24-hour period when the average temperature is 1 degree above the lower developmental threshold of an organism.

density-dependent factors. Factors that have a different effect on population growth when populations are high compared with when they are low.

density-independent factors. Factors that affect population growth similarly regardless of population density; includes disturbances such as floods, drought, fire, other unpredictable environmental conditions, and most pest control actions.

developmental threshold. The temperature below which development of an organism stops.

diapause. A period of physiologically controlled dormancy in insects or other organisms.

dichotomous key. A simple identification key based on a series of sequentially paired statements.

dicots. Commonly called *broadleaves*; plants whose seedlings produce two cotyledons.

disease-suppressive soils. Soils in which disease incidence remains low even though a pathogen, a susceptible host, and environmental conditions that favor disease development are present.

disease triangle. The host plant, causal agent, and favorable environment; when all these elements are present at satisfactory levels, a disease outbreak is likely to occur.

dispersal. The movement of individuals or their offspring.

dormancy. A physiological state in which an organism is inactive or not growing.

dormant season. Time of the year when a plant is not actively growing.

drift. The dispersal through the air of a substance, such as a pesticide, beyond the intended application area.

ecological niche. All the components of the habitat with which an organism or population interacts.

ecology. The study of the interrelationships between organisms and their surrounding environment.

economic injury level. The population level of a pest that will cause economic damage to a crop.

economic threshold. The pest population level at which a control action, such as a pesticide spray, must be taken to prevent economic damage; synonymous with *action threshold* or *treatment threshold*.

ecosystem. The community of organisms in an area and their nonliving environment.

ecotone. A transitional zone between communities.

ecotype. A locally adapted population of a species.

ectoparasitic. A parasite that lives on the outside of its host.

edge effect. The tendency for a greater diversity of species to occur where communities merge. In agricultural fields, many pests move in from adjacent crops and may be concentrated on edges of the field.

efficacy. The ability of a pesticide or other control method to produce a desired effect on the target organism.

ELISA (enzyme-linked immunosorbent assay). A simple test that can be used to confirm the presence of specific proteins; used, for example, to detect certain plant pathogens or pesticide-resistant strains of insects and mites.

emigration. The movement of individuals out of a population.

endoparasite. A parasite that lives inside its host.

entomopathogenic nematodes. Nematode species that infect and kill insects.

epidemic. A rapidly spreading outbreak of disease.

equilibrium position. The mean population level around which a species' characteristic abundance fluctuates.

eradicants. Fungicides that can control pathogens after they have invaded the plant tissue by killing the fungus inside the host or by suppressing sporulation of the fungus.

eradication. Total elimination of a pest from a designated area.

estivation. Summer dormancy.

evapotranspiration. The loss of soil moisture due to evaporation from the soil surface and transpiration by plants.

expert systems. Computer programs designed to simulate the problem-solving behavior of a team of well-trained professionals.

facultative parasites. Parasites, especially plant pathogens, that can survive and grow in the absence of their host, often on decaying matter.

fallowing. Leaving a field unplanted for a season or part of a season.

Federal Insecticide, Fungicide, and Rodenticide Act (FIFRA). The federal law that governs all aspects of agricultural pesticide use in the United States.

field trials. In-field comparisons of different treatments.

food chain. A succession of organisms in a community that constitutes a feeding sequence in which food energy is transferred from one organism to the next as each consumes a lower member and in turn is preyed upon by a higher member. At the bottom of the chain is a photosynthesizing plant, usually followed by an herbivore, a succession of carnivores, and finally decomposers.

food web. A complex of interrelated food chains in a community.

fumigant. A pesticide active ingredient that is a gas under treatment conditions.

fungi. A group of multicellular organisms in the kingdoms Fungi or Strameopila that obtain their nutrients from living or dead plant or animal materials. Although some appear plantlike, they do not contain chlorophyll. The fungus body, or mycelium, normally consists of filamentous strands called hyphae, and reproduction occurs through the production of spores.

fungicide. A pesticide used for control of fungi.

gall. An abnormal swelling of plant tissue typified by rapidly dividing cells and disorganized vascular tissues; often caused by insects, nematodes, or pathogens.

gene. The functional hereditary unit that occupies a fixed location on the chromosomes of an organism.

genetic engineering. Intentional alteration of genetic material.

genotype. The genetic constitution of an organism as determined by the set of genes it carries.

genus. A taxonomic category ranking below the family and above the species. The genus name is used along with a species epithet to form the scientific name of a species.

geographic information systems (GIS). A data management program used in precision farming systems to organize, statistically analyze, and display data about a specific field or orchard. GIS use global positioning systems to precisely identify locations in the field and then link together diverse data such as physical factors, cultural practices, yield, pest problems, and weather collected at these locations.

groundwater. Fresh water trapped in aquifers beneath the surface of the soil; one of the primary sources of water for drinking, irrigation, and manufacturing.

habitat. The environment in which an individual organism lives.

habitat modification. Intentionally changing a managed ecosystem to limit availability of one or more of a pest's survival requirements, thus making the environment less suitable for pest population growth.

half-life. The amount of time it takes for a pesticide to be reduced to half of its original toxicity or effectiveness.

herbicide. A pesticide used to control weeds.

herbivores. Animals that feed on plants.

heterotroph. An organism that cannot synthesize its own food; includes all organisms on the food chain above the green plants (the autotrophs).

hibernation. Winter dormancy.

host. A plant or animal that provides sustenance for another organism.

host plant resistance. The ability of a plant cultivar to ward off or resist attack by pests that damage other cultivars of that plant species.

hot spot. An area of a field or orchard where pest problems tend to develop first.

hyphae. Tiny, tubular filaments that make up the body of a fungus.

identification key. A written and/or illustrated tool that provides a systematic way to identify and distinguish related living organisms.

immigration. The movement of new individuals into a population.

inert ingredients. Materials in a pesticide formulation that are not the active ingredient.

inoculative release. In biological control, a release of a biological control agent in which the agent is expected to reproduce in the field and build up its population so that its progeny provide control for several generations.

inoculum. The form of a pathogen that initiates infection.

insectary. A place where insects are reared.

insect growth regulator (IGR). Insecticide that controls insects by interfering with normal development.

insecticide. Pesticide used to control insects.

integrated pest management (IPM). An ecosystem-based pest management strategy that focuses on long-term prevention of pests or their damage through a combination of techniques such as biological control, habitat manipulation, modification of cultural practices, and use of resistant cultivars. Pesticides are used only after monitoring indicates they are needed according to established guidelines, and treatments are made with the goal of removing only the target organisms. Controls are selected and applied to minimize risk to human health, beneficial and nontarget organisms, and the environment.

intercropping. Growing more than one crop in a field at the same time.

interspecific. Occurring between two or more species.

intraspecific. Occurring within a species.

inundative releases. In biological control, release of biological control agents with the goal of achieving immediate control through the activities of the released individuals. Released agents are not expected to reproduce.

inversion layer. A layer of cool air near the soil surface trapped under a layer of warmer air.

invertebrates. Animals without backbones.

key pest. A pest that causes major damage in a crop on a regular basis unless controlled.

kingdom. In taxonomic classification, the highest division between organisms.

knowledge expectation (KE). A statement defining information that a professional is expected to understand. In the case of pest control advisers, KEs define the basic background information pest control advisers need when starting their first job.

k strategists. Species that demonstrate good competitive abilities when competition for resources is high.

leaching. The process by which a pesticide or other chemical moves in water downward through soil.

limiting factors. Factors that are required for an organism's growth and survival. When any one of these factors is in short supply or overabundant, growth and development will be affected, resulting in reduced growth or reproduction, stress, or death.

metamorphosis. The change in form that takes place as insects grow from immatures to adults.

microbial degradation. The process by which microorganisms break down pesticides.

microbial pesticides. Pesticides that consist of bacteria, fungi, viruses, or other microorganisms used in the control of weeds, invertebrates, or plant pathogens.

migration. The movement of organisms in and out of a population area.

mode of action. The mechanism by which a pesticide kills or controls the target organism.

monitoring. The process of carefully watching the activities, growth, and development of pest organisms or other factors on a regular basis over a period of time, often utilizing specific procedures.

monocots. Plants that produce only a single grasslike leaf in the seedling.

monoculture. A planting system where only one crop species is grown in a field. Most farms in the United States feature crops grown in monoculture.

mortality. Death rate.

MSDS (Material Safety Data Sheet). Information sheets prepared by pesticide manufacturers describing the chemical characteristics, hazards, safe handling instructions, and other important information regarding a specific pesticide.

mulch. A layer of material covering the soil surface.

mummy nut. A nut remaining on a tree long after harvest (also called sticktight); often provides overwintering site for pests.

mutualism. The relationship between two species that have developed a positive, reciprocal dependency, and both populations benefit from this association.

mycelium. The vegetative body of a fungus, consisting of a mass of slender filaments called *hyphae* (plural, *mycelia*).

mycopesticides. Commercially available beneficial microorganisms or their by-products that control plant pathogens.

mycorrhizae. Soil-dwelling fungi that form beneficial associations with the roots of flowering plants, often protecting them from infection by pathogenic fungi; an example of a mutualistic relationship between plant roots and fungi.

natural enemy. An organism that kills, decreases the reproductive potential, or otherwise reduces the numbers of another organism. In pest management, natural enemies are the agents of biological control.

natural selection. The process whereby individuals carrying certain alternative inherited traits survive and reproduce more successfully under stressful conditions, thus increasing the predominance of their genetic traits in the overall population in succeeding generations.

necrosis. The death of plant tissue accompanied by dark brown discoloration.

nematode. An unsegmented roundworm (phylum Nemata) found in soil, water, or plant or animal tissues. Some nematodes are important crop pests; others are used in the biological control of certain insect pests.

nontarget organisms. Organisms that are not the intended target of pesticide applications.

noxious weeds. Weeds that are toxic to livestock.

obligate parasites. Parasites that require living host plants to grow and reproduce.

occasional pests. Pests that are not problems every year in a crop and for which management actions may not be required in many years.

omnivores. Animals that feed on both plants and animals.

organic farming. Farming systems that grow crops without synthetic fertilizers or pesticides.

organic matter. Any material that is derived from living organisms; in soil, this would include decaying plant, microbial, and animal matter.

parasite. A small organism that lives and feeds in or on a larger host organism for all or most of its life.

parasitoid. An insect that parasitizes and kills other insects. Parasitoids are parasitic only in their immature stages, killing the host before emerging as a mature larva or adult. Often referred to as an insect parasite.

pathogen. A microorganism that causes disease.

perennials. Plants that live 3 or more years, repeating the vegetative growth and reproductive cycles each year.

persistence. Amount of time it takes for pesticide to degrade; measured in terms of half-life.

pest. An organism that interferes with the availability, quality, or value of a managed resource.

pest control adviser (PCA). In California, a person licensed to recommend the agricultural use of a pest control product or technique.

pesticide. Any substance or mixture intended for preventing, destroying, repelling, killing, or mitigating problems caused by any insects, rodents, weeds, nematodes, fungi, or other pests; and any other substance or mixture intended for use as a plant growth regulator, defoliant, or desiccant.

pesticide formulation. The pesticide as it comes from its original container, consisting of the active ingredient that controls the target pest blended with the inactive or inert ingredients.

pesticide resistance. The genetically acquired ability of an organism to survive a pesticide application at doses that once killed most individuals of the species.

pest resurgence. The rapid rebound of a pest population after it has been controlled.

pH. A value used to express relative acidity or alkalinity. Lower numbers indicate increasing acidity; higher numbers indicate increasing alkalinity.

phenology model. A mathematical model that can predict when key events in an organism's development will occur, usually based on biological information about the organisms and temperature data.

phenotype. The physical attributes of an individual organism, based on the interaction between the organism's genotype and environmental conditions.

pheromone. A substance secreted by an organism to affect the behavior or development of other organisms of the same species. Insect mating pheromones are used widely in pest management programs.

photoperiod. The amount of daylight in each day.

photosynthesis. Process by which green plants use energy from the sun to convert carbon dioxide and water into carbohydrates.

physical environment. The nonliving factors in an ecosystem, including minerals, water, sunlight, heat, and weather.

phytotoxity. The ability of a material such as a pesticide or fertilizer to cause injury to plants.

pollinators. Agents of pollen transfer, usually bees.

population. A group of individuals of the same species occupying a distinct space and possessing characteristics (such as special adaptations for the habitat) that are unique to the group.

population density. A measure of the number of individuals of a species in a defined area.

predator. An animal that attacks, kills, and feeds on other animals (prey), consuming several to many prey individuals in its lifetime.

presence-absence sampling. A sampling scheme in which the sampler notes only whether or not a designated plant part has any pest individuals on it rather than counting each individual.

preventive methods. Management methods that discourage damaging populations from developing, such as planting weed- and disease-free seeds or growing varieties of plants that are resistant to diseases or insects.

prey. An organism that is attacked and killed by a predator.

primary inoculum. The initial source of a pathogen that starts disease development in a given location.

primary producers. Green plants.

progeny. Offspring.

protectants. Fungicides that prevent spores from germinating but cannot prevent disease once infection occurs; they must be applied before the disease is apparent on the plant.

quarantine. A period of enforced isolation that is required to prevent entry of undesirable organisms.

random distribution. A distribution of individuals in a population that is completely by chance; the distribution is not uniform and cannot be predicted by any biological or environmental factors.

random sample. A sampling scheme that provides an unbiased estimate of the population, with every sample unit having an equal chance of selection.

relative sample. Any sampling method that gives an estimate of the total population or potential damage by sampling only a limited portion of the population. Includes sweep net, beating sheet, sampling frame, or visual counts per plant part samples.

replication. The repetition of treatments in an experiment.

reservoir. A site in which a pest population or quantity of inoculum can survive in the absence of a host crop, and from which a new crop may be invaded.

rhizomes. An underground, horizontal stem, often capable of growing roots at the nodes to produce new plants.

rhizomorph. A rootlike structure produced by certain fungi such as *Armillaria mellea*, which can grow from the root of an infected host plant to the root of an uninfected host plant.

rootstock. The underground portion of a plant such as a root or rhizome; also, on fruit trees and vines, the lower portion of a graft that develops into the root system.

r strategists. Species that have high rates of reproduction and growth and are able to colonize new areas rapidly.

runoff. The movement of water and dissolved or suspended matter such as pesticides or fertilizers from the area of application into surface water or onto neighboring land.

sample. A set of measurements taken from part of a population or a subset of the physical features of an environment, which can then be interpolated to evaluate the presence or impact of that factor in a whole field or area.

sampling pattern. The path followed when collecting samples.

sampling unit. The smallest unit sampled in the sampling universe.

sampling universe. The area that will be covered in a single sampling program.

sanitation. In pest management, removing or destroying pest breeding, refuge, and overwintering sites, as well as pest food sources or the pest species themselves. Also, any activity that reduces the spread of pathogen inoculum, such as removal and destruction of infected plant parts or cleaning of tools and field equipment.

saprophytes. Organisms that derive their nourishment from decaying organic matter.

scientific name. The unique two-word Latin binomial (genus and specific epithet) scientists assign to each organism; name recognized by scientists around the world.

scion. On a grafted fruit tree, the portion above a graft that becomes the trunk, branch, and tree top; determines the cultivar or variety of the fruit.

sclerotia. A compact mass of hardened mycelium that serves as a dormant stage in some fungi.

scouting. Collecting monitoring information in the field. Often used to refer to sampling by field scouts, who are student trainees, interns, or other staff working under the supervision of a pest control adviser.

secondary pest outbreak. A sudden increase in the population of a secondary pest (a pest that is normally at low or nondamaging levels in the crop) caused by the destruction of its natural enemies by a nonselective pesticide applied to control a primary pest.

selectivity. The range of organisms and life stages of organisms affected by a pesticide; a selective pesticide is toxic to only a single pest or group of closely related pests.

sequential sampling. A method for rapidly focusing sampling effort on situations where decisions are not easy to make. Instead of taking a fixed number of samples at each visit, samples are taken until the results confirm that the pest population is either above or below the treatment threshold as determined by a predefined guideline.

signs. In plant disease diagnosis, physical evidence of the presence of a pathogen on the host, such as spores, mycelia, or fruiting structures; as opposed to *symptoms*, which are the plant's expression of disease.

soil solarization. The practice of heating soil to levels lethal to pests through application of clear plastic to the soil surface for 4 to 6 weeks during sunny, warm weather.

species. The basic unit of taxonomic classification of organisms, designated in scientific nomenclature by a unique Latin binomial with a genus and specific epithet. Generally, organisms within a species classification can interbreed with each other but not with other species.

species diversity. A measure of the number of different species found in a community.

specific epithet. The second word (the first is the genus) in the Latin binomial of a scientific name designating a unique species.

spores. Reproductive structures produced by certain fungi and other organisms, capable of growing into a new individual under proper conditions.

stolon. A stem that grows horizontally along the surface of the ground.

stomach poisons. A pesticide that must be ingested to affect the pest.

stratified sampling. A sampling program that independently samples and evaluates areas of a field or orchard that vary significantly in character.

suppressive pest control methods. Pest control methods aimed at reducing existing pest populations to tolerable levels; in contrast to eradication methods, which are aimed at eliminating all individuals of a species in an area.

surface water. Water in aboveground waterways, including lakes, rivers, streams, and creeks, as opposed to groundwater.

sustainable agriculture. Farming systems that can maintain their productivity indefinitely. Sustainable agricultural systems must be resource conserving, environmentally compatible, socially supportive, and commercially competitive.

symptoms. In plant disease diagnosis, changes in the appearance of the infected plant due to the activities of a pathogen; as opposed to *signs*. Typical symptoms of plant disease include appearance of lesions, cankers, or discoloration of leaves.

synthetic organic pesticides. Manufactured pesticides produced from petroleum and containing largely carbon and hydrogen atoms in their basic structure.

systemic pesticide. A pesticide that is taken up by a plant or domestic animal and moves, after application, to other tissues in the plant or animal.

taxonomy. The theory, principles, and process of classifying organisms in categories.

thorax. The second of three major divisions in the body of an insect and the one bearing the legs and wings.

tissue culture. Techniques that allow breeders to propagate entire plants from a single cell or group of cells (*clonal propagation*).

tolerable injury level. The pest density (or damage level) at which the cost of pest control is less than the cost of the damage that the pest infestation causes; synonymous with *economic injury level*.

tolerance. The ability to endure the presence of a pest with little or no long-term damage.

toxicity. The capacity of a chemical to cause injury to organisms.

toxicology. The science that deals with poisons and their effect on living organisms.

transgenic. Having genes that originated from more than one species.

translocated herbicides. Herbicides that are absorbed by the roots or aboveground plant tissue and move throughout plant tissues.

treatment threshold. The level of pest population at which a pesticide or other control measure is needed to prevent eventual economic injury to the crop; also called *economic threshold* or *action threshold*.

trigger. A biological or physical event with a direct relationship to crop or pest development that is used to determine when to initiate a sampling program.

trophic structure. The series of links in a food web that describes the transfer of energy from one nutritional level to the next.

true leaves. Leaves on a plant produced after the cotyledons.

true resistance. Plants that support few or no pest individuals that infest other varieties; as opposed to *tolerance*.

tuber. An enlarged, fleshy underground stem.

uniform distribution. Distribution of individuals in a population so that the individuals are evenly spaced throughout the sampling universe, as in a field where every leaf on every plant has two aphids.

vapor pressure. The pressure exerted by a material in its gaseous form.

variables. Factors that, if not properly controlled, can influence the results of a field trial.

variety. Naturally occurring variants within a subspecies; also, plant strains *(cultivars)* produced through breeding programs.

vector. An organism able to transport and transmit a pathogen to a host.

vegetative growth. In plants, growth of stems and leaves.

viroid. A low-weight nucleic acid that can infect plant cells, replicate itself, and cause plant disease. It has no protein coat and contains only RNA.

virus. An obligate parasite of submicroscopic size that is composed of genetic material and surrounded by a layer of protein; viruses multiply only in living cells.

volatilization. The process by which a pesticide changes from a liquid form to a vapor.

weather. The state of the atmosphere (temperature, humidity, precipitation, wind conditions) over a short period of time (a day or a week) at a specific site.

weed. A plant that interferes with human activities, results in economic loss, or is otherwise undesirable.

winter annuals. Annuals that germinate from late summer to early winter, mature and set seed in late winter or early spring, and die in early summer.

Index

Photographs and illustrations are indicated by page numbers in *italic* type. Latin and common names are included for invertebrate and weed species. For plant diseases, the disease name and causal agent (in Latin) are provided.